# Wireless Networks

**Series Editor**

Xuemin Sherman Shen, University of Waterloo, Waterloo, Canada

The purpose of Springer's Wireless Networks book series is to establish the state of the art and set the course for future research and development in wireless communication networks. The scope of this series includes not only all aspects of wireless networks (including cellular networks, WiFi, sensor networks, and vehicular networks), but related areas such as cloud computing and big data. The series serves as a central source of references for wireless networks research and development. It aims to publish thorough and cohesive overviews on specific topics in wireless networks, as well as works that are larger in scope than survey articles and that contain more detailed background information. The series also provides coverage of advanced and timely topics worthy of monographs, contributed volumes, textbooks and handbooks.

Tho Le-Ngoc • Yuanzhe Gong • Mobeen Mahmood

# Half-Duplex/Full-Duplex MU-mMIMO

## Metamaterial-Based Large-Scale Array & Hybrid Beamforming Design

Springer

Tho Le-Ngoc
Department of Electrical and Computer Engineering
McGill University
Montreal, QC, Canada

Yuanzhe Gong
Department of Electrical and Computer Engineering
McGill University
Montreal, QC, Canada

Mobeen Mahmood
Department of Electrical and Computer Engineering
McGill University
Montreal, QC, Canada

ISSN 2366-1186 ISSN 2366-1445 (electronic)
Wireless Networks
ISBN 978-3-032-09194-9 ISBN 978-3-032-09195-6 (eBook)
https://doi.org/10.1007/978-3-032-09195-6



This Springer imprint is published by the registered company Springer Nature Switzerland AG
The registered company address is: Gewerbestrasse 11, 6330 Cham, Switzerland

# Preface

Massive multiple-input multiple-output (mMIMO) technology has emerged as a key enabler for future wireless systems, providing substantial gains in capacity and coverage through large-scale antenna arrays. With strategically designed beamforming algorithms, these arrays can precisely direct radiation toward intended users while effectively suppressing interference, thereby boosting spectral efficiency and signal quality. Metamaterial-based structures with tunable electromagnetic (EM) properties present promising avenues for further improving large-scale array performance by facilitating miniaturized, high-isolation antenna elements and enabling novel interference mitigation strategies. Moreover, hybrid beamforming (HBF) schemes that integrate radio-frequency (RF) beamformers with digital baseband precoders/combiners using fewer RF chains effectively reduce both hardware and computational complexity while maintaining performance through the joint design of the two stages. This monograph investigates and develops metamaterial-based large-scale antenna arrays alongside beamforming optimization techniques for both half-duplex (HD) multi-user mMIMO (MU-mMIMO) and full-duplex (FD) mMIMO systems. Extensive full-wave simulations and experimental measurements evaluate the performance of various prototype arrays, considering factors such as array aperture size, sub-array placement, and surrounding environments.

In the first part, various HBF architectures are developed while accounting for practical impairments in beamformer components to enhance the capacity of mMIMO systems. Both 2D and 3D array structures are investigated to evaluate their spectral and energy efficiencies, as well as their ability to sustain stable achievable rates across different user angles. Then, nulling control beamforming (NCB) algorithms are proposed to optimize the spatial isolation offered by the RF beamformer, leveraging both non-iterative and nature-inspired approaches while accounting for realistic EM characteristics such as heterogeneous element radiation patterns. These algorithms target far-field, angular-separation-based interference scenarios and near-field, location-differential cases involving spherical wavefronts in extra-large arrays. Full-wave simulations illustrate how beam radiation patterns transition from near-field to far-field propagation with increasing distance. Through simulations and experimental measurements conducted in an anechoic chamber with the proposed array prototypes, the NCB techniques effectively create deep, accurately aligned nulls that reduce residual multi-user interference (MUI) to noise levels while maintaining the desired signal gain.

In the second part, the monograph explores FD mMIMO communications, where concurrent transmission and reception further enhance spectral efficiency. Metamaterial-based large-scale arrays offer additional degrees of freedom for suppressing self-interference (SI), a critical challenge in FD operations. A multi-stage transmit-receive isolation framework is proposed, encompassing: (1) antenna isolation techniques, (2) joint uplink/downlink beamforming optimization, and (3) integration of metamaterial absorber structures. Experimental results demonstrate the effectiveness of these methods in collectively mitigating SI, preventing receiver front-end saturation and nonlinear distortion. Through experimental measurements, the study examines the correlation between beam radiation patterns and beam-level mutual coupling, as well as the effectiveness of various isolation techniques, offering valuable and practical insights into the development of beam-level isolation strategies. By combining these isolation measures with self-interference cancellation techniques, residual SI is reduced to the noise floor, thereby ensuring reliable FD mMIMO communication performance.

Montreal, QC, Canada Tho Le-Ngoc
Montreal, QC, Canada Yuanzhe Gong
Ottawa, ON, Canada Mobeen Mahmood
August 2025

# Declaration

**Competing Interests** The authors have no competing interests to declare that are relevant to the content of this manuscript.

# Contents

# List of Acronyms

| | |
|---|---|
| 1D | One-dimensional |
| 2D | Two-dimensional |
| 3D | Three-dimensional |
| 3GPP | Third-generation partnership project |
| 5G | Fifth-generation |
| 6G | Sixth-generation |
| ABF | Analog beamforming |
| ADC | Analog-to-digital converter |
| AoA | Angle-of-arrival |
| AoD | Angle-of-departure |
| AQNM | Additive quantization noise model |
| AI | Artificial intelligence |
| AR | Axial ratio |
| BB | Baseband |
| BFN | Beamforming network |
| bps | Bits per second |
| BS | Base station |
| CM | Constant-modulus |
| CP | Circularly polarized |
| DAC | Digital-to-analog converter |
| CSI | Channel state information |
| dB | Decibel |
| dBm | Decibel milliwatts |
| DBF | Digital beamforming |
| DGS | Defected ground plane |
| DNN | Deep neural network |
| DRA | Dielectric resonator antenna |
| EBG | Electromagnetic band gap |
| eMBB | Enhanced mobile broadband |
| EVM | Error vector magnitude |
| EM | Electromagnetic |

| | |
|---|---|
| ERP | Element radiation pattern |
| FD | Full-duplex |
| FDD | Frequency-division duplex |
| FEM | Finite element method |
| FIR | Finite-impulse-response |
| FSS | Frequency selective surface |
| HBF | Hybrid beamforming |
| HD | Half-duplex |
| IoT | Internet of things |
| ISI | Inter-symbol interference |
| ITU | International Telecommunication Union |
| IUI | Inter-user interference |
| KKT | Karush-Kuhn-Tucker |
| LCMV | Linearly constrained minimum variance |
| LDMA | Location division multiple access |
| LoS | Line-of-sight |
| LNA | Low noise amplifier |
| LTE | Long-term evolution |
| MC | Mutual coupling |
| MDB | Maximum directivity beamforming |
| MIMO | Multiple input multiple output |
| mMIMO | Massive multiple input multiple output |
| mMTC | Massive machine type communications |
| MMSE | Minimum mean square error |
| mmWave | Millimeter-wave |
| MSE | Mean square error |
| MU | Multi-user |
| MU-mMIMO | Multi-user massive multiple input multiple output |
| MUI | Multi-user interference |
| NCB | Nulling control beamforming |
| NLoS | Non-line-of-sight |
| NSB | Nulling steering beamforming |
| OMP | Orthogonal matching pursuit |
| OTA | Over-the-air |
| PCB | Printed circuit board |
| PNCB | Perturbation-based nulling control beamforming |
| PSD | Power spectral density |
| PSO | Particle swarm optimization |
| RIS | Reconfigurable intelligent surfaces |
| RF | Radio frequency |
| RL | Return loss |
| Rx | Receiver |
| RZF | Regularized zero-forcing |
| SA | Spectrum analyzer |
| SI | Self interference |

| | |
|---|---|
| SIC | Self interference cancellation |
| SINR | Signal-to-interference-plus-noise ratio |
| SIR | Signal-to-interference ratio |
| SISO | Single-input single-output |
| SPI | Serial peripheral interface |
| SNR | Signal-to-noise ratio |
| SV | Steering vector |
| SVD | Singular value decomposition |
| TDD | Time division duplex |
| TDL | Tapped-delay-line |
| Tx | Transmitter |
| UE | User equipment |
| UCA | Uniform circular array |
| ULA | Uniform linear array |
| URA | Uniform rectangular array |
| URLLC | Ultra-reliable and low-latency communications |
| VNA | Vector network analyzer |
| VSA | Vector signal analyzer |
| VSG | Vector signal generator |
| XL | Extra-large |
| XL-MIMO | Extra-large multiple input multiple output |
| ZF | Zero-forcing |

# List of Key Symbols

| | |
|---|---|
| $\boldsymbol{A}$ | Antenna element radiation pattern. |
| $\mathbf{B}$ | Digital baseband precoder. |
| $\beta$ | Propagation phase constant. |
| $D$ | Aperture size of the antenna array. |
| $\boldsymbol{\delta}$ | Shifting factor in the optimization. |
| $\epsilon_r$ | Dielectric constant. |
| $\mathbf{F}$ | Analog RF beamformer. |
| $G_r$ | Free-space power path loss. |
| $\mathbf{h}$ | Array response vector. |
| $\mathbf{H}_{\mathrm{SI}}$ | SI channel matrix |
| $K$ | Number of users. |
| $M$ | Number of antennas. |
| $N_{RF}$ | Number of RF chains. |
| $N_n$ | Number of iterations in the PSO algorithm. |
| $N_p$ | Number of particles in the PSO algorithm. |
| $\Omega$ | Objective (cost) function to be optimized. |
| $\boldsymbol{p}$ | Location vector of the field observation point. |
| $P_T$ | Transmit power. |
| $Q$ | Penalty factor in the optimization. |
| $\mathbf{R}$ | Covariance matrix of the received signal. |
| $r$, $\boldsymbol{v}$ | Distance and position vector (relative to array reference center). |
| $S$ | Scattering parameter. |
| $\boldsymbol{s}$, $s$ | Data signal vector, symbol. |
| $\sigma_n^2$ | Noise power. |
| $\theta, \phi, r$ | Azimuth angle, elevation angle, and radial distance in spherical coordinates. |
| $\boldsymbol{\Theta}$ | Uplink and downlink searching vector of the nulling angles in optimization. |
| $\lambda$ | Wavelength. |
| $\boldsymbol{w}$, $w_m$ | Beam steering vector, beam steering coefficient of $m^{\text{th}}$ element. |
| $\boldsymbol{\zeta}$ | Search vector in the optimization. |

# List of Key Symbols

- Bold upper/lower case letters denote matrices/vectors.
- $\mathbb{C}$ and $\mathbb{R}$ represent the set of complex and real numbers, respectively.
- $(\cdot)^*$, $(\cdot)^T$, $(\cdot)^H$, and $\|\cdot\|_2$, represent the complex conjugate, the transpose, the conjugate transpose, and the 2-norm, respectively.
- $\mathbf{I}_k$, $\mathbf{0}_k$, and $\mathbb{E}\{.\}$ denote $k \times k$ identity matrix, zero vector of length $k$, and the expectation operator, respectively.
- $\mathbf{A}(m, n)$ denotes the element at the intersection of $m$th row and $n$th column of the matrix $\mathbf{A}$.
- [.] and (.) are used to represent vectors and matrices and denote the functional form of variables.
- $\mathbf{x} > \mathbf{y}$ represents vector inequality, which means $x_n > y_n, \forall n$, where $x_n$ and $y_n$ are the $n$th elements of the vectors $\mathbf{x}$ and $\mathbf{y}$, respectively.
- $\mathbf{X} \otimes \mathbf{Y}$ is the Kronecker product of two matrices $\mathbf{X}$ and $\mathbf{Y}$.
- $\mathbf{X} \circ \mathbf{Y}$ is the Hadamard product of two matrices $\mathbf{X}$ and $\mathbf{Y}$.
- $\text{Span}(\mathbf{X})$ denotes the space spanned by the columns of $\mathbf{X}$.
- $\text{clip}\,(x, [a, b]) = \max\,(a, \min\,(x, b))$ denotes a clipping function with $a \preceq b$.
- $x \sim \mathcal{CN}\left(0, \sigma^2\right)$ denotes a complex Gaussian random variable with zero-mean and variance $\sigma^2$.
- ang(.) and mag(.) denote functions that return the phase(s) and magnitude(s), respectively, of a given complex value or vector.

# Chapter 1
# Introduction

## 1.1 Overview of 5G and Beyond

Wireless service providers face a dynamic technological landscape characterized by escalating demands for faster data speeds, improved energy efficiency, expanded capacity, and more reliable communication links. Across multiple sectors, including agriculture, automotive, manufacturing, healthcare, transportation, entertainment, and media, wireless communication technologies have become integral to operations and innovation [1, 2]. This widespread adoption has led to growing data traffic requirements and a corresponding need for continuous advancement in wireless communication systems.

Fifth-generation (5G) wireless networks are central to enabling these transformative innovations such as extended reality (XR), the metaverse, autonomous vehicles, smart cities, and Industry 4.0. By delivering ultra-reliable, low-latency connections at unprecedented speeds, 5G can support millions of devices per square kilometer, creating unprecedented opportunities for connectivity [1]. Rising data usage also underscores the growing global reliance on wireless networks. As presented in Fig. 1.1, the International Telecommunication Union (ITU) projects a 55% annual increase in data traffic between 2020 and 2030, with global monthly traffic expected to increase from 607 exabytes in 2022 to 5016 exabytes by 2030 [3]. According to ITU data, by 2024, about 5.5 billion people, equivalent to 68% of the world's population, are online. The annual growth rate of new internet users also rose from 2.7% in 2023 to 3.4% in 2024. Furthermore, the Ericsson Mobility Report also forecasts that 5G subscriptions will reach 4.4 billion by 2027, covering 71% of the global population. By 2030, an estimated 29.4 billion Internet of Things (IoT) devices will further expand the scope of wireless connectivity.

To shape the evolution of future wireless systems and meet diverse application requirements, three key 5G use cases have been summarized [4], as shown in Fig. 1.2



T. Le-Ngoc et al., *Half-Duplex/Full-Duplex MU-mMIMO*, Wireless Networks,
https://doi.org/10.1007/978-3-032-09195-6_1

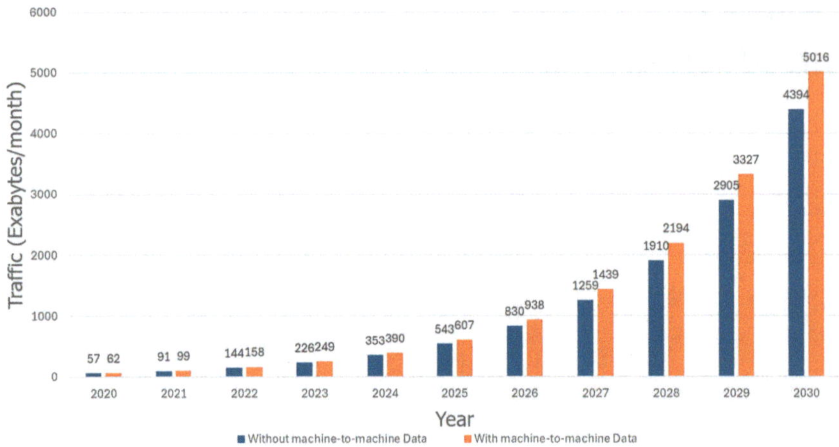


**Fig. 1.1** ITU Global mobile data traffic prediction

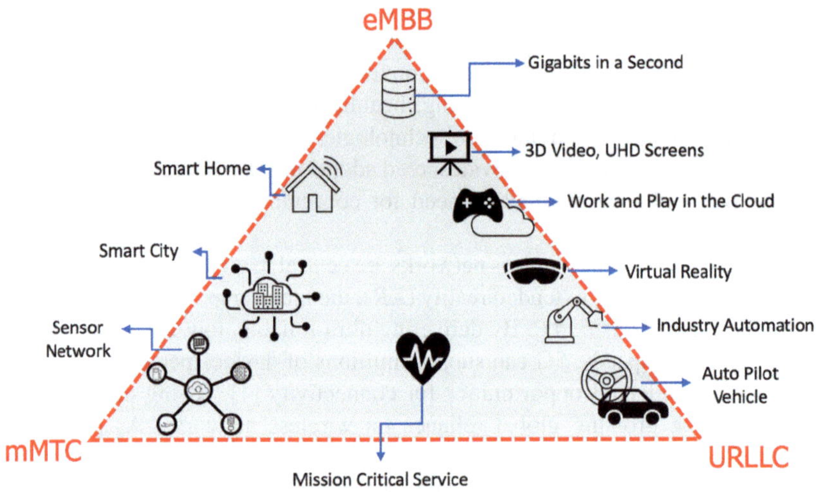


**Fig. 1.2** Three use cases and some important applications for 5G and beyond

- **Enhanced Mobile Broadband (eMBB):** Focused on high data rates and large traffic demands, eMBB supports applications such as augmented and virtual reality, 4K video streaming, and online work or gaming.
- **Massive Machine-Type Communications (mMTC):** Designed for massive device connectivity, mMTC enables applications like IoT, smart cities, and smart homes by emphasizing energy efficiency, extended battery life, and high connection density.
- **Ultra-Reliable and Low-Latency Communications (URLLC):** Targeting mission-critical applications, URLLC provides ultra-high reliability and minimal

latency for use cases such as autonomous vehicles, remote surgery, and industrial automation.

Furthermore, looking beyond 5G, sixth-generation (6G) systems will further integrate human-centric applications, such as XR, haptic feedback, and holographic displays, while substantially increasing traffic demands and imposing even stricter latency and reliability requirements [5]. As IoT devices and wireless sensors continue to proliferate, artificial intelligence (AI) and big data analytics will become essential to improve network throughput and automate system management. In parallel, industrial IoT applications will demand deterministic latency, high reliability, and continuous availability. The 6G networks will focus more on sustainability, automation, and energy efficiency. The resulting ecosystem, shaped by pervasive intelligence, large-scale AI integration, and digital twins, will continue to catalyze technological breakthroughs in wireless communications and underscore the importance of continually improving the wireless communication infrastructure [5].

When considering the fundamental factors that determine the capacity of the wireless communication system, Shannon's channel capacity formula [6] reveals that both available bandwidth and signal-to-noise ratio (SNR) play critical roles. The capacity can be expressed as

$$\text{Capacity [bps]} = \text{Bandwidth [Hz]} \times \underbrace{\text{Spectral Efficiency [bps/Hz]}}_{\log_2(1+\text{SNR})}, \tag{1.1}$$

where spectral efficiency represents the amount of information that can be transmitted over a given bandwidth in the presence of noise, measured in bits per second per Hertz (bps/Hz). Higher spectral efficiency means more data can be transmitted within the same bandwidth, which is essential for maximizing the utilization of limited spectrum resources. Power efficiency measures how effectively transmitted power is used to achieve a certain data rate. Because increasing transmit power is often impractical due to regulatory and power constraints, optimizing power efficiency is essential to maintain desired performance while reducing power consumption and costs. In scenarios where interference becomes a critical factor that significantly limits the achievable capacity, such as concurrent multi-user (MU) operations or full-duplex (FD) communication, the signal-to-interference-plus-noise ratio (SINR) can be used to replace the SNR in the equation to estimate the spectral efficiency. In these cases, interference management becomes essential to maintain high spectral and power efficiency, ensuring reliable communication and maximizing network capacity.

To meet the ever-growing demands of wireless communications while enhancing spectral and power efficiency, several promising technologies and strategies have been proposed and employed:

- **Metamaterial Large-scale Array:** Leverage controllable and spatially isolated beams with miniaturized and inter-element isolated structures to achieve higher power efficiency, improved coverage, and better interference management.
- **Advanced Transmission Techniques:** Implement massive multiple-input multiple-output (mMIMO) systems and advanced beamforming techniques to increase MU capacity and maximize spectral efficiency.
- **Efficient Use of Limited Bandwidth Resources:** Adopt FD communication technologies to improve spectral efficiency by enabling simultaneous transmission and reception on the same frequency band.

## 1.2 Metamaterial-Based Large-Scale Array

Due to the escalating demand for higher data rates and spectral efficiency, large-scale antenna arrays have become indispensable in enabling efficient signal transmission and reception. They provide the critical infrastructure necessary to support advanced communication technologies such as mMIMO, millimeter-wave (mmWave) communications, and FD capabilities, and meet the performance expectations of next-generation networks [7]. In mMIMO systems, the deployment of large-scale antenna arrays allows for the generation of steerable and directional beams characterized by high radiation gain and narrow beamwidth. By dynamically configuring sub-arrays to form multiple programmable beams, the base station (BS) can serve multiple users from various directions concurrently, greatly enhancing spectral efficiency and system throughput [8, 9]. Moreover, precise control over concentrated radiation patterns improves power efficiency and minimizes leakage power toward unintended users, effectively boosting the SNR and expanding coverage [10]. The ability to control and confine radiation patterns introduces new degrees of freedom, enhancing interference management and system performance through nulling and advanced beamforming techniques [11]. Additionally, the use of multiple antennas provides diversity gain, which enhances signal reliability in challenging environments by mitigating the effects of fading and interference [12]. In mmWave communications, the shorter wavelengths allow more antenna elements to be packed into a smaller area, enabling the deployment of larger antenna arrays. Conversely, large-scale antenna arrays are essential at these high frequencies to compensate for higher path losses by providing substantial array gains through beamforming [13]. As the aperture size of antenna arrays increases, near-field beam focusing using spherical wavefronts further enhances system capacity by enabling location-division multiplexing [14]. This approach is particularly relevant for future 6G communications, where extremely large-scale MIMO (XL-MIMO) is poised to deliver substantial performance gains, offering up to a tenfold increase in spectral efficiency compared to existing technologies [15].

Meanwhile, metamaterial-based structures, composed of periodic sub-wavelength unit cells, have gained significant attention in antenna array design. These artificial materials facilitate the development and optimization of antenna arrays, delivering enhanced isolation and improved radiation characteristics. By altering the layout and structure of the unit metacells, the effective dielectric permittivity and magnetic permeability of the metamaterial can be modulated. This capability paves the way to overcome some potential bottlenecks in large-scale antenna array design by facilitating antenna element size miniaturization and interference suppression. For instance, enhancing the apparent permeability of the dielectric substrate allows for antenna element size reduction [16, 17]. Studies in [18, 19] introduced electrically small antenna designs that leverage slow-wave propagation effectively within multi-layer metamaterial patch structures. By manipulating coupling and reflection waves to achieve constructive interference within the periodic structures, the slow-wave effect demonstrates more than a 60% size reduction in patch antennas and shows great adaptability in various array designs [18–20].

When packing antenna elements into arrays, the limited separation results in high inter-element mutual coupling (MC), degrading radiation efficiency and distorting radiation patterns. To enhance inter-element isolation, metamaterial-based decoupling components [21, 22] or frequency-selective structures (FSS) [23, 24], can be implemented between antenna elements. These structures effectively excite constructive interference or act as band-stop filters, suppressing surface leakage waves between elements [25–27]. Although the additional structures can be sizable compared to the limited space on the array, a joint design of multi-layer metamaterials that simultaneously reduces the effective wavelength and improves isolation can strategically overcome this bottleneck. Moreover, metamaterial absorbers can be designed to absorb specific frequencies of electromagnetic (EM) waves, thereby reducing unwanted reflections and EM interference in wireless communication systems, without requiring modifications to the original antenna array structure [28]. These sub-wavelength metacells offer high adaptability and flexibility for implementation in various systems.

Moreover, due to their controllable surface impedance and EM characteristics, metamaterials also play a crucial role in designing reconfigurable intelligent surfaces (RIS) and reconfigurable antennas. RIS are programmable surfaces that manipulate EM wave propagation to enhance signal coverage, capacity, and energy efficiency in wireless networks [29]. By incorporating tunable elements like varactors or MEMS switches into metamaterial structures, antennas achieve agile beam steering and frequency reconfigurability, essential for adaptive wireless communication systems [30]. Other important applications of metamaterials-based structure design include manipulating the phase, amplitude, and polarization of EM waves; shaping radiation wavefronts; improving antenna radiation efficiency and bandwidth; and enabling functions like anomalous reflection and refraction, beam focusing, and holographic beamforming in wireless communications [31–33].

## 1.3 From MIMO to mMIMO

Multiple-input multiple-output (MIMO) technology enhances the capacity and performance of wireless communication systems by employing multiple transmit and receive antennas [12]. Unlike traditional single-antenna systems, MIMO exploits the differences in channel conditions between each transmit-receive antenna pair, providing increased degrees of freedom. This approach mitigates adverse effects of the wireless propagation environment, such as path loss and multipath fading, by leveraging the inherent multipath characteristics of the channels. MIMO has become integral to modern wireless standards, including Long-Term Evolution (LTE) and Wi-Fi [7, 9, 34–37]. LTE-Advanced incorporates MIMO configurations up to 8Tx-to-8Rx, utilizing eight transmit and eight receive antennas [7, 38, 39]. MIMO technology offers primary advancements in communication systems [40]:

- Spatial Multiplexing Gain: Enables the transmission and reception of multiple parallel data streams without increasing the occupied bandwidth, thereby facilitating higher data rates and improving spectral efficiency;
- Diversity Gain: Enhances throughput and reliability performance by exploiting different channel conditions between antenna pairs, leading to improved SNR.
- Array Gain: Achieves coherent combination and concentration of signals across multiple antennas to boost the effective transmit or receive power and extend coverage, thus improving SNR and range without additional power consumption.

However, due to the limited number of antennas employed, there is usually a trade-off between these features, known as the diversity-multiplexing trade-off.

By employing large-scale antenna arrays, mMIMO was proposed more than a decade ago as a transformative shift in wireless communications [8, 12, 36]. By equipping the BS with a large number of antennas, mMIMO significantly boosts spectral efficiency and network capacity, as illustrated in Fig. 1.3. Recognizing its

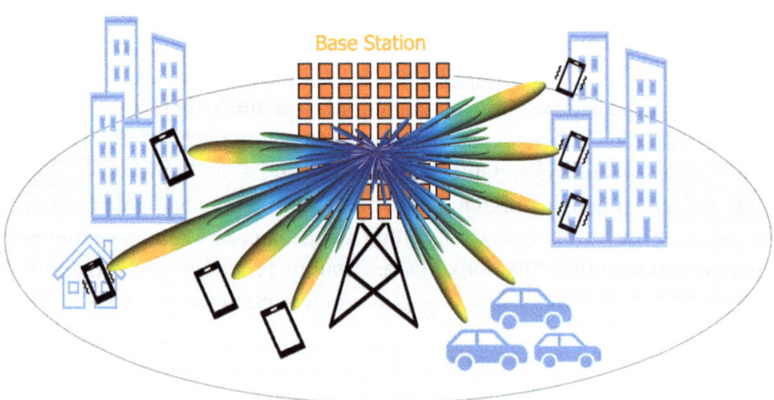


**Fig. 1.3** mMIMO with the large-scale antenna array

potential, the Third Generation Partnership Project (3GPP) standardized deployments of BSs with 64 to 256 antennas in Release 17 [41]. As such, mMIMO is considered a key technology for 5G and beyond 5G (B5G) networks [8, 42–44]. In MU-mMIMO systems, these large antenna arrays allow BS to serve multiple users over the same frequency band at the same time. By creating focused and steerable signal beams, spatial separation between users is improved, and signal strength is enhanced. This results in significant improvements in signal quality and reduces interference when multiple data streams are active. With more antennas, the system gains additional flexibility to manage signals, allowing advanced techniques like strategic beamforming and signal processing to further reduce interference among users. As the size of the antenna arrays increases, the number of simultaneous data streams that can be supported also increases, leading to even higher spectral efficiency and network capacity. The significant beamforming gains provided by large antenna arrays also offer users a more consistent experience across the network, especially benefiting those at the edges of cells where signals are typically weaker [45]. mMIMO systems can thus extend network coverage and improve signal reception in areas with weak signals. The focused beams also help reduce interference between neighboring cells in the network. Furthermore, the increased diversity from using more antennas enhances error performance and reliability of communication links, especially in challenging wireless environments. These advantages enable 5G and future networks to support a wide range of emerging services, particularly those requiring high data rates like eMBB and those involving a large number of connected devices like mMTC.

However, when mMIMO is implemented in practice, challenges such as high hardware costs, increased system complexity, and the overhead associated with estimating channel conditions remain. Key challenges include:

- **High Hardware Costs and Complexity:** One of the most significant obstacles is the high cost and complexity of hardware. Traditional MIMO systems typically rely on fully digital beamforming (DBF), where each antenna element is associated with its own radio frequency (RF) chain, digital-to-analog converters (DACs) for transmission, analog-to-digital converters (ADCs) for reception, mixers, and other components. In mMIMO systems, the large number of antenna elements dramatically increases hardware costs and power consumption. While substantial beamforming gains can reduce transmit power without compromising spectral efficiency, the high power consumption of these hardware components can significantly impact energy efficiency.
- **Substantial Channel Estimation Overhead:** In mMIMO systems using traditional DBF techniques, estimating the high-dimensional channel state information (CSI) is crucial for effective beamforming. The increased number of antennas leads to higher channel estimation overhead. Given a fixed coherence time (the period during which the channel remains stable), acquiring extensive CSI requires longer pilot transmission times, which in turn reduces the time available for data transmission. This increased overhead can negatively affect effective spectral efficiency. For example, as the number of antennas approaches

infinity, the required pilot transmission length could surpass any finite coherence time, resulting in zero effective spectral efficiency. Therefore, it is essential to develop new beamforming methods that require less CSI overhead for mMIMO systems.

- **High Computational Complexity:** As the size of the antenna array increases, so does the size of the channel matrix. Applying conventional MIMO signal processing algorithms to tasks such as beamforming, channel estimation, and signal detection involves large-scale matrix computations (e.g., matrix inversions, multiplications, singular value decompositions (SVD), and determinants), leading to higher computational complexity.

To address these issues, it is essential to develop advanced signal-processing techniques that are efficient and have low computational complexity. Hybrid beamforming (HBF) has been proposed as a promising solution to these challenges in mMIMO systems [46–48]. Unlike DBF, HBF splits the beamforming process into two stages: an analog stage that processes signals at radio frequencies, and a digital stage that processes signals at baseband (BB) frequencies, connected by a reduced number of radio frequency (RF) chains. This architecture reduces hardware costs and computational complexity, as well as the overhead for estimating channel conditions, making HBF mMIMO systems more practical for deployment [49].

## 1.4 From Half-Duplex to Full-Duplex Communications

Half-duplex (HD) mMIMO systems have demonstrated significant potential for capacity enhancement in wireless communications. Utilizing spatially isolated beams, these systems can concurrently serve multiple users in either the uplink or downlink direction. This approach improves spectrum efficiency and boosts overall network performance through spatial multiplexing. To prevent interference between local transmitters (Tx) and receivers (Rx), HD mMIMO systems transmit downlink signals and receive uplink signals using orthogonal time or frequency resources, as illustrated in Fig. 1.4. Downlink and uplink transmissions occur either in different frequency bands, as in Frequency Division Duplexing (FDD), or in different time slots, as in Time Division Duplexing (TDD). While this separation ensures that transmission and reception do not interfere with each other, it inherently limits maximum data rates and spectral efficiency [50].

In TDD systems, the device alternates between transmitting and receiving over the same frequency band at different times. This arrangement can lead to underutilization of the spectrum, especially when traffic demands are asymmetric between uplink and downlink, and also an increase in communication latency. In contrast, FDD systems allow simultaneous transmission and reception but require separate frequency bands for each direction, doubling the required spectral resources. These operational constraints prevent the full utilization of available spectral resources and

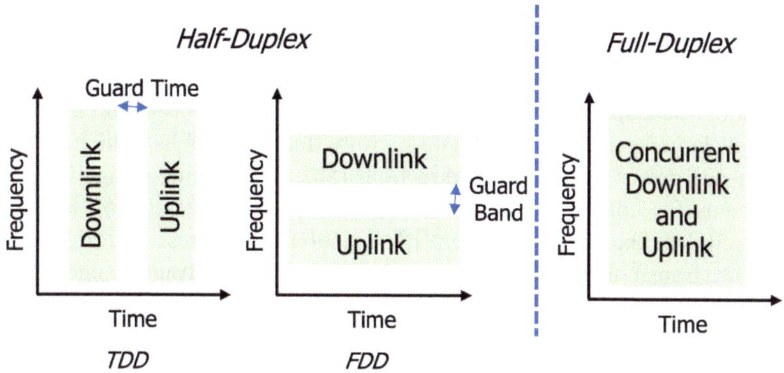


**Fig. 1.4** Half-duplex and full-duplex communication schemes

hardware capabilities, resulting in suboptimal system performance in terms of data throughput and efficiency.

On the other hand, FD communication enables simultaneous transmission and reception of signals on the same frequency band, potentially doubling spectral efficiency compared to conventional HD systems [51, 52]. By facilitating concurrent uplink and downlink services, FD communication optimizes the utilization of available bandwidth, an essential advancement to address the escalating demand for higher data rates and lower latency in wireless networks with limited spectrum resources. FD communication plays a crucial role in enabling 5G and beyond networks, where stringent requirements for capacity, spectral efficiency, and latency must be meticulously addressed across diverse application scenarios. Furthermore, integrating FD communication with mMIMO systems can further enhance energy efficiency, increase network capacity, reduce latency, and improve spectral efficiency [52–54].

Despite the advantages of FD communication, a significant obstacle to its widespread adoption is the strong self-interference (SI) generated by a node's own transmissions [51, 55, 56]. This SI can overload the Rx because the transmitted leakage signal power is usually substantially higher than the desired signal power from the remote user. The problem arises because the receive antenna is typically much closer to the local transmit antenna than the remote user, resulting in minimal path loss for the SI relative to the intended signal. The excessive SI power can significantly degrade the received SINR, limiting channel throughput [57]. To prevent performance degradation, it is essential to suppress the residual SI to or below the receiver noise floor.

While the use of large-scale antenna arrays in FD communications increases the complexity of signal processing and hardware costs, it also enables a multi-stage SI isolation and cancellation architecture, which is crucial for realizing FD mMIMO technology. Additionally, to leverage the advantages of large-scale arrays for SI management while reducing hardware complexity, HBF has gained traction for FD joint uplink and downlink beamformer design [46]. SI mitigation relies on two primary approaches: RF isolation and SI cancellation (SIC) [11]. RF isolation separates transmit and receive signals, preventing or minimizing leakage between local Tx and Rx components. Techniques such as antenna separation, cross-polarization, decoupling structures, and metamaterial absorbers can be employed. Moreover, large-scale arrays facilitate narrow, steerable beams, which help minimize leakage power in unintended directions [58]. Beamformer design then refines these beams for uplink and downlink interactions, further reducing interference [59–61]. SIC techniques complement RF isolation by suppressing residual SI that arises when transmitted signals leak into the receive path. Advanced algorithms and signal processing methods address this residual interference [62]. Recent progress in large-scale array and metamaterial design, alongside strategic signal processing algorithms, has made FD mMIMO systems increasingly practical through multi-stage isolation and cancellation [51, 54–56, 58, 63, 64].

## 1.5 Book Organization

Massive MIMO technology stands out as a promising advancement for future wireless systems. The deployment of large-scale arrays marks a key milestone, enabling networks to efficiently accommodate the ever-growing demand for data. With programmable radiation patterns, arrays equipped with strategically designed beamforming algorithms can effectively focus power in specific directions, thereby substantially boosting signal strength and mitigating interference. Meanwhile, metamaterial-based structures with tunable EM properties can further enhance large-scale antenna array design and create new opportunities for improved interference suppression.

In this book, we investigate and develop metamaterial-based large-scale antenna arrays alongside various beamforming optimization techniques for both half-duplex (HD) in Part I (Chaps. 4 to 8) and full-duplex (FD) mMIMO systems in Part II (Chaps. 9 to 11). Full-wave simulations and experimental measurements have been carried out to evaluate and illustrate the performance of these designs under practical considerations. Collectively, the results demonstrate the potential of metamaterial-based large-scale arrays to meet evolving demands by effectively boosting desired signal gain and mitigating interference in HD MU-/FD mMIMO systems. An overview of the book outline, highlighting the key topics covered in each chapter, is provided in Fig. 1.5. The remainder of this book is organized as follows.

In Chap. 3, two antenna element structures operating around 3.5 GHz are proposed. The first design utilizes a dual-layer metamaterial-based design that achieves patch antenna miniaturization through a slow-wave effect and enhances inter-element isolation via an electromagnetic bandgap (EBG) structure. A strategically designed L-slotted radiating patch enables circular polarization using a simple probe-feeding mechanism. The second design features a dielectric resonator antenna (DRA) structure with dual-polarization, achieved through two distinct feeding

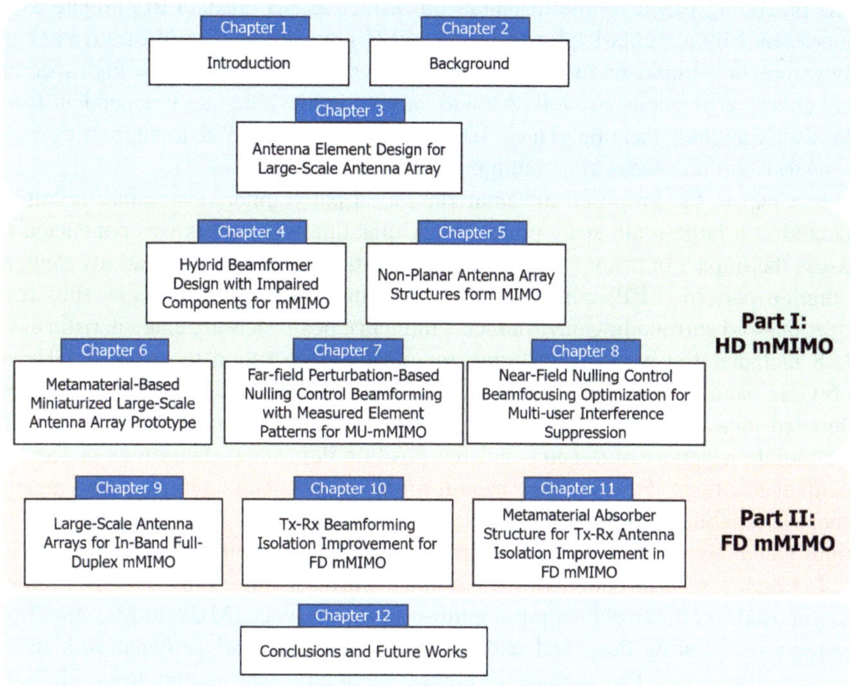


**Fig. 1.5** Book outline

mechanisms. Full-wave simulations are performed to optimize the structure and investigate the achieved radiation characteristics.

Chapter 4 investigates different 2D array configurations to be used at BS for enhanced capacity in mMIMO systems. We design three HBF architectures: (i) HBF using full-resolution PSs and DACs, with a BB transfer block for a constant-modulus RF beamformer, (ii) HBF using b-bit PSs and full-resolution DACs, with an orthogonal matching pursuit (OMP) based algorithm that can approach the optimal unconstrained RF beamformer, and (iii) HBF using b-bit PSs and q-bit DACs, taking into account also DAC quantization noise. The proposed HBF schemes not only reduce the hardware cost/complexity by utilizing a limited number of RF chains but also lower the channel state information (CSI) overhead size by designing the RF stage via slow time-varying CSI (i.e., angular information). Therefore, the reduced-size effective CSI is utilized during the BB-stage design. We compare the spatial, spectral and energy efficiencies of HBF designs for different antenna arrays. In the illustrative results, we show that the proposed HBF schemes achieve higher sum-rate and energy efficiency than other HBF benchmarks.

Chapter 5 compares the performance of various 3D array structures in the design of HBF for mmWave MU mMIMO systems. The proposed HBF scheme involves two stages: RF beamforming based on SVD of the channel matrix, and BB

MU precoding based on the instantaneous effective BB channel to mitigate MU-interference by a regularized zero-forcing (RZF) technique. The illustrative results show that, in comparison to 2D arrays, 3D array structures can provide high spectral and energy efficiencies as well as a non-varying achievable rate independent from the user's angular location. Thus, 3D structures can be useful in supporting both terrestrial and non-terrestrial communications.

In Chap. 6, the proposed metamaterial-based patch antenna structure is implemented in a large-scale array prototype. A quantitative analysis was conducted to assess the impact of array length on beam radiation patterns under various element radiation pattern (ERP) considerations. The study also examined how sub-array locations and surrounding environments influence beam steering characteristics. An $8\times8$ metamaterial-based patch array for mMIMO operation in the 3.35 GHz to 3.6 GHz band was designed and experimentally studied in the anechoic chamber. Detailed measurements of the ERPs across all elements revealed both the inherent radiation pattern symmetry and the position-dependent distortions at various element locations. Furthermore, beamforming experiments confirmed the array's steering capabilities, investigated the influence of array length on steerability, and highlighted the potential impact of surrounding EM environments on performance.

In Chap. 7, a Perturbation-Based Nulling Control Beamforming (PNCB) scheme is proposed to effectively suppress multi-user interference (MUI) in MU-mMIMO systems by creating deep and wide radiation nulls precisely aligned with interference directions. The scheme leverages symmetry information from ERPs to streamline optimization by reusing optimized beamforming vectors for symmetric sub-arrays. The PNCB approach begins with weights generated by the Linearly Constrained Minimum Variance (LCMV) method. Subsequently, the weights are refined through perturbations to identify optimal beamforming vectors that account for variations caused by heterogeneous ERPs. Particle Swarm Optimization (PSO) algorithm is employed to enhance the search efficiency. Moreover, the effects of attenuator and phase-shifter quantization on nulling depth are investigated, providing valuable insights into achievable performance under different hardware configurations. Experimental validations confirm the PNCB's effectiveness, resulting in considerable improvements in Error Vector Magnitude (EVM) and Signal-to-Interference Ratio (SIR), which successfully bring the MUI to the noise floor.

As the array aperture size continues to grow, Chap. 8 explores the opportunities provided by extra-large (XL) arrays for near-field beamfocusing in MU-mMIMO, accounting for the spherical wavefront characteristics. Comprehensive full-wave simulation analyses are conducted to investigate near-field beam radiation pattern transitions with changing distances. Two nulling control beamforming (NCB) algorithms are developed to effectively mitigate MUI by exploiting distance-dependent beam patterns in near-field communication. The first algorithm employs an LCMV-based NCB scheme. The second scheme introduces a constant-modulus beamfocusing scheme, based on a perturbation-based NCB algorithm that utilizes PSO and relies solely on phase shifters. The approaches demonstrate their effectiveness by achieving significant differences in focusing gain between desired and interfering locations in both colinear and non-colinear user scenarios. Variations in achievable beamfocusing gain differences across various colinear user locations have been explored.

To further enhance spectral efficiency, FD technologies have emerged as a promising advancement, enabling simultaneous transmission and reception over the same bandwidth. In Chap. 9, a cross-polarized 8×8Tx/8×8Rx large-scale array prototype was designed, fabricated, and characterized to evaluate its effectiveness in achieving Tx-Rx antenna isolation for FD mMIMO applications around 3.5 GHz. Building on the dual-layer metamaterial-based antenna design introduced in Chap. 3, the prototype employs mirrored L-slotted radiating patches to form cross-polarized CP Tx/Rx pairs. The efficacy of the cross-polarized configuration and strategic element separation in achieving high Tx-Rx isolation is studied through full-wave simulations. Experimental measurements across all 64Tx-to-64Rx SI channels are conducted and the spatial isolation between various Tx and Rx element positions is investigated. The comprehensive dataset of measured SI channels offers a realistic foundation for developing and evaluating beam-level isolation and cancellation algorithms.

Moreover, in Chap. 10, joint uplink/downlink beamforming optimization schemes are implemented to further suppress SI by optimizing beam-level spatial isolation while maintaining gain toward desired directions. The chapter highlights the necessity and advantages of multi-stage SI isolation and cancellation techniques in FD mMIMO systems. The relationship between array size, beamwidth, and beam-level MC is analyzed. Two beam-level isolation optimization algorithms are proposed. The first algorithm employs an angular perturbation search to identify optimal uplink and downlink steering angles with a tradeoff gain degradation for desired users. The second algorithm integrates a Null Steering Beamforming (NSB) approach with a PSO algorithm to determine optimal nulling angles for enhanced isolation. The effectiveness of these beam-level optimization algorithms in achieving robust SI mitigation is studied using the 8×8Tx/8×8Rx antenna array prototype with quantized beamforming networks in the anechoic chamber. Measurement results also illustrate the correlation between Tx/Rx beam radiation patterns and achieved isolation, offering valuable insights for the development of advanced beam-level isolation algorithms. Additional SI mitigation through SIC techniques successfully reduces the residual SI to levels comparable to thermal noise.

To achieve more uniform Tx-Rx isolation performance across all uplink and downlink user locations, Chap. 11 presents the design and evaluation of a metamaterial absorber structure aimed at enhancing Tx-Rx RF isolation in FD mMIMO systems without requiring modifications to the existing antenna array design. The chapter details the absorber unit element design, its placement, and its influence on beamforming radiation patterns. Using the 8×8Tx/8×8Rx FD mMIMO antenna array prototype, the addition of the absorber significantly improves Tx-Rx element- and beam-level isolation, with negligible impact on the original beam radiation patterns. The study further explores the correlation between near-field and far-field dominant SI scenarios and the alignment of downlink and uplink beam directions.

Findings emphasize the critical role of the absorber in mitigating near-field SI, which is essential for maintaining effective beamforming isolation performance.

Finally, Chap. 12 presents concluding remarks, summarizes the key findings of this book, and discusses potential directions for future research.

## References

1. 5G Americas, Becoming 5G-advanced: the 3GPP 2025 roadmap. White paper, 5G Americas (2022)
2. Ericsson, The 5G business potential, pp. 1–8 (2017). https://www.5gsummit.org/greece/slides/Panel_08_Maria%20Boura.pdf
3. I. T. U. (ITU), IMT traffic estimates for the years 2020 to 2030. Tech. Rep. ITU-R M.2370-0 (2015)
4. Huawei, 5G wireless network planning solution white paper. Tech. Rep. (2018)
5. W. Tong, P. Zhu (Eds.), *6G: The Next Horizon: From Connected People and Things to Connected Intelligence* (Cambridge University Press, Cambridge, 2021)
6. D. H. Morais, *Key 5G physical layer technologies* (Springer, Berlin, 2020)
7. L. Lu, G. Y. Li, A.L. Swindlehurst, A. Ashikhmin, R. Zhang, An overview of massive MIMO: benefits and challenges. IEEE J. Sel. Topics Signal Process. **8**(5), 742–758 (2014)
8. T.L. Marzetta, Noncooperative cellular wireless with unlimited numbers of base station antennas. IEEE Trans. Wireless Commun. **9**(11), 3590–3600 (2010)
9. E.G. Larsson, O. Edfors, F. Tufvesson, T.L. Marzetta, Massive MIMO for next generation wireless systems. IEEE Commun. Mag. **52**(2), 186–195 (2014)
10. H.Q. Ngo, E.G. Larsson, T.L. Marzetta, Energy and spectral efficiency of very large multiuser MIMO systems. IEEE Trans. Commun. **61**(4), 1436–1449 (2013)
11. Y. Gong, A. Yaseen, T. Zhang, R. Morawski, T. Le-Ngoc, Perturbation-based nulling control beamforming with measured element radiation patterns for MU-mMIMO. IEEE Open J. Veh. Technol. **5**, 1273–1293 (2024)
12. F. Rusek, D. Persson, B.K. Lau, E.G. Larsson, T.L. Marzetta, O. Edfors, F. Tufvesson, Scaling up MIMO: opportunities and challenges with very large arrays. IEEE Signal Process. Mag. **30**(1), 40–60 (2012)
13. Z. Pi, F. Khan, An introduction to millimeter-wave mobile broadband systems. IEEE Commun. Mag. **49**(6), 101–107 (2011)
14. Z. Wu, L. Dai, Multiple access for near-field communications: SDMA or LDMA? IEEE J. Sel. Areas Commun. **41**(6), 1918–1935 (2023)
15. T.S. Rappaport, Y. Xing, O. Kanhere, S. Ju, A. Madanayake, S. Mandal, A. Alkhateeb, G.C. Trichopoulos, Wireless communications and applications above 100 GHz: opportunities and challenges for 6G and beyond. IEEE Access **7**, 78729–78757 (2019)
16. P. Mookiah, K.R. Dandekar, Metamaterial-substrate antenna array for MIMO communication system. IEEE Trans. Antennas Propag. **57**(10), 3283–3292 (2009)
17. R.O. Ouedraogo, E.J. Rothwell, A.R. Diaz, K. Fuchi, A. Temme, Miniaturization of patch antennas using a metamaterial-inspired technique. IEEE Trans. Antennas Propag. **60**(5), 2175–2182 (2012)
18. E. Korkontzila, D. Papafilippou, D. Chrissoulidis, Miniaturization of microstrip patch antenna for wireless applications by use of multilayered electromagnetic band gap substrate, in *2006 First European Conference on Antennas and Propagation* (IEEE, Piscataway, 2006), pp. 1–6
19. S. Ghosh, T.N. Tran, T. Le-Ngoc, Dual-layer EBG-based miniaturized multi-element antenna for MIMO systems. IEEE Trans. Antennas Propag. **62**(8), 3985–3997 (2014)

20. Y. Gong, R. Morawski, H.H. Lee, T. Le-Ngoc, A miniaturized 8x8 dual-layer EBG slotted circularly polarized patch antenna array for mMIMO, in *GLOBECOM 2022 - 2022 IEEE Global Communications Conference* (2022), pp. 6511–6516
21. K.-L. Wu, C. Wei, X. Mei, Z.Y. Zhang, Array-antenna decoupling surface. IEEE Trans. Antennas Propag. **65**(12), 6728–6738 (2017)
22. J. OuYang, F. Yang, Z. Wang, Reducing mutual coupling of closely spaced microstrip MIMO antennas for WLAN application. IEEE Antennas Wireless Propag. Lett. **10**, 310–313 (2011)
23. C.X. Mao, S.S. Gao, Y. Wang, J.T.S. Sumantyo, Compact broadband dual-sense circularly polarized microstrip antenna/array with enhanced isolation. IEEE Trans. Antennas Propag. **65**(12), 7073–7082 (2017)
24. G. Zhai, Z.N. Chen, X. Qing, Enhanced isolation of a closely spaced four-element MIMO antenna system using metamaterial mushroom. IEEE Trans. Antennas Propag. **63**(8), 3362–3370 (2015)
25. Z. Iluz, R. Shavit, R. Bauer, Microstrip antenna phased array with electromagnetic bandgap substrate. IEEE Trans. Antennas Propag. **52**(6), 1446–1453 (2004)
26. H.S. Farahani, M. Veysi, M. Kamyab, A. Tadjalli, Mutual coupling reduction in patch antenna arrays using a UC-EBG superstrate. IEEE Antennas Wireless Propag. Lett. **9**, 57–59 (2010)
27. K. Payandehjoo, R. Abhari, Employing EBG structures in multiantenna systems for improving isolation and diversity gain. IEEE Antennas Wireless Propag. Lett. **8**, 1162–1165 (2009)
28. N.I. Landy, S. Sajuyigbe, J.J. Mock, D.R. Smith, W.J. Padilla, Perfect metamaterial absorber. Phys. Rev. Lett. **100**(20), 207402 (2008)
29. C. Huang, S. Hu, G.C. Alexandropoulos, A. Zappone, C. Yuen, R. Zhang, M. Di Renzo, M. Debbah, Holographic MIMO surfaces for 6G wireless networks: opportunities, challenges, and trends. IEEE Wireless Commun. **27**(5), 118–125 (2020)
30. S. Bang, J. Kim, G. Yoon, T. Tanaka, J. Rho, Recent advances in tunable and reconfigurable metamaterials. Micromachines **9**(11), 560 (2018)
31. H.T. Chen, A.J. Taylor, N. Yu, A review of metasurfaces: physics and applications. Rep. Prog. Phys. **79**(7), 076401 (2016)
32. A. Alù, N. Engheta, Multifrequency optical invisibility cloak with layered plasmonic shells. Phys. Rev. Lett. **100**(11), 113901 (2008)
33. J.B. Pendry, Negative refraction makes a perfect lens. Phys. Rev. Lett. **85**(18), 3966 (2000)
34. A. Gupta, R.K. Jha, A survey of 5G network: architecture and emerging technologies IEEE Access **3**, 1206–1232 (2015)
35. P.D. Arapoglou, P. Burzigotti, M. Bertinelli, A.B. Alamanac, R. De Gaudenzi, To MIMO or not to MIMO in mobile satellite broadcasting systems. IEEE Trans. Wireless Commun. **10**(9), 2807–2811 (2011)
36. E. Bjornson, L. Van der Perre, S. Buzzi, E.G. Larsson, Massive MIMO in sub-6 GHz and mmWave: physical, practical, and use-case differences. IEEE Wireless Commun. **26**(2), 100–108 (2019)
37. J. Lee, J.K. Han, J. Zhang, MIMO technologies in 3GPP LTE and LTE-advanced. EURASIP J. Wireless Commun. Network. **2009**, 1–10 (2009)
38. T.T. Tran, Y. Shin, O.S. Shin, Overview of enabling technologies for 3GPP LTE-advanced. EURASIP J. Wireless Commun. Netw. **2012**, 1–12 (2012)
39. D. Schoolar, Massive MIMO comes of age. Samsung, White Paper (2017), pp. 1–13
40. S. Chatzinotas, B. Ottersten, R. De Gaudenzi, *Cooperative and cognitive satellite systems* (Academic Press, Cambridge, 2015)
41. 3GPP TR 38.913, 5G: study on scenarios and requirements for next generation access technologies. Tech. Rep. Ver. 17.0.0 (2022)
42. M. Agiwal, A. Roy, N. Saxena, Next generation 5G wireless networks: a comprehensive survey. IEEE Commun. Surv. Tutor. **18**(3), 1617–1655 (2016)
43. M. Shafi, A.F. Molisch, P.J. Smith, T. Haustein, P. Zhu, P. De Silva, F. Tufvesson, A. Benjebbour, G. Wunder, 5G: a tutorial overview of standards, trials, challenges, deployment, and practice. IEEE J. Sel. Areas Commun. **35**(6), 1201–1221 (2017)

44. S.A. Busari, K.M.S. Huq, S. Mumtaz, L. Dai, J. Rodriguez, Millimeter-wave massive MIMO communication for future wireless systems: a survey. IEEE Commun. Surv. Tutor. **20**(2), 836–869 (2017)
45. E. Björnson, J. Hoydis, L. Sanguinetti, et al., Massive MIMO networks: spectral, energy, and hardware efficiency. Found. Trends Signal Process. **11**(3–4), 154–655 (2017)
46. M. Rihan, T.A. Soliman, C. Xu, L. Huang, M.I. Dessouky, Taxonomy and performance evaluation of hybrid beamforming for 5G and beyond systems. IEEE Access **8**, 74605–74626 (2020)
47. F. Sohrabi, W. Yu, Hybrid digital and analog beamforming design for large-scale antenna arrays. IEEE J. Sel. Topics Signal Process. **10**(3), 501–513 (2016)
48. O. El Ayach, S. Rajagopal, S. Abu-Surra, Z. Pi, R.W. Heath, Spatially sparse precoding in millimeter wave MIMO systems. IEEE Trans. Wireless Commun. **13**(3), 1499–1513 (2014)
49. I. Ahmed, H. Khammari, A. Shahid, A. Musa, K.S. Kim, E. De Poorter, I. Moerman, A survey on hybrid beamforming techniques in 5G: architecture and system model perspectives. IEEE Commun. Surv. Tutor. **20**(4), 3060–3097 (2018)
50. J. Flordelis, F. Rusek, F. Tufvesson, E.G. Larsson, O. Edfors, Massive MIMO performance–TDD versus FDD: what do measurements say? IEEE Trans. Wireless Commun. **17**(4), 2247–2261 (2018)
51. Z. Zhang, K. Long, A.V. Vasilakos, L. Hanzo, Full-duplex wireless communications: challenges, solutions, and future research directions. Proc. IEEE **104**,(7), 1369–1409 (2016)
52. A. Sabharwal, P. Schniter, D. Guo, D.W. Bliss, S. Rangarajan, R. Wichman, In-band full-duplex wireless: challenges and opportunities. IEEE J. Sel. Areas Commun. **32**(9), 1637–1652 (2014)
53. D. Bharadia, E. McMilin, S. Katti, Full duplex radios, in *Proceedings of the ACM SIGCOMM 2013 Conference on SIGCOMM* (2013), pp. 375–386
54. M. Duarte, A. Sabharwal, V. Aggarwal, R. Jana, K.K. Ramakrishnan, C.W. Rice, N. Shankaranarayanan, Design and characterization of a full-duplex multiantenna system for WiFi networks. IEEE Trans. Veh. Technol. **63**(3), 1160–1177 (2013)
55. D. Kim, H. Lee, D. Hong, A survey of in-band full-duplex transmission: from the perspective of PHY and MAC layers. IEEE Commun. Surv. Tutor. **17**(4), 2017–2046 (2015)
56. K.E. Kolodziej, B.T. Perry, J.S. Herd, In-band full-duplex technology: techniques and systems survey. IEEE Trans. Microw. Theory Technol. **67**(7), 3025–3041 (2019)
57. F. Chen, H.H. Lee, R. Morawski, T. Le-Ngoc, RF/analog self-interference canceller for 2×2 MIMO full-duplex transceiver, in *2017 IEEE International Conference on Communications (ICC)* (IEEE, Piscataway, 2017), pp. 1–6
58. E. Everett, A. Sahai, A. Sabharwal, Passive self-interference suppression for full-duplex infrastructure nodes. IEEE Trans. Wireless Commun. **13**(2), 680–694 (2014)
59. A. Koc, T. Le-Ngoc, Intelligent non-orthogonal beamforming with large self-interference cancellation capability for full-duplex multiuser massive MIMO systems. IEEE Access **10**, 51771–51791 (2022)
60. Y. Gong, M. Mahmood, R. Morawski, T. Le-Ngoc, Dual-layer metamaterial rectangular antenna arrays for in-band full-duplex massive MIMO. IEEE Access **11**, 135708–135727 (2023)
61. Y. Gong, R. Morawski, T. Le-Ngoc, Metamaterial absorber structure for tx-rx antenna isolation improvement in full-duplex massive MIMO. *IEEE Access* **12**, 64571–64588 (2024)
62. T. Le-Ngoc, A. Masmoudi, *Full-Duplex Wireless Communications Systems*. Wireless Networks (Springer, Berlin, 2017)
63. G. Liu, F.R. Yu, H. Ji, V.C. Leung, X. Li, In-band full-duplex relaying: a survey, research issues and challenges. IEEE Commun. Surv. Tutor. **17**(2), 500–524 (2015)
64. I.P. Roberts, J.G. Andrews, H.B. Jain, S. Vishwanath, Millimeter-wave full duplex radios: new challenges and techniques. IEEE Wireless Commun. **28**(1), 36–43 (2021)

# Chapter 2
# Background

This chapter provides an overview of metamaterial-based design techniques for enhancing the performance of antenna elements and arrays. It focuses on antenna miniaturization, improved radiation characteristics, and enhanced isolation. It then examines and compares various beamforming architectures for mMIMO systems, including DBF, analog beamforming (ABF), and HBF schemes. To further boost the SINR in MU-mMIMO systems, the chapter reviews different NCB algorithms to suppress MUI. As array apertures grow, the discussion expands to recent research and future opportunities in near-field location-differential beamfocusing for XL-arrays in wireless communication. Finally, the chapter addresses SI isolation and cancellation approaches in FD systems. A multi-stage SI mitigation strategy is highlighted as an effective solution to enhance overall FD communication efficiency.

## 2.1 Metamaterial-Based Antenna Improvement in Array Design

### *2.1.1 Metamaterial for Electrically Small Antennas*

Metamaterials, composed of periodic sub-wavelength unit cells with tunable EM properties, have emerged as a transformative tool in antenna design. By modifying the layout and structure of these meta-cells, the effective dielectric permittivity and magnetic permeability of the material can be tailored, enabling the development of resonant structures that operate at frequencies lower than typically dictated by their physical dimensions. One prominent application of metamaterials is the realization of electrically small antennas. For instance, in [1], metamaterials are engineered to have negative or near-zero apparent permittivity and permeability and a metamaterial shell is designed to compress the EM wavelength around the antenna. Embedding this negative refractive-index metamaterial shell enhances the



T. Le-Ngoc et al., *Half-Duplex/Full-Duplex MU-mMIMO*, Wireless Networks,
https://doi.org/10.1007/978-3-032-09195-6_2

antenna structure radiation and resonance performance and makes it appear electrically larger without increasing its physical size. Additionally, artificial magnetic conductors (AMCs) can be designed to create surfaces that mimic perfect magnetic conductors over a specific frequency range. Integrating AMCs into antenna designs allows antennas to be placed very close to a ground plane without suffering from destructive interference caused by image currents. This proximity reduces the antenna's profile and overall size while maintaining radiation performance. In [2], low-profile microstrip antennas are integrated with EBG structures acting as AMCs, resulting in considerable antenna height reduction without degrading radiation patterns or impedance matching.

Loading antennas with metamaterial elements such as split-ring resonators (SRRs) introduces additional inductance or capacitance, effectively lowering the antenna's resonant frequency and allowing for size reduction without compromising radiation performance. In [3], the use of concentric SRR metamaterials enhances the effective permittivity of the medium, subsequently reducing the resonant frequency of the patch. By manipulating the near-field reactive components through metamaterial loading, these antennas can achieve resonance and radiation efficiency that surpasses the limitations imposed by the Chu-Harrington limit for small antennas. A metamaterial-inspired electrically small antenna using near-field resonant parasitic elements design has been proposed in [4]. These antennas effectively couple energy to the far-field while maintaining a small physical footprint, demonstrating that metamaterials can overcome traditional size-efficiency trade-offs.

While metamaterial-based techniques effectively facilitate antenna miniaturization, they often necessitate the integration of meta-cells into dielectric substrates or the incorporation of meta-shells around the antenna structure. These modifications increase manufacturing complexity and can sometimes lead to bulkier designs. To overcome these challenges, dual-layer slow-wave structures have been proposed as an alternative in patch antenna designs, offering enhanced manufacturing simplicity and adaptability in controlling the effective wavelength. The structures are engineered to reduce the phase velocity of EM waves within the antenna, thereby effectively increasing its electrical length without altering its physical dimensions [5, 6]. The slow-wave effect is achieved by manipulating wave coupling and reflection within periodic structures. Constructive interference between incident and reflected waves in the multi-layer structures reduces the phase velocity compared to its velocity in free space. This reduction in phase velocity shortens the effective wavelength, enabling a compact antenna design while maintaining desired resonant frequencies. Studies in [5–7] showcase miniaturized patch antenna designs that exploit slow-wave propagation within multi-layer metamaterial structures. The prototypes, fabricated using multi-layer printed circuit board (PCB) processes, demonstrate the ability to achieve compact, low-frequency antennas while preserving robust radiation performance. This highlights the practical utility of slow-wave structures in advancing modern antenna engineering.

### 2.1.2 Metamaterial Structure for Inter-Element Isolation Enhancement

Maintaining an inter-element spacing below half a wavelength is critical to prevent spatial aliasing in antenna arrays. When element spacing exceeds this threshold, coherent summation of the array patterns occurs in unintended directions, leading to grating lobes. These unwanted lobes degrade the desired main lobe gain and introduce strong interference in undesired directions [8, 9]. However, placing antenna elements within a limited separation can lead to high inter-element MC. This coupling stems from spatial and surface leakage waves propagating within or near the dielectric substrate, ultimately degrading radiation efficiency and distorting the antenna's radiation pattern [10].

To enhance inter-element isolation and mitigate MC, a variety of metamaterial-based designs have been investigated. One common strategy involves introducing metamaterial decoupling or FSS structures between antenna elements [2, 11, 12]. Among these, EBG structures have gained popularity due to their ease of fabrication and effectiveness. By exhibiting high surface impedance at resonant frequencies, EBG structures act as band-stop filters that suppress the surface leakage wave propagation between elements, which is a main source to the MC [13–19]. Moreover, Defected Ground Structures (DGS) offer another method for mitigating unwanted coupling [20, 21]. Etching specific patterns or defects into the ground plane beneath antennas disrupts surface currents, effectively filtering out coupling paths. This technique can substantially improve isolation without increasing the array's physical footprint, making it attractive for compact designs. Destructive interference can also be introduced through decoupling networks or neutralization lines, which incorporate specific circuit elements such as reactive components or transmission lines, to produce canceling fields that counteract the leakage waves, thereby suppressing MC [12, 22, 23]. However, the presence of a non-uniform ground plane may adversely affect the antenna radiation patterns. In addition, parasitic elements and metasurfaces can be strategically positioned to dissipate, absorb, or redirect coupling energy, thereby enhancing antenna performance and isolation [24, 25]. However, incorporating these parasitic structures can be challenging and may add excessive complexity, particularly in large-scale antenna arrays where space is severely limited.

Although incorporating structures like EBGs typically necessitates substantial spacing between antenna elements, multi-layer metamaterial designs help overcome these constraints by reducing the effective wavelength in the design. This approach enables the creation of electrically small patch antennas, preserving sufficient space for additional structures and allowing the isolation-enhancement elements themselves to be miniaturized, thereby reducing their overall footprint. By integrating multi-layer metamaterials with top-layer FSS, designers can further improve inter-element isolation and enhance overall radiation performance in large-scale arrays.

### 2.1.3 Metamaterial-Based Absorber for Isolation Improvement

The absorber technology presents a promising solution to enhance the isolation performance in communication systems without requiring modifications to the original large-scale antenna array configuration. Research in [26] demonstrates that adding RF absorbers can achieve more than 10 dB improvement in isolation by effectively blocking and absorbing line-of-sight (LoS) coupling waves. However, there is often a tradeoff between the thickness and the height of the conventional RF absorber and its absorption effectiveness. The large size of RF absorbers limits their widespread use in various applications.

First proposed in [27], metamaterial absorbers have gained more and more popularity in research due to their advantages of adjustable effective permittivity and permeability, low profile, easy fabrication, and low reflectivity. The metamaterial absorber structures typically consist of periodic top metallic patch circuits and ground metal planes separated by dielectric substrates. Therefore, the fabrication can take advantage of the PCB manufacturing process. The interaction between the homogenized structures creates the electric and magnetic response. The separation between the top conductor circuit and the ground plane can achieve an electric Drude response from the ground plane plus a Drude-Lorentz response from the top metallic circuit [28]. The magnetic response transfers the incident coupling waves to the surface current on the top-layer metallic circuit. Adjusting the geometry of the metallic circuits and the spacing of the two metallic structures changes the resonant frequency position and strength of a Lorentz resonance and the magnetic response.

When the effective relative permittivity ($\widetilde{\epsilon}_r$) and effective relative permeability ($\widetilde{\mu}_r$) of the homogenized structure are adjusted to be equal to each other, i.e., $\widetilde{\epsilon}_r/\widetilde{\mu}_r = 1$, the structure's input impedance closely matches the impedance of the air, and low reflection at the intersections of the air and the absorber surface can be realized. This design allows a significant portion of the coupling wave to be absorbed and dissipated within the dielectric substrate. Over 90% power absorption can usually be achieved [29–31], and over 8 dB extra isolation can be realized [16, 31]. A typical metamaterial is seen to have elements with sizes smaller than $0.2\lambda_0 \times 0.2\lambda_0$ and a thickness less than $0.04\lambda_0$. Therefore, metamaterial absorbers offer greater adaptability and flexibility due to their compact size, allowing seamless integration into various antenna systems without altering the overall system dimensions. With their sub-wavelength structure and minimal reflection interference to the original radiation waves, metamaterial absorbers have been widely adopted in antenna design to enhance isolation [16, 31–35]. In [31], a double-layer metamaterial absorber is employed in a four-element FD antenna array. Experimental findings indicate the absorber contributes to an SI suppression around 8 dB, resulting in a total isolation of 22 dB. Additionally, the study in [36] demonstrates a reduction of MC by 8 dB between sub-arrays in a large $3\times8$ array MIMO, achieved through the orthogonal placement of a metamaterial absorber between neighbouring sub-arrays. Table 2.1 summarizes and compares

**Table 2.1** Comparison of absorber structures in the literature

| Ref. | Half-/full-duplex | Operating frequency | Absorber type | Mounted position | Absorber element size length × width (×thickness) ($\lambda_0$) | Absorber isolation improvement | Total isolation level |
|---|---|---|---|---|---|---|---|
| [26] | FD | 2.45 GHz | Conventional ABS | Vertical | | 10.1 dB | 73.8 dB |
| [16] | HD | 2.4 GHz | Metamaterial | Vertical | 0.17×0.16×0.013 | 16 dB | 42 dB |
| [32] | HD | 3.4 GHz | Metamaterial | Horizontal | 0.12×0.25 | 27 dB | 52 dB |
| [33] | HD | 5.1 GHz | Metamaterial | Horizontal | 0.08×0.06 | 10 dB | 29 dB |
| [34] | HD | 5.1 GHz | Metamaterial | Horizontal | 0.20×0.20 | 10.5 dB | 33 dB |
| [31] | FD | 1.27 GHz | Metamaterial | Vertical | 0.19×0.19×0.03 | 8 dB | 22 dB |

recent absorber research and the achieved isolation improvement in the HD and FD communication systems.

## 2.2 Beamforming Architectures

To harness the benefits of antenna arrays in enhancing wireless communication performance, beamforming serves as a key technique in array control and signal processing. It ensures reliable transmission and reception quality in mMIMO systems. By dynamically adjusting the amplitude and phase of signals fed to the antenna elements, beamforming algorithms focus energy toward the desired user directions while suppressing radiation toward potential sources of interference. This process leverages constructive and destructive interference among transmitted signals and also ensures receiving signals are coherently combined to maximize performance metrics such as the SINR and achievable rate. By producing more concentrated beams, an increased array size offers greater beamforming gain, enhanced steerability, and higher beam resolution, thereby improving overall beamforming effectiveness [8, 37–39]. Distinguished by the domain in which signal processing occurs and the hardware implementation, beamforming architectures can be generally divided into three schemes as shown in Fig. 2.1

- fully digital beamforming (DBF);
- fully analog beamforming (ABF);
- two-stage hybrid beamforming (HBF).

The following sections delve into the advantages and limitations of each approach.

### 2.2.1 *Digital Beamforming*

DBF is a widely adopted technique in conventional MIMO systems with relatively small antenna arrays [40], providing extensive degrees of freedom for advanced signal processing methods that boost spectral efficiency and support multiple simul-

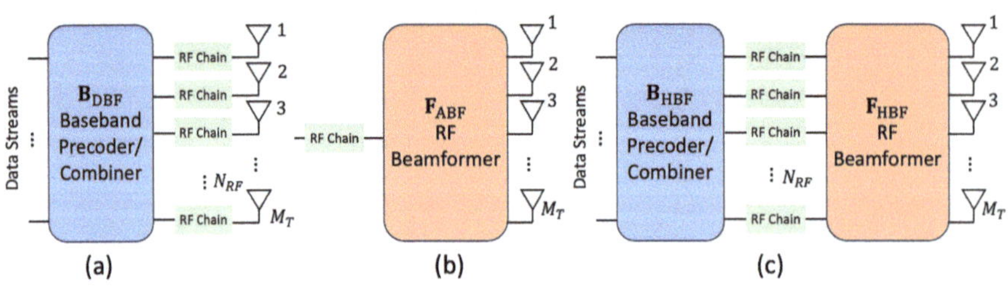


**Fig. 2.1** Beamforming scheme comparison (**a**) DBF, (**b**) ABF, (**c**) HBF

taneous data streams [41, 42]. Operating at the digital baseband level, DBF readily adapts to dynamic channel conditions and enhances overall system performance [43]. Manipulating the phase and amplitude of transmitted signals, DBF improves the system performance by effectively suppressing interference, such as inter-symbol interference (ISI) in point-to-point links to inter-user interference (IUI) or MUI in multi-user scenarios [41].

As illustrated in Fig. 2.1a, DBF employs a single-stage digital precoder or combiner at the baseband. This design offers substantial flexibility, as beamforming weights are implemented digitally for each antenna element. For an $M_T$-element array, DBF supports $K$ data streams (where $K \leq M_T$) via a beamforming matrix $\mathbf{B}_{\text{DBF}} \in \mathbb{C}^{M_T \times K}$. However, DBF typically mandates one RF chain per antenna element, thus $N_{RF} = M_T$. Each RF chain consists of power-hungry components such as analog-to-digital converters, up/down-conversion mixers, amplifiers, and filters, which together facilitate signal processing between the FD domain and digital baseband. As arrays scale to mMIMO, deploying a dedicated RF chain for every element becomes prohibitively expensive, power-intensive, and operationally complex [44, 45]. For instance, 3GPP Release 17 specifies up to 256 antennas at a BS, where implementing 256 RF chains is economically and practically infeasible.

Moreover, DBF's computational overhead escalates with the size of the beamforming matrix $\mathbf{B}_{\text{DBF}}$, complicating real-time optimization in large-scale systems [45]. While DBF can theoretically attain optimal spectral efficiency for mMIMO [46], the exponential growth in hardware and processing requirements motivates the exploration of alternative beamforming architectures. These approaches strive to achieve near-optimal performance with reduced complexity and cost, thereby ensuring scalability for next-generation communication networks [47, 48].

### 2.2.2 Analog Beamforming

Unlike DBF, ABF can maintain a fixed number of RF chains regardless of the number of antennas, providing a hardware- and power-efficient strategy for mMIMO systems. As depicted in Fig. 2.1b, a single RF chain is commonly employed to support $M_T$ antennas. The beamformer stage can rely on $M_T$ low-cost phase shifters, which manage constructive and destructive interference among signals transmitted or received across the antenna elements [49, 50].

Despite this advantage, ABF faces notable drawbacks. For example, under constant-modulus (CM) constraints, the analog beamforming vector $\mathbf{F}_{\text{ABF}}$ is restricted to phase-only adjustments, thereby significantly reducing the available degrees of freedom compared to DBF scheme. As a result, ABF often delivers lower beamforming performance compared to DBF [49, 50]. Furthermore, ABF generally supports only a single data stream, whereas DBF can simultaneously handle multiple streams. This restriction reduces spectral efficiency and partially offsets the benefits that mMIMO systems aim to provide.

### 2.2.3 Hybrid Beamforming

As discussed in the contexts of DBF and ABF, selecting an appropriate beamforming architecture entails balancing spectral efficiency, flexibility, hardware complexity, and computational overhead. While DBF can achieve high spectral efficiency and support multiple data streams, it requires a dedicated RF chain per antenna element. As array sizes grow, this approach becomes prohibitively expensive, power-intensive, and computationally demanding. In contrast, ABF utilizes a single RF chain for all antennas, thus reducing hardware costs and complexity. However, because ABF offers limited flexibility in controlling beamforming weights and can support only a single data stream, it severely reduces degrees of freedom and diminishes spectral efficiency.

To leverage the advantages of both DBF and ABF, a two-stage HBF architecture has emerged as a promising solution for mMIMO systems [43, 48, 51]. As illustrated in Fig. 2.1c, HBF divides the beamforming process into a combination of a digital baseband stage and an analog RF stage. The large-scale antenna array can be divided into $K$ sub-arrays to serve up to $K$ concurrent data streams. A reduced number of $N_{RF}$ RF chains connects two stages, which satisfy $K \leq N_{RF} \ll M_T$. The digital stage employs a reduced-size precoder/combiner $\mathbf{B}_{\text{HBF}} \in \mathbb{C}^{N_{\text{RF}} \times K}$, while the RF beamforming stage is represented as $\mathbf{F}_{\text{HBF}} \in \mathbb{C}^{M_T \times N_{\text{RF}}}$. By using fewer RF chains and still accommodating multiple data streams, HBF significantly lowers hardware costs, power consumption, and computational complexity while maintaining near-optimal spectral efficiency and offering considerable flexibility [52]. HBF architectures can be further classified into fully-connected and sub-connected configurations [51]. In the fully-connected design, each RF chain is connected to all $M_T$ antennas. Although this configuration offers greater beamforming flexibility and higher performance, it increases hardware costs compared to the sub-connected design, where each RF chain only serves a subset of antennas.

By jointly optimizing the RF and baseband stages, the HBF architecture enhances the signal quality and overall system performance by leveraging both the analog domain's capabilities for spatial beam control and the flexible, adaptive processing offered by the digital domain. Studies have shown that, with appropriate design, HBF can closely approximate the spectral efficiency of DBF while significantly outperforming ABF [43, 51]. The two-stage design also offers substantial flexibility, facilitating the development of low-complexity beamforming algorithms [53]. Various HBF approaches, both for precoding/combining in downlink/uplink directions, are examined for HD transmission in [52, 54–58] and FD transmission in [59–66]. The comparison between the three beamforming architectures is summarized in Table 2.2.

**Table 2.2** Comparison of beamforming architectures

| | DBF | ABF | HBF |
|---|---|---|---|
| Number of RF chains | $N_{RF} = M_T$ | 1 | $N_{RF} < M_T$ |
| Number of RF beamforming modules | – | $M_T$ | $M_T$ |
| Number of data streams | Multiple | One | Multiple |
| Spectral efficiency | High | Low | High |
| Energy efficiency | Low | High | High |
| Hardware complexity | High | Low | Low |

## 2.3 Beamforming and Beamfocusing for Multi-User mMIMO

The deployment of large-scale antenna arrays in mMIMO systems enables the formation of steerable, high-gain, and narrow-beamwidth directional beams. These focused radiation patterns improve channel conditions by leveraging spatial diversity, as users positioned at different locations experience distinct channel characteristics, leading to higher channel rank. The beamforming capability significantly enhances spectral efficiency and MU capacity by enabling the simultaneous directional transmission and reception of multiple data streams [67]. Dividing the array into sub-arrays allows the BS to communicate with multiple users in different directions concurrently, further improving overall throughput and spectral efficiency [42]. However, with the advancement of MU operations in mMIMO systems, MUI emerges as a critical challenge. Sharing the same frequency resources among multiple users introduces significant interference, which degrades the SIR. A lower SIR leads to higher bit error rates and reduces system throughput, emphasizing the need for effective interference management strategies. To address these challenges, appropriate beamforming techniques must be designed for transmissions, ensuring improved interference suppression and enhanced overall MU-mMIMO system performance.

### *2.3.1 Nulling Control Beamforming*

NCB plays a pivotal role in enhancing signal reception quality. By strategically adjusting the magnitudes and phase shifts of feedings within the antenna array, NCB acts as a spatial filter, maximizing the beam gain in the target user directions and minimizing the radiation power in the potential interference directions. Consequently, the array beam main lobe can be focused toward the intended signal's direction of arrival and radiation nulls are created concurrently to eliminate unwanted interference. This dual functionality not only maintains signal strength from the designated user but also efficiently suppresses interference from other users, significantly enhancing the efficacy of the MU-mMIMO system [68, 69].

Various NCB techniques are widely studied in different large-scale antenna arrays to mitigate the MUI. In general, popular beamforming algorithms can be divided into two main categories, which are the non-iterative scheme and the iterative optimization scheme.

For the non-iterative beamforming approach, the feeding weights can be derived from instant calculations based on the estimation of the inverse covariance matrix of the desired and interference channels. For instance, in [70], null steering in antenna arrays is approached as a sparse recovery problem, introducing a compressed-sensing method that perturbs minimal elements for efficient wide null generation. The simulation results illustrate an interference suppression improvement by 10 to 20 dB with a reduced hardware complexity. In [71], an NCB using the Minimum Variance Distortionless Response (MVDR) algorithm was explored. The method optimizes beam outputs via a second-order-cone programming approach, enhancing beam steering and null placement with rapid convergence. In [72], the study introduces a computationally efficient null widening technique through a covariance matrix taper (CMT)-based method, which simulates fictitious interference in snapshots to broaden sharp nulls. By adapting Mailloux's methodology with added random disturbances to the covariance matrix and mutual correlation vector, this approach significantly reduces computational demands as demonstrated by simulation results. In [73], the authors propose a flexible approach to CMT that allows for the creation of nulls with controllable widths and asymmetries, tailored to the specific spatial distribution of interference. This method optimizes the use of degrees of freedom by avoiding the need for uniformly wide nulls, as confirmed by simulations demonstrating the technique's effectiveness and efficiency. A dual Least-Mean-Square (LMS) algorithm, which interposes an array-image factor between two successive LMS filtering stages, is proposed to enhance convergence speed and interference suppression in [74]. Simulation results demonstrate the algorithm's enhanced robustness and faster convergence rate in scenarios with interference angle variation, as well as its stability amidst additive white Gaussian noise (AWGN) and Rayleigh fading. These improvements are evidenced by better EVM outcomes and scatter plots compared to previous LMS schemes. A novel NCB algorithm utilizing real-valued coefficients is introduced, leveraging a preprocessing transformation matrix within a ULA framework [75]. The approach, adaptable for either minimizing the mean square error (MSE) or maximizing the output SIR, varies the transformation-matrix key parameter based on the beamforming context. Other popular beamforming algorithms include MVDR [76–78], LCMV [79–81], and the recursive least squares constant modulus algorithm [82–84].

In non-iterative beamforming techniques, performance is critically dependent on the precision of the channel's statistical data, particularly the estimated steering vector (SV). However, real-world conditions introduce challenges like wavefront distortions, variations in ERPs, directional inaccuracies, and incoherent local scattering, all of which can cause notable discrepancies in the SV estimation. Additionally, collecting realistic ERPs presents significant challenges. The simulation and evaluations of the proposed algorithms [70–84] predominantly assume isotropic or uniform ERPs, where the variations of the ERPs across a large-scale antenna array are ignored. The theoretical isotropic SV encapsulates solely the phase differences induced by an incoming signal at the inputs of isotropic sources, attributable to their distinct spatial positions. Consequently, the non-iterative methods usually generate suboptimal beamforming vectors, characterized by insufficient null depths and imprecise null positioning.

The iterative beamforming scheme is a search and optimization strategy that continuously adjusts feeding weights while evaluating the resulting objective metrics such as nulling depth, SIR, or achievable data rate. By introducing strategic perturbations to the beamforming weights, the scheme progressively enhances overall beamforming performance. This approach often leverages techniques such as successive fast Fourier transformations (FFT) [85, 86] or sophisticated optimization algorithms like PSO [87–90] and genetic algorithms (GA) [88, 91, 92]. These techniques aim to refine the accuracy and efficiency of the tuning and search process within the predefined solution space. For example, a PSO-based beamforming algorithm for linear array synthesis, aimed at sidelobe suppression and strategic null placement, is proposed in [87]. The method optimizes array geometry alongside the magnitude and phase of element feedings, demonstrating successful beam pattern control. Similarly, [91] integrates GA with the Schelkunoff nulling method for efficient beam steering and null placement in circular antenna arrays with a minimal number of elements. Additionally, [92] presents a GA-based nulling steering algorithm that optimizes element positions to minimize power at null locations and maintain the main lobe pattern. The approach demonstrates enhanced accuracy in null placement and nulling depth, confirmed by simulations that align with theoretical predictions for effective interference suppression in wireless systems.

Recently, the application of intelligent neural networks (NN) for predicting beamforming vectors has gained popularity [93–96]. In [94], a beamforming approach is developed using a recurrent neural network (RNN) based on the gated recurrent unit architecture. This RNN is trained on an extensive dataset generated through an NCB technique, which takes into account the effects of MC within a microstrip linear antenna array. An improved method for solving wide nulling problems in adaptive digital beamforming arrays by combining the Bat algorithm with the general regression neural network to generate and estimate complex weights efficiently is proposed in [96]. Enhanced time efficiency and acceptable performance are observed through numerical experiments with a 32-element ULA. While NN implementations offer the advantage of immediate prediction of feeding weights, these approaches generally rely on either non-iterative or iterative search schemes for the initial generation of their extensive training sets. These sets, comprising feeding weights and the corresponding beam characteristics, necessitate thorough optimization simulations or measurements. Hence, designing an optimization scheme that incorporates realistic considerations of ERPs and ensures robust control over beam patterns remains the foundational step.

### 2.3.2 XL-Array Near-field and Far-Field

To enhance spatial resolution and meet the ever-growing demands for greater spectral efficiency and increased capacity, antenna arrays are being designed with increasingly larger apertures [97–99]. As the antenna aperture size increases, the Rayleigh distance ($r_{FF}$), which can be the boundary between the near-field and far-field operating regions, also increases quadratically, leading to a wider radiative near-field region characterized by a spherical wavefront [100, 101]. This boundary can be estimated by $r_{FF} = 2D^2/\lambda$, where $D$ is the antenna aperture size and $\lambda$ is the operating wavelength [100, 102]. Although spherical wavefronts present challenges in channel modelling and very large-scale antenna arrays increase the hardware complexity in the beamforming, near-field propagation offers appealing opportunities [97, 103, 104]. Compared to conventional far-field beamforming, which focuses on shaping array radiation patterns in only the angular domain, near-field beamfocusing adds an additional degree of freedom by shaping beam patterns in both angular and distance domains [105, 106]. As illustrated in [105], near-field Location Division Multiple Access (LDMA) can be realized by leveraging beamfocusing properties and employing a spherical-domain sampling method. This approach significantly enhances interference mitigation and spectral efficiency in future 6G extremely large antenna array (ELAA) systems, potentially contributing to a tenfold increase in spectral efficiency [107]. Consequently, many recent research efforts have focused on exploiting near-field ELAA processing for advanced wireless communication systems.

In [97], a polar-domain representation capturing both angular and distance information of users is introduced to estimate the near-field channel in ELAA systems. This approach exploits the channel's sparsity, overcoming the limitations of traditional angular-domain methods under near-field spherical-wavefront conditions. Simulation results confirm the accuracy of the proposed on-grid and off-grid near-field channel estimation schemes. To maximize the achievable sum rate, researchers in [103] investigate the potential of near-field beamfocusing in downlink MU-mMIMO systems by developing and analyzing various precoding schemes, including fully digital, hybrid phase shifter-based precoders, and dynamic metasurface antennas (DMAs). The designed focused beams exploit both angle and distance domains to mitigate interference, illustrating new degrees of freedom beyond conventional far-field beam steering and enabling reliable communications even for users located in the same angular direction. In [108], MU-mMIMO systems are modeled by considering both spherical wavefront phases and variations in signal amplitude and projected aperture across array elements. Various beamforming schemes, including maximal-ratio combining (MRC), zero-forcing (ZF), and minimum mean-square error (MMSE) beamforming, are developed. The achievable SINR reveals that inter-user interference can be suppressed through differential user angles and distances. Moreover, to maximize spectral efficiency and effectively mitigate the spatial wideband effect in the near-field, [109] proposes two wideband beamforming optimization approaches: the fully digital approximation and heuristic two-stage methods. Simulation results demonstrate that an ELAA employing the proposed sub-connected HBF architecture based on true-time delayers achieves high energy efficiency and reduced hardware requirements while serving multiple users. In [110], analytical expressions for the gain and beam depth are presented and validated to better estimate the separation between users and the gain variation in near-field beamfocusing. These expressions estimate how the beam gain varies with the antenna aperture size and the beamfocusing locations, providing valuable insights on the correlation between the achievable beamfocusing gain differences and user relative distances. Furthermore, [111] presents the impact of planar array antenna size, inter-element distance, and focal distance on near-field focusing performance. A multi-focus antenna array system capable of targeting multiple focal points within the near-field region is demonstrated.

## 2.4 Self-Interference Mitigation Techniques for Full-Duplex Communications

Traditional HD systems, such as TDD or FDD, inherently limit maximum data rates and spectral efficiency. In contrast, when Tx and Rx operate in FD mode, the system can simultaneously transmit and receive on the same frequency band, theoretically doubling the spectral efficiency [112–114]. Integrating mMIMO with FD is expected to substantially increase spectral efficiency and system capacity [115]. However, strong SI from the local Tx can severely degrade Rx performance, posing a major obstacle in the FD design. The SI power often greatly surpasses the received signals from intended users, as Rx antennas are much closer to the local Tx antennas than remote users. This discrepancy is attributed to the rapid growth of path loss with distance, with Rx antennas being situated much closer to the local Tx antennas compared to those of remote users. This imbalance leads to a significant reduction in SINR and ultimately constrains FD communication performance [116]. To realize the capacity gains promised by FD, the SI power must be reduced to or below the noise floor.

To illustrate the level of SI mitigation required in an FD system, let us consider an example with an ideal receiver noise figure of 0 dB and a noise power spectrum density (PSD) of − 174 dBm/Hz [117]. Over a transmission bandwidth of 20 MHz, the corresponding noise power can be estimated as

$$\sigma_n^2 = -174\,\text{dBm} + 10\log_{10}(20 \times 10^6\,\text{Hz}) = -101\,\text{dBm}. \tag{2.1}$$

In FD operation, co-located Tx and Rx antennas result in strong in-band SI at the receiver due to high leakage coupling transmit power. To achieve detection performance comparable to HD systems, the residual SI must ideally be suppressed below the thermal noise floor. The required SI suppression level is determined by the difference between the transmit power $P_T$ and the noise power $\sigma_n^2$. For example

[118], assuming a transmit power of $P_T = 30\,\text{dBm}$, the required SI suppression for a 20 MHz bandwidth can be calculated as

$$\text{Required SI Mitigation} \geq P_T - \sigma_n^2 = 30\,\text{dBm} - (-101\,\text{dBm}) = 131\,\text{dB}. \quad (2.2)$$

Achieving such a substantial level of SI suppression presents a significant technical challenge in FD mMIMO system design. Relying solely on single-stage isolation or cancellation techniques is either practically infeasible or results in prohibitive costs [119]. To maintain the FD performance under such conditions, advanced joint Tx-Rx isolation and SIC techniques are pivotal for mitigating the SI strength to near or below the noise floor [113].

### *2.4.1 Antenna Design for Isolation Improvement*

Tx-Rx isolation refers to a system's ability to effectively separate the transmit and receive signals. It is achieved by preventing and minimizing RF leakage between the local Tx and Rx, thereby ensuring that the receiver's low noise amplifier (LNA) and analog-to-digital converter (ADC) remain within their linear range. To enable optimal FD communications, numerous antenna isolation techniques can be implemented. The most direct method is increasing the separation distance between the Tx and the local Rx antennas to increase the path loss. Experiments show that a 65 dB to 85 dB attenuation can be reached with a separation distance of 3.5m to 5m [120, 121]. However, the compactness requirements of the overall system often impose limitations on the permissible separation distance between antennas in practical FD design. Furthermore, cross-polarization increases the polarization mismatch between the Tx and the local Rx antenna while decreasing the strength of leakage MC waves being captured by its local Rx antenna. It is demonstrated that orthogonal polarization can increase isolation by 10 to 18 dB [122, 123].

In [124], a proposed 2×2 cross-polarized FD prototype achieving a Tx-Rx isolation exceeding 45 dB under various test scenarios is demonstrated. Furthermore, by integrating Tx and Rx cross-polarization with both antenna port spacing and grounding via shorting pins techniques, an isolation level of 52 dB can be realized, as illustrated in [125]. Additionally, [126, 127] present two 4-element FD antenna designs equipped with feeding phase control, designed for out-of-phase near-field coupling wave cancellation. Experimental measurement outcomes have reported a Tx-to-Rx isolation more than 40 dB. However, the addition of signal cancellation mechanisms within the feeding networks can potentially lead to excessive implementation complexity in the large-scale antenna array design.

Moreover, the integration of various structures, such as FSS, DGS, and absorber structures, can effectively diminish, cancel out, or dissipate the coupling waves [6, 16, 18, 128, 129]. Nevertheless, these structures may impact the design's adaptability and the original beam pattern. It has been demonstrated that the RF isolation stage plays an important role in FD designs with more than 60 dB isolation achievable through the combination of various strategies [26].

However, most antenna isolation designs focus on a very limited number of antennas because of the challenges associated with simulation, manufacturing, and measurement [16, 18, 26, 122, 124, 126–129]. The experimental studies of Tx-to-Rx elements/sub-arrays between large-scale antenna arrays are still missing. Therefore, it is still essential to examine the antenna isolation approaches in large-scale array prototypes in order to discover potential opportunities, and advance the research and development of FD mMIMO systems in practice.

### 2.4.2 RF Beamforming Isolation

The employment of large-scale arrays introduces additional degrees of freedom for beam-level isolation design, enabling improved spatial isolation through narrow, steerable beams. By focusing signal energy within confined scattering areas, the strength of leakage waves in undesirable directions can be significantly reduced. Additionally, by steering and changing the interactions between the Tx and Rx beams, the beamformer design can play an essential role in the FD design. Beamforming isolation electronically steers beams toward the radiation nulls of interfering antennas, effectively reducing SI intensity [63, 130]. A well-designed combination of antenna and beamforming isolation ensures residual SI at the Rx RF chain remains sufficiently low, avoiding front-end ADC or LNA overloading or saturation.

To harness the advantages of large-scale arrays while managing hardware and computational complexity, the two-stage HBF design has gained popularity to enhance the Tx-Rx isolation in recent FD research [50, 59–66]. HBF splits the beamforming architecture into two segments: RF Tx/Rx beamforming and digital baseband precoding/combining. Joint uplink and downlink beamformer designs in FD systems optimize Tx/Rx spatial isolation, leveraging both stages to maximize system performance. A hybrid precoding/combining (HPC) technique for mmWave FD mMIMO systems is proposed in [59] for FD mMIMO, which uses the angle-of-departure (AoD) and angle-of-arrival (AoA) information to suppress SI while also reducing the number of RF chains. The authors show that the power of the far-field (near-field) SI channel is canceled by 81.5 dB (44.5 dB) in point-to-point FD mMIMO systems. In [60], separate and joint FD beamforming algorithms based on sequential convex programming are presented to maximize the sum rate for both single-user (SU-) or MU-MIMO FD systems. In [61], an FD mmWave relaying system is considered to suppress SI while maximizing spectral efficiency using an orthogonal matching pursuit-based SIC algorithm. Similarly, an HPC design for FD amplify-and-forward (AF) mmWave relay systems is considered in [62], introducing a zero-space SIC method based on correlated mmWave channel estimation errors. A FD mmWave HBF scheme is introduced in [63] for a MU-mMIMO system, where the non-orthogonal beams are generated to serve multiple users while suppressing the near-field component of the SI channel and maximizing the sum-rate capacity. The authors show that the proposed HBF scheme can achieve 78.1 dB of SIC in FD mMIMO systems. In [64], the authors proposed an iterative FD HBF scheme to mitigate SI by jointly designing the transmit and receive RF beamformer weights, the precoder, and combiner matrices and achieving SI suppression by up to 30 dB. The authors in [65] designed the analog and digital beamformers of FD mMIMO systems based on the practical constraints of RF chains, channel knowledge, beam alignment, and a limited receive dynamic range. A two-timescale HBF scheme for FD mmWave multiple-relay transmission is investigated in [66], where the analog and digital beams are updated based on channel samples and real-time low-dimensional effective Channel State Information (CSI) matrices, respectively. By employing the instantaneous SI channel knowledge during the HBF design, the FD MU-mMIMO systems are designed to simultaneously support a single downlink UE and a single uplink UE over the same frequency band [131], and multiple downlink/uplink UEs [132, 133], where both employ the instantaneous SI channel knowledge during the HBF design are proposed.

### 2.4.3 Baseband SI-Cancellation

The residual SI occurs when the transmitted signal leaks into the receive path of the same device or system. SIC techniques aim to suppress or cancel out this interference, allowing the device or system to effectively transmit and receive signals simultaneously. These techniques involve sophisticated algorithms and signal processing methods, which aim to estimate the residual SI for reconstructing its counterpart and canceling it from the received signal [113, 134]. To reconstruct the residual SI signal, the baseband sampled Tx signals before the Tx RF chains can be processed by the set of adaptive finite impulse response (FIR) filters and the summed/superimposed outputs are subtracted from the corresponding Rx signals. Most commonly used structures for digital baseband-SIC are based on the transversal symbol-spaced (SS) FIR filter, whose tap-delay is set to the signal's symbol interval, as shown in studies [135] and [136] for FD single-input single-output (SISO) and studies [137–139] for FD MIMO. By implementing different design algorithms, such as least-square, MMSE, and maximum-likelihood, a set of FIR filters with tap coefficients and digitally controlled delay elements can be used to estimate the SI channel [113, 134, 137, 138, 140]. A 16-tap baseband canceller in [135] further suppressed a 400 MHz bandwidth SI by around 50 dB after an RF-SIC stage. The work in [136] implemented an SIC stage directly after the antenna isolation to reduce the SI by around 30 dB for a low-power or short-range, single-antenna FD communication system. In [139], a baseband delay-based FIR filtering SI canceller for $2\times2$ FD MIMO has been implemented and up to 35/45 dB average SIC across 40 and 20 MHz bandwidth has been achieved.

## 2.5 Concluding Remarks

In this chapter, a background and literature review of metamaterial-based structures for improving antenna design has been presented. The potentials of metamaterials in antenna miniaturization and isolation improvement are highlighted, and they will be further investigated in our metamaterial-based antenna array prototype development in Chap. 3.

Various beamforming architectures for mMIMO systems are reviewed and compared in terms of spectral efficiency, as well as hardware and computational complexity. Due to its favorable trade-off among spectral efficiency, available degrees of freedom, and moderate hardware and computational complexity, the HBF scheme is adopted as the primary beamforming strategy in this thesis. Later on, the existing research on interference mitigation for MU-mMIMO is reviewed, including far-field angular-based beamforming techniques and near-field location-based beamforming technologies, which will be further illustrated in Part I.

Finally, to further improve spectrum efficiency, the motivation and potential techniques for SI mitigation in FD mMIMO are presented. Various isolation and cancellation technologies, along with their achievable performance, are discussed. Detailed studies for each SI suppression stage will be presented in Part II.

## References

1. R.W. Ziolkowski, A.D. Kipple, Application of double negative materials to increase the power radiated by electrically small antennas. IEEE Trans. Antennas Propag. **51**(10), 2626–2640 (2003)
2. F. Yang, Y. Rahmat-Samii, Microstrip antennas integrated with electromagnetic band-gap (EBG) structures: a low mutual coupling design for array applications. IEEE Trans. Antennas Propag. **51**(10), 2936–2946 (2003)
3. R.O. Ouedraogo, E.J. Rothwell, A.R. Diaz, K. Fuchi, A. Temme, Miniaturization of patch antennas using a metamaterial-inspired technique. IEEE Trans. Antennas Propag. **60**(5), 2175–2182 (2012)
4. A. Erentok, R.W. Ziolkowski, Metamaterial-inspired efficient electrically small antennas. IEEE Trans. Antennas Propag. **56**(3), 691–707 (2008)
5. E. Korkontzila, D. Papafilippou, D. Chrissoulidis, Miniaturization of microstrip patch antenna for wireless applications by use of multilayered electromagnetic band gap substrate, in *2006 First European Conference on Antennas and Propagation* (IEEE, Piscataway, 2006), pp. 1–6
6. S. Ghosh, T.N. Tran, T. Le-Ngoc, Dual-layer EBG-based miniaturized multi-element antenna for MIMO systems. IEEE Trans. Antennas Propag. **62**(8), 3985–3997 (2014)
7. Y. Gong, R. Morawski, H.H. Lee, T. Le-Ngoc, A miniaturized 8x8 dual-layer EBG slotted circularly polarized patch antenna array for mMIMO, in *GLOBECOM 2022 - 2022 IEEE Global Communications Conference* (2022), pp. 6511–6516
8. C.A. Balanis, *Antenna Theory: Analysis and Design* (Wiley, Hoboken, 2015)
9. R.J. Mailloux, *Phased Array Antenna Handbook* (Artech House, Norwood, 2017)

10. D.M. Pozar, The active element pattern. IEEE Trans. Antennas Propag. **42**(8), 1176–1178 (1994)
11. L.K. Lin, C.C. Yu, Y.T. Chang, M.H. Huang, A novel electromagnetic bandgap (EBG) structure and its application for mutual coupling reduction, in *IEICE Proceedings Series*, vol. 35, no. 3B07-2 (2008)
12. K.L. Wu, C. Wei, X. Mei, Z.Y. Zhang, Array-antenna decoupling surface. IEEE Trans. Antennas Propag. **65**(12), 6728–6738 (2017)
13. F. Yang, Y. Rahmat-Samii, Mutual coupling reduction of microstrip antennas using electromagnetic band-gap structure, in *IEEE Antennas and Propagation Society International Symposium. 2001 Digest. Held in Conjunction With: USNC/URSI National Radio Science Meeting (Cat. No. 01CH37229)*, vol. 2 (IEEE, Piscataway, 2001), pp. 478–481
14. J.Y. Lee, S.H. Kim, J.H. Jang, Reduction of mutual coupling in planar multiple antenna by using 1-D EBG and SRR structures. IEEE Trans. Antennas Propag. **63**(9), 4194–4198 (2015)
15. E. Rajo-Iglesias, O. Quevedo-Teruel, L. Inclan-Sanchez, Mutual coupling reduction in patch antenna arrays by using a planar EBG structure and a multilayer dielectric substrate. IEEE Trans. Antennas Propag. **56**(6), 1648–1655 (2008)
16. G. Zhai, Z.N. Chen, X. Qing, Enhanced isolation of a closely spaced four-element MIMO antenna system using metamaterial mushroom. IEEE Trans. Antennas Propag. **63**(8), 3362–3370 (2015)
17. Z. Iluz, R. Shavit, R. Bauer, Microstrip antenna phased array with electromagnetic bandgap substrate. IEEE Trans. Antennas Propag. **52**(6), 1446–1453 (2004)
18. H.S. Farahani, M. Veysi, M. Kamyab, A. Tadjalli, Mutual coupling reduction in patch antenna arrays using a UC-EBG superstrate. IEEE Antennas Wireless Propag. Lett. **9**, 57–59 (2010)
19. K. Payandehjoo, R. Abhari, Employing EBG structures in multiantenna systems for improving isolation and diversity gain. IEEE Antennas Wireless Propag. Lett. **8**, 1162–1165 (2009)
20. K. Wei, J.Y. Li, L. Wang, Z.J. Xing, R. Xu, Mutual coupling reduction by novel fractal defected ground structure bandgap filter. IEEE Trans. Antennas Propag. **64**(10), 4328–4335 (2016)
21. J. OuYang, F. Yang, Z. Wang, Reducing mutual coupling of closely spaced microstrip MIMO antennas for WLAN application. IEEE Antennas Wireless Propag. Lett. **10**, 310–313 (2011)
22. S. Zhang, P. Zetterberg, S. He, Printed MIMO antenna system of four closely-spaced elements with large bandwidth and high isolation. Electron. Lett. **46**(15), 1052–1053 (2010)
23. M. Li, L. Jiang, K.L. Yeung, A general and systematic method to design neutralization lines for isolation enhancement in MIMO antenna arrays. IEEE Trans. Veh. Technol. **69**(6), 6242–6253 (2020)
24. S. Roy, U. Chakraborty, Mutual coupling reduction in a multi-band MIMO antenna using meta-inspired decoupling network. Wireless Pers. Commun. **114**(4), 3231–3246 (2020)
25. Y.F. Cheng, X. Ding, W. Shao, B.Z. Wang, Reduction of mutual coupling between patch antennas using a polarization-conversion isolator. IEEE Antennas Wireless Propag. Lett. **16**, 1257–1260 (2016)
26. E. Everett, A. Sahai, A. Sabharwal, Passive self-interference suppression for full-duplex infrastructure nodes. IEEE Trans. Wireless Commun. **13**(2), 680–694 (2014)
27. N.I. Landy, S. Sajuyigbe, J.J. Mock, D.R. Smith, W.J. Padilla, Perfect metamaterial absorber. Phys. Rev. Lett. **100**(20), 207402 (2008)
28. C.M. Watts, X. Liu, W.J. Padilla, Metamaterial electromagnetic wave absorbers. Adv. Mater. **24**(23), OP98–OP120 (2012)
29. F. Tofigh, M. Amiri, N. Shariati, J. Lipman, M. Abolhasan, Polarization-insensitive metamaterial absorber for crowd estimation based on electromagnetic energy measurements. IEEE Trans. Antennas Propag. **68**(3), 1458–1467 (2019)
30. A. Jafargholi, A. Jafargholi, J.H. Choi, Mutual coupling reduction in an array of patch antennas using CLL metamaterial superstrate for MIMO applications. IEEE Trans. Antennas Propag. **67**(1), 179–189 (2018)

31. J. Zhang, J. Li, J. Chen, Mutual coupling reduction of a circularly polarized four-element antenna array using metamaterial absorber for unmanned vehicles. IEEE Access **7**, 57469–57475 (2019)
32. Z. Qamar, U. Naeem, S.A. Khan, M. Chongcheawchamnan, M.F. Shafique, Mutual coupling reduction for high-performance densely packed patch antenna arrays on finite substrate. IEEE Trans. Antennas Propag. **64**(5), 1653–1660 (2016)
33. Q.L. Zhang, Y.T. Jin, J.Q. Feng, X. Lv, L.M. Si, Mutual coupling reduction of microstrip antenna array using metamaterial absorber, in *2015 IEEE MTT-S International Microwave Workshop Series on Advanced Materials and Processes for RF and THz Applications (IMWS-AMP)* (IEEE, Piscataway, 2015), pp. 1–3
34. M. Agarwal, M.K. Meshram, Isolation improvement of 5 GHz WLAN antenna array using metamaterial absorber, in *2016 URSI Asia-Pacific Radio Science Conference (URSI AP-RASC)* (IEEE, Piscataway, 2016), pp. 1050–1053
35. Y. Gong, R. Morawski, T. Le-Ngoc, Metamaterial absorber structure for Tx-Rx antenna isolation improvement in full-duplex massive MIMO. IEEE Access **12**, 64571–64588 (2024)
36. D.H.T. Hoang, V.X.T. Nguyen, T.T. Do, D.N. Nguyen, T.A. Vu, Q.N. Tran, Mutual coupling reduction of MIMO antenna using metamaterial absorber for 5G base stations, in *2022 Asia-Pacific Microwave Conference (APMC)* (IEEE, Piscataway, 2022), pp. 767–769
37. S. Kutty, D. Sen, Beamforming for millimeter wave communications: an inclusive survey. IEEE Commun. Surv. Tutor. **18**(2), 949–973 (2015)
38. W. Hong, K.H. Baek, Y. Lee, Y. Kim, S.T. Ko, Study and prototyping of practically large-scale mmwave antenna systems for 5G cellular devices. IEEE Commun. Mag. **52**(9), 63–69 (2014)
39. S.A. Busari, K.M.S. Huq, S. Mumtaz, L. Dai, J. Rodriguez, Millimeter-wave massive MIMO communication for future wireless systems: a survey. IEEE Commun. Surv. Tutor. **20**(2), 836–869 (2017)
40. A. Goldsmith, *Wireless Communications* (Cambridge University Press, Piscataway, 2005)
41. R.W. Heath, N. Gonzalez-Prelcic, S. Rangan, W. Roh, A.M. Sayeed, An overview of signal processing techniques for millimeter wave MIMO systems. IEEE J. Sel. Topics Signal Process. **10**(3), 436–453 (2016)
42. E.G. Larsson, O. Edfors, F. Tufvesson, T.L. Marzetta, Massive MIMO for next generation wireless systems. IEEE Commun. Mag. **52**(2), 186–195 (2014)
43. A.F. Molisch, V.V. Ratnam, S. Han, Z. Li, S.L.H. Nguyen, L. Li, K. Haneda, Hybrid beamforming for massive MIMO: a survey. IEEE Commun. Mag. **55**(9), 134–141 (2017)
44. A. Alkhateeb, G. Leus, R.W. Heath, Limited feedback hybrid precoding for multi-user millimeter wave systems. IEEE Trans. Wireless Commun. **14**(11), 6481–6494 (2015)
45. F. Sohrabi, W. Yu, Hybrid digital and analog beamforming design for large-scale antenna arrays. IEEE J. Sel. Topics Signal Process. **10**(3), 501–513 (2016)
46. H.Q. Ngo, E.G. Larsson, T.L. Marzetta, Energy and spectral efficiency of very large multiuser MIMO systems. IEEE Trans. Commun. **61**(4), 1436–1449 (2013)
47. T.E. Bogale, L.B. Le, Massive MIMO and mmWave for 5G wireless HetNet: potential benefits and challenges. IEEE Veh. Technol. Mag. **11**(1), 64–75 (2016)
48. X. Gao, L. Dai, S. Han, I. Chih-Lin, R.W. Heath, Energy-efficient hybrid analog and digital precoding for mmWave MIMO systems with large antenna arrays. IEEE J. Sel. Areas Commun. **34**(4), 998–1009 (2016)
49. S.S. Ioushua, Y.C. Eldar, A family of hybrid analog–digital beamforming methods for massive MIMO systems. IEEE Trans. Signal Process. **67**(12), 3243–3257 (2019)
50. M. Rihan, T.A. Soliman, C. Xu, L. Huang, M.I. Dessouky, Taxonomy and performance evaluation of hybrid beamforming for 5G and beyond systems. IEEE Access **8**, 74605–74626 (2020)
51. I. Ahmed, H. Khammari, A. Shahid, A. Musa, K.S. Kim, E. De Poorter, I. Moerman, A survey on hybrid beamforming techniques in 5G: architecture and system model perspectives. IEEE Commun. Surv. Tutor. **20**(4), 3060–3097 (2018)

52. X. Wu, D. Liu, F. Yin, Hybrid beamforming for multi-user massive MIMO systems. IEEE Trans. Commun. **66**(9), 3879–3891 (2018)
53. L. Liang, W. Xu, X. Dong, Low-complexity hybrid precoding in massive multiuser MIMO systems. IEEE Commun. Lett. **3**(6), 653–656 (2014)
54. M. Mahmood, A. Koc, T. Le-Ngoc, Energy-efficient MU-massive-MIMO hybrid precoder design: low-resolution phase shifters and digital-to-analog converters for 2D antenna array structures. IEEE Open J. Commun. Soc. **2**, 1842–1861 (2021)
55. D. Zhang, Y. Wang, X. Li, W. Xiang, Hybridly connected structure for hybrid beamforming in mmwave massive MIMO systems. IEEE Trans. Commun. **66**(2), 662–674 (2017)
56. M. Mahmood, A. Koc, T. Le-Ngoc, 3-D antenna array structures for millimeter wave multi-user massive MIMO hybrid precoder design: a performance comparison. IEEE Commun. Lett. **26**(6), 1393–1397 (2022)
57. A. Koc, A. Masmoudi, T. Le-Ngoc, 3D angular-based hybrid precoding and user grouping for uniform rectangular arrays in massive MU-MIMO systems. IEEE Access **8**, 84689–84712 (2020)
58. M. Mahmood, A. Koc, T. Le-Ngoc, 2D antenna array structures for hybrid massive MIMO precoding, in *GLOBECOM 2020 - 2020 IEEE Global Communications Conference* (IEEE, 2020), pp. 1–6
59. A. Koc, T. Le-Ngoc, Full-duplex mmWave massive MIMO systems: a joint hybrid precoding/combining and self-interference cancellation design. IEEE Open J. Commun. Soc. **2**, 754–774 (2021)
60. S. Huberman, T. Le-Ngoc, MIMO full-duplex precoding: a joint beamforming and self-interference cancellation structure. IEEE Trans. Wireless Commun. **14**(4), 2205–2217 (2014)
61. Y. Zhang, M. Xiao, S. Han, M. Skoglund, W. Meng, On precoding and energy efficiency of full-duplex millimeter-wave relays. IEEE Trans. Wireless Commun. **18**(3), 1943–1956 (2019)
62. Z. Luo, L. Zhao, L. Tonghui, H. Liu, R. Zhang, Robust hybrid precoding/combining designs for full-duplex millimeter wave relay systems. IEEE Trans. Veh. Technol. **70**(9), 9577–9582 (2021)
63. A. Koc, T. Le-Ngoc, Intelligent non-orthogonal beamforming with large self-interference cancellation capability for full-duplex multiuser massive MIMO systems. IEEE Access **10**, 51771–51791 (2022)
64. K. Satyanarayana, M. El-Hajjar, P.H. Kuo, A. Mourad, L. Hanzo, Hybrid beamforming design for full-duplex millimeter wave communication. IEEE Trans. Veh. Technol. **68**(2), 1394–1404 (2018)
65. I.P. Roberts, J.G. Andrews, S. Vishwanath, Hybrid beamforming for millimeter wave full-duplex under limited receive dynamic range. IEEE Trans. Wireless Commun. **20**(12), 7758–7772 (2021)
66. Y. Cai, K. Xu, A. Liu, M. Zhao, B. Champagne, L. Hanzo, Two-timescale hybrid analog-digital beamforming for mmwave full-duplex MIMO multiple-relay aided systems. IEEE J. Sel. Areas Commun. **38**(9), 2086–2103 (2020)
67. F. Rusek, D. Persson, B.K. Lau, E.G. Larsson, T.L. Marzetta, O. Edfors, F. Tufvesson, Scaling up MIMO: opportunities and challenges with very large arrays. IEEE Signal Process. Mag. **30**(1), 40–60 (2012)
68. I.P. Gravas, Z.D. Zaharis, T.V. Yioultsis, P.I. Lazaridis, T.D. Xenos, Adaptive beamforming with sidelobe suppression by placing extra radiation pattern nulls. IEEE Trans. Antennas Propag. **67**(6), 3853–3862 (2019)
69. Z.D. Zaharis, I.P. Gravas, P.I. Lazaridis, T.V. Yioultsis, T.D. Xenos, Improved beamforming in 3D space applied to realistic planar antenna arrays by using the embedded element patterns. IEEE Trans. Veh. Technol. **71**(6), 6145–6157 (2022)
70. M. Khosravi, M. Fakharzadeh, M.H. Bastani, Large array null steering using compressed sensing. IEEE Signal Process. Lett. **23**(8), 1032–1036 (2016)
71. R. Qian, M. Sellathurai, D. Wilcox, A study on MVDR beamforming applied to an ESPAR antenna. IEEE Signal Process. Lett. **22**(1), 67–70 (2014)

72. Z. Liu, S. Zhao, G. Zhang, B. Jiao, Robust adaptive beamforming for sidelobe canceller with null widening. IEEE Sens. J. **19**(23), 11213–11220 (2019)
73. Z. Liu, S. Zhao, C. Zhang, G. Zhang, Flexible robust adaptive beamforming method with null widening. IEEE Sens. J. **21**(9), 10579–10586 (2021)
74. J.A. Srar, K.S. Chung, A. Mansour, Adaptive array beamforming using a combined LMS-LMS algorithm. IEEE Trans. Antennas Propag. **58**(11), 3545–3557 (2010)
75. L. Zhang, W. Liu, R.J. Langley, Adaptive beamforming with real-valued coefficients based on uniform linear arrays. IEEE Trans. Antennas Propag. **59**(3), 1047–1053 (2010)
76. J. Capon, High-resolution frequency-wavenumber spectrum analysis. Proc. IEEE **57**(8), 1408–1418 (1969)
77. L. Yang, M.R. McKay, R. Couillet, High-dimensional MVDR beamforming: optimized solutions based on spiked random matrix models. IEEE Trans. Signal Process. **66**(7), 1933–1947 (2018)
78. X. Wang, M. Amin, X. Cao, Analysis and design of optimum sparse array configurations for adaptive beamforming. IEEE Trans. Signal Process. **66**(2), 340–351 (2017)
79. O.L. Frost, An algorithm for linearly constrained adaptive array processing. Proc. IEEE **60**(8), 926–935 (1972)
80. B.R. Breed, J. Strauss, A short proof of the equivalence of LCMV and GSC beamforming. IEEE Signal Process. Lett. **9**(6), 168–169 (2002)
81. J. Xiong, H. Hong, H. Zhang, N. Wang, H. Chu, X. Zhu, Multitarget respiration detection with adaptive digital beamforming technique based on SIMO radar. IEEE Trans. Microw. Theory Technol. **68**(11), 4814–4824 (2020)
82. Y. Chen, T. Le-Ngoc, B. Champagne, C. Xu, Recursive least squares constant modulus algorithm for blind adaptive array. IEEE Trans. Signal Process. **52**(5), 1452–1456 (2004)
83. R.C. de Lamare, M. Haardt, R. Sampaio-Neto, Blind adaptive constrained reduced-rank parameter estimation based on constant modulus design for CDMA interference suppression. IEEE Trans. Signal Process. **56**(6), 2470–2482 (2008)
84. Y. Cai, R.C. de Lamare, Low-complexity variable step-size mechanism for code-constrained constant modulus stochastic gradient algorithms applied to CDMA interference suppression. IEEE Trans. Signal Process. **57**(1), 313–323 (2008)
85. G. Mahanti, V.B. Bikkani, S. Mandal, Placement of wide nulls in the radiation pattern of a linear array antenna using iterative fast Fourier transform, in *2012 IEEE International Conference on Signal Processing, Communication and Computing (ICSPCC 2012)* (IEEE, Piscataway, 2012), pp. 552–555
86. W.P. Keizer, Fast low-sidelobe synthesis for large planar array antennas utilizing successive fast Fourier transforms of the array factor. IEEE Trans. Antennas Propag. **55**(3), 715–722 (2007)
87. M.M. Khodier, C.G. Christodoulou, Linear array geometry synthesis with minimum sidelobe level and null control using particle swarm optimization. IEEE Trans. Antennas Propag. **53**(8), 2674–2679 (2005)
88. S.K. Goudos, V. Moysiadou, T. Samaras, K. Siakavara, J.N. Sahalos, Application of a comprehensive learning particle swarm optimizer to unequally spaced linear array synthesis with sidelobe level suppression and null control. IEEE Antennas Wireless Propag. Lett. **9**, 125–129 (2010)
89. Y. Liao, J. Wang, Q.H. Liu, Transmit beampattern synthesis for frequency diverse array with particle swarm frequency offset optimization. IEEE Trans. Antennas Propag. **69**(2), 892–901 (2020)
90. Y. Gong, A. Yaseen, R. Morawski, T. Le-Ngoc, Perturbation-based adaptive beamforming for MU-mMIMO, in *2023 IEEE 98th Vehicular Technology Conference (VTC2023-Fall)* (IEEE, Piscataway, 2023), pp. 1–5
91. A. Hamza, H. Attia, Fast beam steering and null placement in an adaptive circular antenna array. IEEE Antennas Wireless Propag. Lett. **19**(9), 1561–1565 (2020)

92. M.R.R. Khan, V. Tuzlukov, Null steering beamforming for wireless communication system using genetic algorithm, in *2011 IEEE International Conference on Microwave Technology & Computational Electromagnetics* (IEEE, Piscataway, 2011), pp. 289–292
93. A.E. Zooghby, C. Christodoulou, M. Georgiopoulos, Neural network-based adaptive beamforming for one-and two-dimensional antenna arrays. IEEE Trans. Antennas Propag. **46**(12), 1891–1893 (1998)
94. I. Mallioras, Z.D. Zaharis, P.I. Lazaridis, S. Pantelopoulos, A novel realistic approach of adaptive beamforming based on deep neural networks. IEEE Trans. Antennas Propag. **70**(10), 8833–8848 (2022)
95. Z.D. Zaharis, C. Skeberis, T.D. Xenos, P.I. Lazaridis, J. Cosmas, Design of a novel antenna array beamformer using neural networks trained by modified adaptive dispersion invasive weed optimization based data. IEEE Trans. Broadcast. **59**(3), 455–460 (2013)
96. X. Xiao, Y. Lu, Data-based model for wide nulling problem in adaptive digital beamforming antenna array. IEEE Antennas Wireless Propag. Lett. **18**(11), 2249–2253 (2019)
97. M. Cui, L. Dai, Channel estimation for extremely large-scale MIMO: far-field or near-field? IEEE Trans. Commun. **70**(4), 2663–2677 (2022)
98. H. Lei, J. Zhang, Z. Wang, B. Ai, D.W.K. Ng, Hybrid-field channel estimation for XL-MIMO systems with stochastic gradient pursuit algorithm. IEEE Trans. Signal Process. **72**, 2998–3012 (2024)
99. M. Cui, L. Dai, Near-field wideband beamforming for extremely large antenna arrays. IEEE Trans. Wireless Commun. **23**, 13110–13124 (2024)
100. K.T. Selvan, R. Janaswamy, Fraunhofer and fresnel distances: unified derivation for aperture antennas. IEEE Antennas Propag. Mag. **59**(4), 12–15 (2017)
101. M. Cui, Z. Wu, Y. Lu, X. Wei, L. Dai, Near-field MIMO communications for 6G: fundamentals, challenges, potentials, and future directions. IEEE Commun. Mag. **61**(1), 40–46 (2022)
102. J.D. Kraus, R.J. Marhefka, *Antennas for All Applications*. (McGraw-Hill, New York, 2002)
103. H. Zhang, N. Shlezinger, F. Guidi, D. Dardari, M.F. Imani, Y.C. Eldar, Beam focusing for near-field multiuser MIMO communications. IEEE Trans. Wireless Commun. **21**(9), 7476–7490 (2022)
104. J. An, C. Yuen, L. Dai, M. Di Renzo, M. Debbah, L. Hanzo, Near-field communications: research advances, potential, and challenges. IEEE Wireless Commun. **31**(3), 100–107 (2024)
105. Z. Wu, L. Dai, Multiple access for near-field communications: SDMA or LDMA? IEEE J. Sel. Areas Commun. **41**(6), 1918–1935 (2023)
106. P. Nepa, A. Buffi, Near-field-focused microwave antennas: near-field shaping and implementation. IEEE Antennas Propag. Mag. **59**(3), 42–53 (2017)
107. T.S. Rappaport, Y. Xing, O. Kanhere, S. Ju, A. Madanayake, S. Mandal, A. Alkhateeb, G.C. Trichopoulos, Wireless communications and applications above 100 GHz: opportunities and challenges for 6G and beyond. IEEE Access **7**, 78729–78757 (2019)
108. H. Lu, Y. Zeng, Near-field modeling and performance analysis for multi-user extremely large-scale MIMO communication. IEEE Commun. Lett. **26**(2), 277–281 (2021)
109. Z. Wang, X. Mu, Y. Liu, Beamfocusing optimization for near-field wideband multi-user communications. IEEE Trans. Commun. **73**, 555–572 (2024)
110. A. Kosasih, E. Björnson, Finite beam depth analysis for large arrays. IEEE Trans. Wireless Commun. **23**, 10015–10029 (2024)
111. A. Buffi, P. Nepa, G. Manara, Design criteria for near-field-focused planar arrays. IEEE Antennas Propag. Mag. **54**(1), 40–50 (2012)
112. A. Sabharwal, P. Schniter, D. Guo, D.W. Bliss, S. Rangarajan, R. Wichman, In-band full-duplex wireless: challenges and opportunities. IEEE J. Sel. Areas Commun. **32**(9), 1637–1652 (2014)
113. T. Le-Ngoc, A. Masmoudi, *Full-Duplex Wireless Communications Systems*. Wireless Networks (Springer, Berlin, 2017)
114. Z. Zhang, K. Long, A.V. Vasilakos, L. Hanzo, Full-duplex wireless communications: challenges, solutions, and future research directions. Proc. IEEE **104**(7), 1369–1409 (2016)

115. G.C. Alexandropoulos, M.A. Islam, B. Smida, Full-duplex massive multiple-input, multiple-output architectures: recent advances, applications, and future directions. IEEE Veh. Technol. Mag. **17**(4), 83–91 (2022)
116. F. Chen, H.H. Lee, R. Morawski, T. Le-Ngoc, RF/analog self-interference canceller for 2×2 MIMO full-duplex transceiver, in *2017 IEEE International Conference on Communications (ICC)* (IEEE, Piscataway, 2017), pp. 1–6
117. 3GPP TR 36.931, LTE; evolved universal terrestrial radio access (E-UTRA); radio frequency (RF) requirements for LTE pico node B. Tech. Rep. Ver. 17.0.0 (2022)
118. 3GPP TR 38.901, 5G: study on channel model for frequencies from 0.5 to 100 GHz. Tech. Rep. Ver. 17.0.0 (2022)
119. T. Le-Ngoc, Y. Gong, M. Mahmood, A. Koc, R. Morawski, J.G. Griffiths, P. Guillemette, J. Zaid, P. Wang, Full-duplex in massive multiple-input multiple-output. IEEE Open J. Veh. Technol. **5**, 560–576 (2024)
120. M. Duarte, A. Sabharwal, V. Aggarwal, R. Jana, K.K. Ramakrishnan, C.W. Rice, N. Shankaranarayanan, Design and characterization of a full-duplex multiantenna system for WiFi networks. IEEE Trans. Veh. Technol. **63**(3), 1160–1177 (2013)
121. W. Slingsby, J. McGeehan, Antenna isolation measurements for on-frequency radio repeaters, in *1995 Ninth International Conference on Antennas and Propagation, ICAP'95 (Conf. Publ. No. 407)*, vol. 1 (IET, London, 1995), pp. 239–243
122. B. Debaillie, D.J. van den Broek, C. Lavin, B. van Liempd, E.A. Klumperink, C. Palacios, J. Craninckx, B. Nauta, A. Pärssinen, Analog/RF solutions enabling compact full-duplex radios. IEEE J. Sel. Areas Commun. **32**(9), 1662–1673 (2014)
123. H. Nawaz, I. Tekin, Dual-polarized, differential fed microstrip patch antennas with very high interport isolation for full-duplex communication. IEEE Trans. Antennas Propag. **65**(12), 7355–7360 (2017)
124. F. Chen, R. Morawski, T. Le-Ngoc, Self-interference channel characterization for wideband 2× 2 MIMO full-duplex transceivers using dual-polarized antennas. IEEE Trans. Antennas Propag. **66**(4), 1967–1976 (2018)
125. J.H. Chen, L.H. Ye, T. Liu, D.L. Wu, A low-profile dual-polarized patch antenna with simple feed and multiple decoupling techniques. IEEE Antennas Wireless Propag. Lett. **22**(8), 1883–1887 (2023)
126. J. Wu, M. Li, N. Behdad, A wideband, unidirectional circularly polarized antenna for full-duplex applications. IEEE Trans. Antennas Propag. **66**(3), 1559–1563 (2018)
127. D. Wu, Y.X. Sun, B. Wang, R. Lian, A compact, monostatic, co-circularly polarized simultaneous transmit and receive (STAR) antenna with high isolation. IEEE Antennas Wireless Propag. Lett. **19**(7), 1127–1131 (2020)
128. D. Ahn, J.S. Park, C.S. Kim, J. Kim, Y. Qian, T. Itoh, A design of the low-pass filter using the novel microstrip defected ground structure. IEEE Trans. Microw. Theory Technol. **49**(1), 86–93 (2001)
129. A. Dadgarpour, B. Zarghooni, B.S. Virdee, T.A. Denidni, A.A. Kishk, Mutual coupling reduction in dielectric resonator antennas using metasurface shield for 60-GHz MIMO systems. IEEE Antennas Wireless Propag. Lett. **16**, 477–480 (2016)
130. Y. Gong, M. Mahmood, R. Morawski, T. Le-Ngoc, Dual-layer metamaterial rectangular antenna arrays for in-band full-duplex massive MIMO. IEEE Access **11**, 135708–135727 (2023)
131. J.M.B. da Silva, A. Sabharwal, G. Fodor, C. Fischione, 1-bit phase shifters for large-antenna full-duplex mmwave communications. IEEE Trans. Wireless Commun. **19**(10), 6916–6931 (2020)
132. K. Satyanarayana, M. El-Hajjar, A.A. Mourad, L. Hanzo, Multi-user full duplex transceiver design for mmwave systems using learning-aided channel prediction. IEEE Access **7**, 66068–66083 (2019)
133. M. Darabi, A.C. Cirik, L. Lampe, Transceiver design in millimeter wave full-duplex multi-user massive MIMO communication systems. IEEE Access **9**, 165394–165408 (2021)

134. K.E. Kolodziej, B.T. Perry, J.S. Herd, In-band full-duplex technology: techniques and systems survey. IEEE Trans. Microw. Theory Technol. **67**(7), 3025–3041 (2019)
135. H. Luo, M. Holm, T. Ratnarajah, Design and analysis of wideband self-interference cancellation for full-duplex wireless networks, in *2022 IEEE Wireless Communications and Networking Conference (WCNC)* (IEEE, Piscataway, 2022), pp. 2697–2702
136. M.S. Amjad, H. Nawaz, K. Özsoy, Ö. Gürbüz, I. Tekin, A low-complexity full-duplex radio implementation with a single antenna. IEEE Trans. Veh. Technol. **67**(3), 2206–2218 (2017)
137. M.B. Dastjerdi, S. Jain, N. Reiskarimian, A. Natarajan, H. Krishnaswamy, Analysis and design of a full-duplex two-element MIMO circulator-receiver with high TX power handling exploiting MIMO RF and shared-delay baseband self-interference cancellation. IEEE Antennas Propag. Mag. **54**(12), 3525–3540 (2019)
138. G. Shao, R. Morawski, T. Le-Ngoc, Digital baseband self-interference cancellation using fractionally-spaced finite-impulse-response structure for full-duplex MIMO, in *2024 Tenth International Conference on Communications and Electronics (ICCE)* (IEEE, Piscataway, 2024), pp. 48–53
139. M.B. Dastjerdi, S. Jain, N. Reiskarimian, A. Natarajan, H. Krishnaswamy, 28.6 full-duplex $2\times2$ MIMO circulator-receiver with high TX power handling exploiting MIMO RF and shared-delay baseband self-interference cancellation, in *2019 IEEE International Solid-State Circuits Conference-(ISSCC)* (IEEE, Piscataway, 2019), pp. 448–450
140. Z. Gao, L. Dai, D. Mi, Z. Wang, M.A. Imran, M.Z. Shakir, Mmwave massive-MIMO-based wireless backhaul for the 5G ultra-dense network. IEEE Wireless Commun. **22**(5), 13–21 (2015)

# Chapter 3
# Antenna Element Design for Large-Scale Antenna Arrays

## 3.1 Introduction

Employing large-scale antenna arrays enables systems to generate directional beams with high directivity and narrow beamwidth, leading to improved power efficiency and reduced spatial correlation. Beamforming focuses and directs the transmission or reception of signals in specific directions, considerably boosting signal strength and expanding coverage range [1, 2]. However, as the number of elements increases, the large physical size presents challenges in manufacturing, installation, and adaptability. Additionally, inter-element MC between tightly packed antenna elements becomes a significant obstacle, undermining radiation efficiency and resulting in array beam pattern distortion.

Thanks to their adjustable EM characteristics, metamaterial-based structures have emerged as a highly promising technology for overcoming the challenges in conventional antenna array design [3–5]. By altering the layout and structure of the meta-cells, the effective wavelength of the metamaterial can be modulated, enabling the design of electrically small antennas. As presented in Chap. 2, antenna miniaturization can be achieved by enhancing the apparent permittivity or permeability of the dielectric substrate or by leveraging near-field coupling interactions between the metamaterials and the original radiation waves. In [6, 7], the use of concentric SRR metamaterials enhances the effective permittivity of the medium, subsequently reducing the resonant frequency of the patch. Studies in [8, 9] introduced a miniaturized patch antenna design that leverages slow-wave propagation within multi-layer metamaterial patch structures. The fabrication process can be simplified by leveraging the advantages of multi-layer PCB manufacturing techniques. This slow-wave effect is realized by manipulating the coupling and reflection waves to achieve constructive interference within the periodic structures. By controlling this interference between waves in the metamaterial cells, the phase velocity can be reduced, resulting in a shorter effective wavelength [10]. The sub-wavelength

 
T. Le-Ngoc et al., *Half-Duplex/Full-Duplex MU-mMIMO*, Wireless Networks,
https://doi.org/10.1007/978-3-032-09195-6_3

meta-cells in the dual-layer structure demonstrate high adaptability for various array enhancement designs.

When packing antenna elements into arrays, the limited separation between elements can lead to significant inter-element MC driven by strong spatial and surface leakage waves [11]. This coupling degrades radiation efficiency, distorts radiation patterns, and compromises the effectiveness and accuracy of array steerability [12]. Metamaterials can be engineered to effectively suppress surface leakage waves between array elements, leading to considerable improvements in inter-element isolation. For instance, metamaterial-based decoupling structures [13, 14] and FSS [12, 15] are often introduced between antenna elements. Among them, EBG structures have become popular due to their ease of fabrication and good adaptability in various planar structures. The EBG structure introduces high surface impedance at the resonant frequency, suppressing surface leakage wave propagation in the undesired directions [16–18]. Although these additional EBG structures can be relatively large compared to the limited space between elements on the array, a strategic joint design of multi-layer metamaterials, aimed at reducing the effective wavelength and improving isolation, can overcome this bottleneck.

Despite the benefits of easy fabrication and design versatility, the popular patch antenna suffers from drawbacks due to high surface leakage waves that compromise its radiation efficiency, purity of polarization, and beamforming capabilities. Although metamaterial-based designs can mitigate these challenges, they often result in increased structural complexity. DRAs have gained popularity recently due to their pronounced resonance characteristics, which enable the confinement of EM fields within the resonator. This confinement leads to more focused radiation and a substantial reduction of surface leakage waves that typically degrade radiation efficiency and impair polarization purity in conventional patch antennas [19–21]. Moreover, in array configurations, the strong resonance characteristics of DRAs enable more concentrated radiation power towards the target directions and improved beam steerability. Furthermore, the improved inter-element isolation and reduced sensitivity to environmental influences help maintain the integrity of the beam radiation patterns, making DRAs highly appealing for large-scale array designs. Additionally, the strong resonance characteristics of the DRA also mitigate the contribution of near-field SI in FD applications by confining the wave propagation directions and enhancing the Tx-Rx antenna cross-polarization ratio. Although research on FD DRA arrays remains scarce, the few existing studies on FD DRAs involving a very limited number of antenna elements show great potential in DRA designs in enhancing the Tx-Rx isolation. In [22], a DRA structure equipped with a decoupling mechanism is proposed for FD communications, and experiments illustrate a peak gain of 8.9 dB and isolation better than 50 dB. A two-port DRA is studied in [23]; by implementing the method of mode cancellation, an isolation level of 40 dB can be achieved. With these attributes, DRAs emerge as strong candidates for large-scale array applications, offering superior radiation efficiency, enhanced beamforming capabilities, and better isolation.

Motivated by the above observations, in this chapter, two antenna element structures are proposed and developed operating around 3.5 GHz. To begin with, a miniaturized metamaterial-based dual-layer circularly-polarized (CP) patch antenna structure is proposed. The dual-layer slow-wave structure is implemented to reduce the effective wavelength and miniaturize the radiating patch in the design, and a top-layer EBG structure is used to improve the inter-element isolation. A strategically designed L-slot radiation patch achieves high-purity circular polarization with a simple probe-feeding mechanism. Then, a dual-polarized, high-isolation DRA structure is developed. The design integrates cross-polarized dual ports, utilizing two distinct mechanisms, probe feeding and aperture feeding, to enhance inter-port isolation performance. The confinement of EM fields within the dielectric resonators helps the proposed DRA structure eliminate leakage waves and achieve a wider range of beam steerability when implemented in arrays. Additionally, a ring-shaped copper shield is implemented to further improve inter-element isolation. Through ANSYS HFSS full-wave simulations, we comprehensively study the characteristics of the proposed two antenna elements in terms of impedance bandwidth, radiation patterns, inter-element MC, and beamforming performance in different scenarios. These antenna element designs serve as the first steps toward the large-scale antenna array design, which will be presented in the next chapter. The main illustrative results and contributions of this chapter are summarized as follows:

- **Miniaturized metamaterial-based dual-layer EBG-slotted CP patch antenna element:**[1] A dual-layer EBG slotted CP patch antenna is proposed for mMIMO operating at 3.5 GHz. The antenna structure achieves an impedance bandwidth better than 250 MHz (7.1% at 3.5 GHz), a gain of 5.1 dB, a directivity of 6.1 dB, and a 3dB-beamwidth of 94°. Incorporating a dual-layer slow-wave structure reduces the radiating patch size by 65% compared to a conventional half-wavelength patch antenna structure. A strategically designed L-slot radiating patch and a simple probe feeding mechanism achieve high-purity circular polarization with an axial ratio (AR) better than 2 dB. The utilization of a top-layer EBG structure provides high surface impedance and achieves an isolation of 14.3 dB between radiating patches with a separation distance of 0.32 wavelength edge-to-edge. The full-wave simulation using sample $1 \times 8$ array illustrates an effective beam steering range from $-50°$ to $50°$ degrees, with an average directivity and 3 dB-beamwidth of 13.3 dB and 15.5°, respectively.
- **Dual-polarized high-isolation dielectric resonator antenna structure:** A dual-polarized high-isolation DRA structure is proposed operating around 3.5 GHz. The design integrates cross-polarized dual ports, utilizing two distinct mechanisms, probe feeding, and aperture feeding, to enhance isolation performance. The proposed antenna element demonstrates a realized gain of 8.6 dB and a remarkable radiation efficiency of 97.7% at 3.5 GHz. The inter-port isolation remains better than 38.7 dB across a bandwidth from 3.40 to 3.61 GHz. Additionally, the beamforming capability and potential Tx-to-Rx port isolation with monostatic and bistatic array configurations are studied. In sample $1 \times 8$

[1] To maintain consistency throughout our discussions, the illustrative results of the cross-polarized design using the proposed patch antenna for FD communications will be presented in Part II.

array configurations, the design delivers an effective beam steering range from $-60^{\circ}$ to $+60^{\circ}$ degrees, with an average directivity and 3 dB-beamwidth of 14.9 dB and 16.2°, respectively. The proposed structure achieves an average Tx-Rx port isolation of 53.8 and 61.7 dB, when deployed in sample monostatic and bistatic full-duplex arrays, respectively.

The rest of this chapter is organized as follows. Section 3.2 presents the detailed design process of the proposed metamaterial-based dual-layer EBG-slotted CP patch antenna element. Full-wave simulation results are provided to evaluate its performance in terms of impedance bandwidth, radiation patterns, inter-element MC, and beamforming capabilities. Then, Sect. 3.3 details the design process of the proposed dual-polarized, high-isolation DRA structure. The antenna's radiation characteristics and isolation performance for single antenna elements and in array configurations are investigated. Finally, the chapter is concluded in Sect. 3.4 which compares the two antenna element structures and illustrates the decision regarding the choice of antenna elements for fabrication and prototyping in future chapters.

## 3.2 Dual-Layer EBG-Slotted Miniaturized Antenna Element Structure

### *3.2.1 Dual-Layer EBG Structure*

Microstrip patch antennas are inherently prone to significant wave leakage along the dielectric substrate and antenna surface due to the excitation of surface waves. Additionally, in array configurations designed for beamforming, maintaining element separation at less than half a wavelength is essential to prevent the formation of grating lobes. Consequently, this close proximity intensifies inter-element MC, which can lead to reduced radiation efficiency, impedance mismatches, and distortion of the overall radiation pattern.

To address these challenges, EBG structures can be integrated between antenna elements. EBG structures are engineered periodic lattices that exhibit band-stop filtering characteristics for EM waves within specific frequency ranges. By creating regions of high surface impedance, EBG structures effectively suppress the propagation of surface waves and reduce MC between closely spaced antenna elements, thereby enhancing the isolation. Additionally, EBG structures confine EM energy within desired directions and regions, resulting in more focused and efficient antenna radiation. To characterize the relationship between the surface impedance ($Z_S$), the resonant frequency, and the square EBG dimensions, an Effective Medium Model is introduced in [24] with the expression

$$
\begin{aligned}
C &= \frac{W_u \epsilon_0 (1+\epsilon_r)}{\pi} \cosh^{-1}\left(\frac{W_u + g_u}{g_u}\right), \ L = \mu h, \\
Z_S &= \frac{j(2\pi f)L}{1-(2\pi f)^2 LC},
\end{aligned}
\tag{3.1}
$$

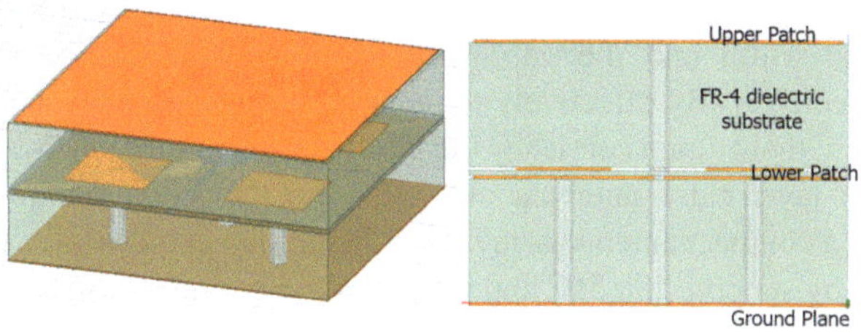


**Fig. 3.1** Top and side views of the dual-layer unit EBG cell

where $g_u$ and $W_u$ are the gap width between neighboring upper EBG patches and the width of upper EBG patches, respectively. $h$ is the dielectric substrate thickness. $\epsilon_r$ is the dielectric constant of the substrate and $\epsilon_0$ is the permittivity of free space. The edge capacitance ($C$) is generated by the gaps between the conducting surfaces of the square EBG patches, acting like capacitors due to the electric fields across these gaps. The inductance ($L$) arises from the currents flowing along the paths of adjacent EBG cells, which create magnetic fields. Together, these capacitance and inductance components form a resonant circuit within the EBG structure, enabling it to exhibit high surface impedance and effectively suppress the propagation of surface waves at specific frequencies.

Moreover to integrate this periodic EBG lattice between the half-wave spacing patch antenna elements, the miniaturization of both the patch antenna and the EBG patches is crucial. This miniaturization is achieved through a strategic dual-layer patch design that incorporates a slow-wave structure. As illustrated in Fig. 3.1, periodic square patches are embedded between two dielectric substrates in a sandwich-like arrangement. The EBG patches, which enhance isolation, are located on top of the upper dielectric substrate, while the ground plane is at the bottom of the lower dielectric substrate. Each middle-layer square patch is connected to the ground plane with a through-hole via, forming a slow-wave structure that interacts with the top radiating patch. The interaction between the upper and lower conductive patches induces near-field coupling and reflective interference, modifying both the wave reflection phase and the dispersion relationship [10]. This interaction can be modelled as a parallel plate waveguide. When incident and reflected waves within this structure constructively interfere, the phase velocity of the wave travelling through the structure decreases relative to its velocity in free space. By adjusting the sizes and gaps between the top and middle-layer square patches, a reduced effective wavelength ($\lambda_{eff}$) is achieved, enabling a reduction in the physical size of the antenna. The relationship between the effective short wavelength $\lambda_{eff}$ and the relative dimensions of the lower and the upper square patches can be estimated as [9]

$$\lambda_{eff} = \frac{2\pi(W_l + g_l)}{\psi}, \tag{3.2}$$

where $\psi$ is the phase difference between the side walls of the lattice. $W_l$ and $g_l$ denote the width and the gap width between the middle layer slow-wave patches, respectively. By effectively reducing $\lambda_{eff}$, electrically small antennas can be achieved without compromising the resonant and radiation performance. The smaller upper-layer patch antennas also allow for the accommodation of the EBG lattice within a compact antenna array. This integration enhances the overall antenna performance by suppressing MC and improving radiation efficiency.

### 3.2.2 L-Slot CP Radiating Patch Element

CP antennas are widely used in wireless systems due to their resilience to multipath distortion, polarization mismatch losses, and interference [25, 26]. However, CP patch antennas often suffer from elliptical polarization because of the ease of surface wave excitation and the difficulty in generating identical orthogonal wave paths.

As illustrated in Fig. 3.2, an L-slot patch antenna on the top layer of a dual-layer metamaterial structure is developed in our design. The antenna features a central square patch surrounded by copper strips, interconnected by two copper bridges. These bridges create two orthogonal current paths essential for generating the EM fields associated with the antenna's operation. Currents flowing through these paths produce EM waves in two orthogonal directions. The resulting fields are perpendicular to each other and to the direction of wave propagation, with the wave's polarization determined by the interaction and rotation of these field components as they propagate. The direction of the simple circular polarization excitation using an L-slot mechanism can be easily modified by mirroring the relative positions of the probe feed and cutting slots. This allows for the creation of a cross-polarized antenna structure with identical radiation characteristics, providing great potential

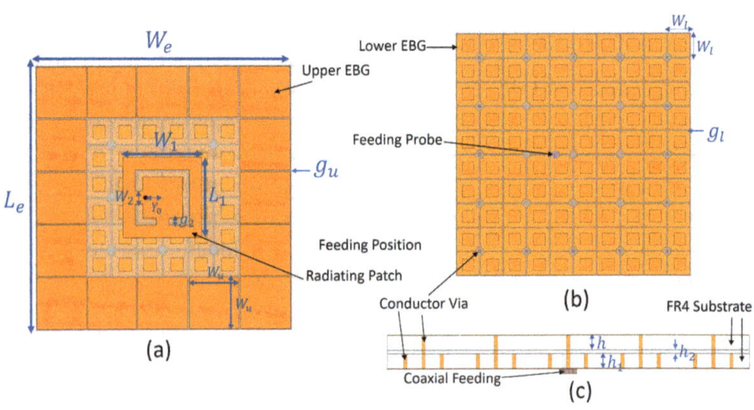


**Fig. 3.2** (**a**) Dual-layer L-slotted CP antenna element top view; (**b**) top view of the element medium layer; (**c**) side view of the multi-layer structure

for FD isolation improvement. The orthogonal and equal-magnitude fields ensure the excitation of pure CP with a simple probe-feeding mechanism. The performance details will be illustrated in Chap. 7.

The surface capacitance is controlled by the gaps between and within the top patch's metallic circuits, as well as by the dielectric material sandwiched between the circuits and the ground plane. Reducing the gap between conductors increases capacitance, enhancing the structure's ability to store and interact with electric fields. The inductance is determined by the geometry of the conducting strips in the patch antenna, which governs the magnetic field linkage within the structure. The positioning of the feed relative to the patch geometry determines the initial phase and orientation of the emitted fields. A feeding probe placed 3 mm away from the center of the radiating patch is employed to achieve good antenna impedance matching.

The ANSYS High-Frequency Structural Simulator (HFSS) with the Eigen Solver and Optimetrics Analyzer is utilized to optimize the effective wavelength, the upper-layer EBGs, and the dimensions of the antenna element structure, as well as to evaluate performance through full-wave simulations. The software employs the finite element method (FEM) to solve Maxwell's equations, accurately simulating real-world electromagnetic interactions. In the design process, to ensure practical manufacturability with standard PCB fabrication techniques, parameters such as the dielectric substrate thickness, via radius, and minimum cutting width are selected according to the standard options provided by PCB manufacturers.

The final dimensions of the proposed patch antenna structure are detailed in Table 3.1. Regular patch size on FR-4 PCB material ($\epsilon_r = 4.34$) at 3.5 GHz would be $20.5 \times 20.5\,\text{mm}^2$. With the EBGs underneath, the element size is reduced to $12.2 \times 12.2\,\text{mm}^2$, which is about a 40% reduction per side and a 65% reduction in area.

**Table 3.1** Parameters of the proposed antenna element

| Parameter | Description | Value (mm) |
|---|---|---|
| $W_e$ | Element width | 40 |
| $L_e$ | Element length | 40 |
| $W_1$ | Patch width | 12.2 |
| $L_1$ | Patch length | 12.2 |
| $W_2$ | Bridge width | 2.0 |
| $Y_0$ | Feeding point from the center | 3.0 |
| $g_u$ | Upper EBG gap | 0.2 |
| $g_l$ | Lower EBG gap | 0.2 |
| $g_2$ | Center patch slot gap | 1.1 |
| $W_l$ | Lower EBG size | 3.8 |
| $W_u$ | Upper EBG size | 7.8 |
| $h$ | Upper FR-4 substrate thickness | 1.6 |
| $h_1$ | Lower FR-4 substrate thickness | 1.6 |
| $h_2$ | Prepreg thickness | 0.2 |

## 3.2.3 Full-Wave Simulation Results of the Proposed Dual-Layer CP Patch Antenna Element

### 3.2.3.1 EBG Unit Cell Surface-Impedance

The surface impedance of our proposed EBG unit, shown in Fig. 3.1, was simulated in HFSS with Floquet Port setup over the operating frequency range from 3 to 4 GHz. The simulation results are presented in Fig. 3.3. It can be observed that the surface impedance exceeds 400 Ω between 3.14 and 3.85 GHz. In our target band from 3.35 to 3.6 GHz, the surface impedance is above 900 Ω. At the center frequency of 3.5 GHz, a surface impedance of 4476.4 Ω is achieved. Within this frequency range, the high surface impedance effectively suppresses the propagation of leakage EM waves along the conductor patch surface, thereby reducing MC between neighbouring elements. As a result, the element separation distance can be reduced while maintaining a satisfactory level of isolation between elements.

### 3.2.3.2 Antenna Element Impedance Bandwidth

In the ANSYS HFSS full-wave simulation, we enclose the model in an airbox with perfectly matched layer (PML) boundary conditions to eliminate reflections at its edges. This conventional configuration for antenna radiation-pattern simulations mimics a low-reflection environment, such as measurements in an anechoic chamber, ensuring that we observe only the antenna's intrinsic electromagnetic behaviour without interference from surrounding structures.

The impedance bandwidth of an antenna element is defined as the continuous frequency interval over which its input impedance remains adequately matched

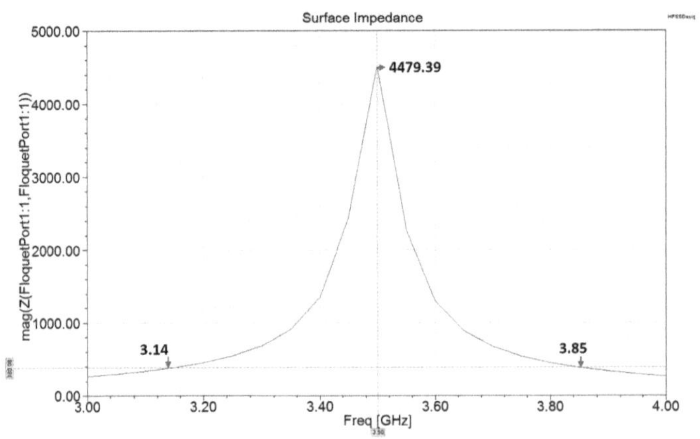


**Fig. 3.3** Surface impedance of the unit EBG cell (simulation results)

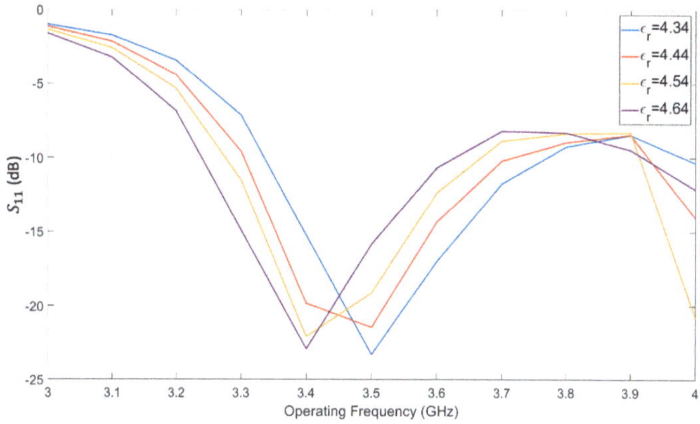


**Fig. 3.4** Return loss versus operating frequency (simulation results)

to the feeding network, typically characterized by a reflection coefficient $|S_{11}| \leq -10$ dB. Here, $S_{11}$ denotes the reflection coefficient, which is defined by the fraction of incident power reflected back to the source, and thus identifies the frequency range over which the antenna can transmit and receive electromagnetic energy with minimal reflection loss. The economically attractive FR-4 PCB material has a dielectric constant $\epsilon_r$ that can potentially vary with operating frequency and differ between PCB manufacturers, ranging from 4.34 to 4.64 within the target bandwidth of 3.35 to 3.6 GHz. This variation can affect the return loss (RL) and shift the resonant frequency, influencing both the radiation efficiency and the effective bandwidth of the antenna. To assess this effect, HFSS simulations were conducted to illustrate the proposed antenna structure return loss across the entire range of $\epsilon_r$ variation, incrementing from 4.34 to 4.64 in steps of 0.1.

As shown in Fig. 3.4, the RL of the proposed antenna element with different substrate dielectric constants demonstrates a similar manner of change with operating frequency over the range from 3.00 to 4.00 GHz. The center frequency, characterized by minimal reflection (lowest $S_{11}$ values), shifts horizontally in response to different dielectric constant ($\epsilon_r$) values. Notably, the $S_{11}$ remains below $-10$ dB from 3.34 to 3.62 GHz for all $\epsilon_r$ values, providing an effective bandwidth of 280 MHz, or approximately 8.0% relative to the center frequency of 3.5 GHz.

#### 3.2.3.3 Antenna Element Radiation Pattern

The radiation pattern of the proposed patch antenna element was simulated at 3.5 GHz with a dielectric constant of $\epsilon_r = 4.34$, and the results are presented in Fig. 3.5, which displays both the 3D element pattern and the 2D cut plot along the $\phi = 0°$ and $\theta = 0°$. The surrounding EBG patches around the center radiating patch suppress the propagation of surface waves and confine EM energy within desired

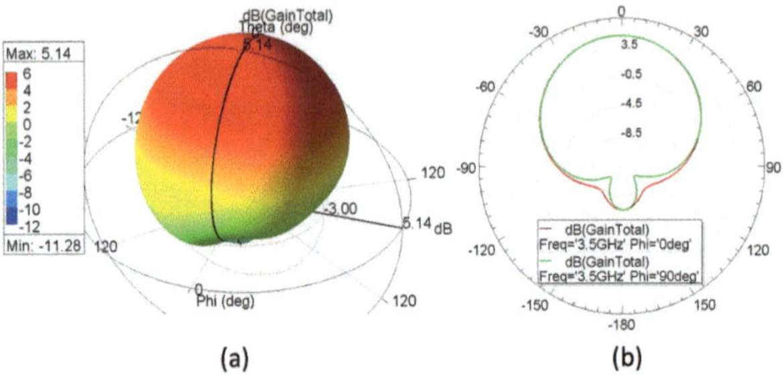


**Fig. 3.5** Antenna element (**a**) 3D radiation pattern at 3.5 GHz (gain in dB); (**b**) 2D cut radiation patterns (simulation results)

directions and regions, resulting in more focused and symmetric antenna radiation. The antenna achieves a maximum gain of 5.1 dB and a maximum directivity of 6.1 dB, with the difference primarily attributed to the dielectric loss of the FR-4 substrate. Consequently, the radiation efficiency is approximately $-1$ dB or 79.4%. The radiation pattern is symmetric along both the $\phi = 0°$ and $\phi = 90°$ planes. The 3dB-beamwidth is about 94°, and the front-to-back ratio is 11.7 dB, thanks to the ground plane, which contributes to the elimination of back radiation. Additionally, the side radiation level at $\theta = 90°$ is approximately 9 dB lower than the magnitude in the main lobe direction.

#### 3.2.3.4 Antenna Circular-Polarization Purity

High CP purity is crucial for antennas to effectively transmit or receive the desired polarization. In scenarios involving multipath propagation, high CP purity ensures that the signal retains its polarization characteristics, minimizing signal degradation and improving communication performance. The surface current distribution and E-field excitation results from the HFSS simulation of the proposed L-slot patch antenna structure are illustrated in Fig. 3.6. The simulation reveals that the proposed structure introduces two symmetrically rotating current paths with closely matched magnitudes. As a result, these uniform orthogonal currents facilitate the consistent excitation of the E-field in two perpendicular directions, significantly contributing to high circular polarization purity in the radiated wave. The orthogonal and equal magnitude fields ensure the excitation of pure CP with a simple probe-feeding mechanism.

Defined as the ratio of the major axis to the minor axis of the polarization ellipse formed by the electric field components, the AR is a key parameter for assessing circular polarization purity, quantifying how closely an antenna's radiation approaches ideal circular polarization. Figure 3.7a illustrates the HFSS simulated AR as a function of operating frequency for the antenna element. An AR of 2 dB is achieved at 3.42 GHz, and the antenna maintains an AR below 3 dB across the

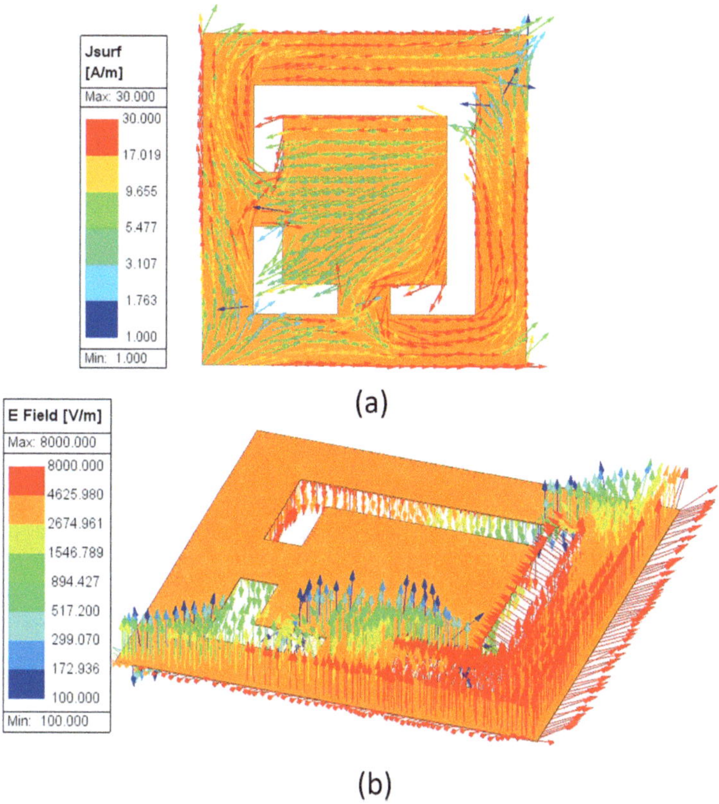


**Fig. 3.6** (**a**) Surface current; (**b**) Electric field excitation of the proposed L-slot patch antenna structure (simulation results)

operating bandwidth of 3.38 to 3.46 GHz, demonstrating high polarization purity within this range. Over a bandwidth of 0.18 GHz, from 3.34 to 3.52 GHz, the AR remains better than 6 dB. Beyond this frequency range, the AR rapidly increases, indicating a deterioration in circular polarization purity due to the degradation of the antenna's radiation characteristics. Figure 3.7b presents the AR as a function of radiation angles in the $\theta$ cuts at the operating frequency of 3.42 GHz. Within the 3 dB-beamwidth of the antenna radiation pattern, the AR consistently stays below (better than) 2.6 dB.

#### 3.2.3.5 Antenna Element Inter-Element Isolation

To evaluate the MC levels between antenna elements in different relative positions, a sample 3 × 3 antenna array was simulated in HFSS. The radiating patch antennas are separated 4 cm (approximately 0.47 wavelength) center-to-center or 2.8 cm (0.32

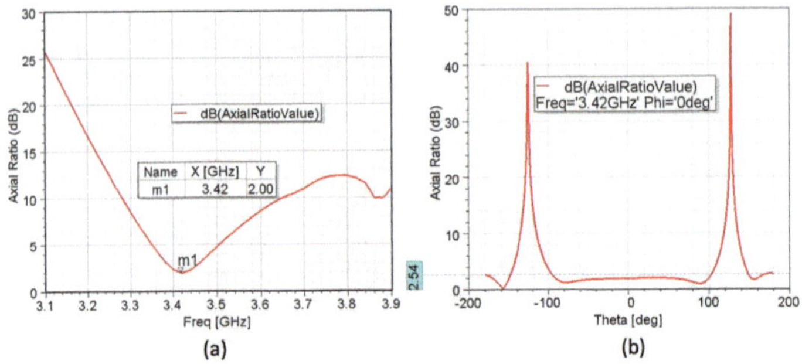


**Fig. 3.7** Axial ratio of the proposed antenna structure (**a**) versus the operating frequency, and (**b**) at 3.42 GHz $\theta$ and $\phi$ cuts (simulation results)

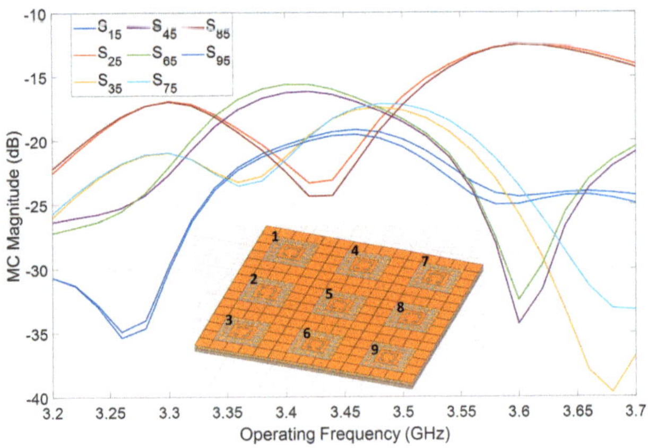


**Fig. 3.8** Inter-element MC (in dB) from surrounding elements to the center element (simulation results)

wavelength) edge-to-edge. Figure 3.8 illustrates the MC between the center element and each of the surrounding elements, along with their corresponding element indices. It can be observed that antenna elements positioned diagonally symmetric to the center exhibit very similar MC magnitudes. For example, the MC between elements 1 and 5, and between elements 9 and 5, are consistent due to the symmetry of the EM surrounding environments.

Across all eight coupling combinations, the MC remains below − 15 dB in the frequency band from 3.20 to 3.52 GHz. Among the eight directions, the MC between neighboring elements on the same row (the same axis as the probe feeding mechanism) exhibits a relatively higher MC level. Specifically, the MC between elements 2 and 5, and between elements 8 and 5, exceeds − 15 dB when the frequency is greater than 3.52 GHz.

This observation also applies to all $9 \times 8 = 72$ MC combinations in the $3 \times 3$ array. The worst (highest) MC magnitudes within the operating frequency range (3.2

**Table 3.2** Worst MC magnitude between antenna elements in frequency range 3.2–3.7 GHz

| Element | 1 | 2 | 3 | 4 | 5 | 6 | 7 | 8 | 9 |
|---|---|---|---|---|---|---|---|---|---|
| 1 | | −17.3 | −16.1 | −14.8 | −19.2 | −21.5 | −15.2 | −21.0 | −21.8 |
| 2 | −17.3 | | −15.6 | −17.6 | −12.5 | −19.2 | −21.7 | −16.7 | −21.4 |
| 3 | −16.1 | −15.6 | | −21.6 | −17.4 | −14.4 | −24.5 | −21.7 | −15.0 |
| 4 | −14.8 | −17.6 | −21.6 | | −16.2 | −16.4 | −15.6 | −19.0 | −21.8 |
| 5 | −19.2 | −12.5 | −17.4 | −16.2 | | −15.6 | −17.1 | −12.5 | −19.5 |
| 6 | −21.5 | −19.2 | −14.4 | −16.4 | −15.6 | | −22.0 | −16.8 | −14.7 |
| 7 | −15.2 | −21.7 | −24.5 | −15.6 | −17.1 | −22.0 | | −16.3 | −15.7 |
| 8 | −21.0 | −16.7 | −21.7 | −19.0 | −12.5 | −16.8 | −16.3 | | −17.6 |
| 9 | −21.8 | −21.4 | −15.0 | −21.8 | −19.5 | −14.7 | −15.7 | −17.6 | |

to 3.7 GHz) are summarized in Table 3.2. It can be noted that combinations between two neighboring elements on the same row have relatively high MC magnitudes. The positions of the elements, whether at the center or edge, do not significantly influence the MC magnitude. Except for the combinations between elements 2 and 5, and between elements 8 and 5, all other combinations have MC levels lower than − 14.4 dB. Furthermore, except for the same-row neighboring combinations, all other combinations have MC levels better than − 15 dB.

### 3.2.4 *Sample 1×8 Patch Antenna Array Radiation Patterns*

To evaluate the beamforming capability of the proposed metamaterial-based patch antenna structure, a sample 1×8 uniform linear array (ULA) is built in HFSS and simulated over a steering angle range $\{-60^\circ : 10^\circ : +60^\circ\}$, as shown in Fig. 3.9. The maximum directivity beamforming (MDB) vector, which is calculated by aligning the phase difference across the array element, is used in the simulation. Figure 3.10 depicts the 2D beamforming cuts along $\phi = 90^\circ$ (narrow axis), with the characteristics of the beam pattern summarized in Table 3.3. At a steering angle 0°, an array directivity of 13.9 dB is achieved, with a 3 dB- and 10 dB-beamwidths (BW) 13 and 26°, respectively.

The array achieves a uniform and concentrated beam pattern across the steering range from − 50° to 50°. Within the steering range, an average directivity of 13.3 dB is attained, with the 3 dB- and 10 dB-beamwidths averaging at 15.5 and 26.6°, respectively. The normalized sidelobe (SLL) level (w.r.t. the beam peak) is below (better) −9.5 dB. The difference between the achievable peak radiation direction and the desired steering angle is within 3°. The directivity and 3 dB-beamwidth degradation are limited to less than 1.7 dB and 4°, respectively, ensuring high array gain in the desired user direction and consistent beam resolution across the steering range. These simulation results demonstrate the sample 1×8 patch array's capability to effectively focus radiation power across a 100° range.

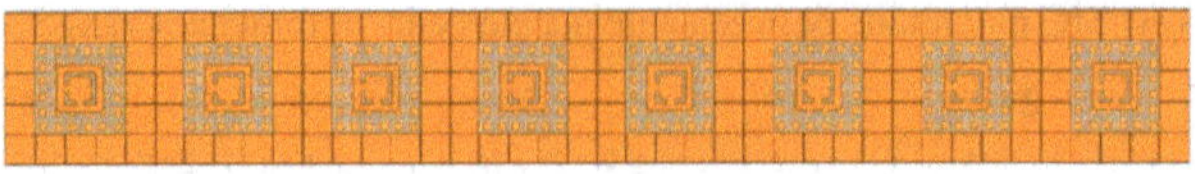

**Fig. 3.9** Sample 1×8 metamaterial-based patch antenna array beamforming capability simulation setup

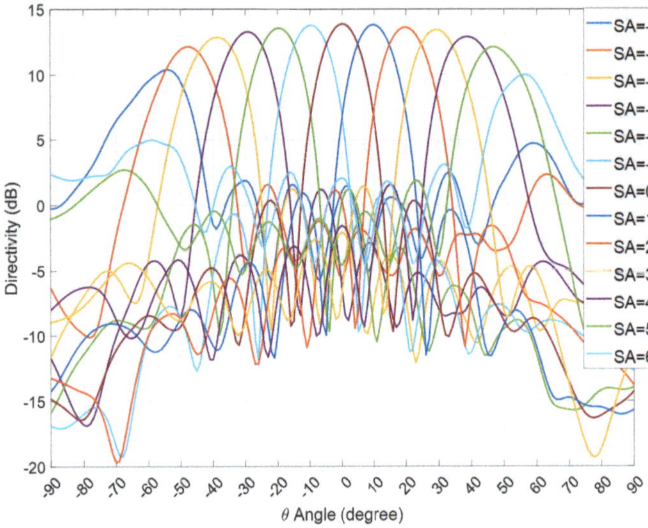


**Fig. 3.10** Sample 1×8 Patch array beamforming: 2D cuts along $\phi = 90°$ with steering angle (SA) from $-60°$ to $60°$ at 3.5 GHz (simulation results)

**Table 3.3** Tx 1×8 patch array beam steering HFSS simulation results

| Steering angle (deg) | Achieved angle (deg) | Directivity (dB) | 3 dB BW(deg) | 10 dB BW (deg) | SLL (dB) |
|---|---|---|---|---|---|
| −60 | −54 | 10.6 | 21.0 | 44.0 | −5.9 |
| −50 | −48 | 12.3 | 17.0 | 32.0 | −10.0 |
| −40 | −39 | 13.0 | 17.0 | 30.0 | −11.8 |
| −30 | −29 | 13.4 | 16.0 | 26.0 | −12.2 |
| −20 | −20 | 13.7 | 15.0 | 24.0 | −12.5 |
| −10 | −10 | 13.8 | 14.0 | 23.0 | −13.2 |
| 0 | 0 | 13.9 | 13.0 | 23.0 | −13.6 |
| 10 | 10 | 13.8 | 13.0 | 23.0 | −12.9 |
| 20 | 20 | 13.6 | 15.0 | 24.0 | −12.5 |
| 30 | 29 | 13.4 | 15.0 | 26.0 | −12.0 |
| 40 | 39 | 13.0 | 17.0 | 30.0 | −11.4 |
| 50 | 47 | 12.2 | 18.0 | 31.0 | −9.5 |
| 60 | 56 | 10.0 | 18.0 | 66.0 | −5.0 |

When the steering angle exceeds the range of $-50°$ to $+50°$, a directivity degradation of more than 3 dB is observed, along with a rapid increase in sidelobe levels and beamwidth. These observations indicate a rapid degradation in beam resolution and concentration. This degradation at large steering angles is due to the inherent limitations of a finite-length antenna array in terms of steerability. This limitation arises because the beam's steering capability is intrinsically tied to the array's geometry and the phase differences that can be created between individual antenna elements. As the desired steering angle increases, the required phase shift between elements also increases. For a finite-length antenna array, there comes a point where it is not possible to introduce sufficient phase differences to achieve the desired steerability, especially towards the edges of the frontal radiation range. Additionally, grating lobes appear in the radiation pattern when attempting to steer the beam at large angles. Grating lobes are unwanted lobes that occur when the antenna elements are effectively spaced more than half a wavelength apart. As the steering angle increases, the effective spacing between elements, in terms of wavelength, also increases due to the projection of element spacing along the direction of the beam. This can result in the effective element spacing exceeding half a wavelength, leading to the formation of grating lobes and restricting the system's steerability. Finally, realistic antenna elements exhibit variations in gain across different angles. As the steering angle increases, the effective radiation from these elements–which contributes to the main lobe of the array's overall pattern–may decrease due to the natural roll-off in gain at extreme angles within the individual elements' radiation patterns. This effect can lead to a reduction in the array's overall directivity in the desired steering direction, thereby constraining the array's effective steering range.

## 3.3 Dielectric Resonator Antenna Element Design

### *3.3.1 Dual-Polarization Feeding Mechanism*

Thanks to their simple dielectric resonator geometry, strong resonant performance, and versatile feed placement options, DRAs offer remarkable flexibility in feeding strategies, including coaxial probe feeds, microstrip lines, slot coupling, and waveguide feeds [27, 28]. By precisely tuning parameters such as resonator shape, feed location, and coupling aperture dimensions, DRAs can be optimized to achieve effective impedance matching and robust radiation performance across a wide range of feeding configurations. Figure 3.11 depicts the feeding mechanisms employed in the proposed DRA element. This mechanism utilizes a tri-layered sandwich structure designed to support dual-polarization through two distinct feeding ports. The first port incorporates probe feeding to excite the DRA with a vertically polarized wave, while the second utilizes aperture feeding to induce a horizontally

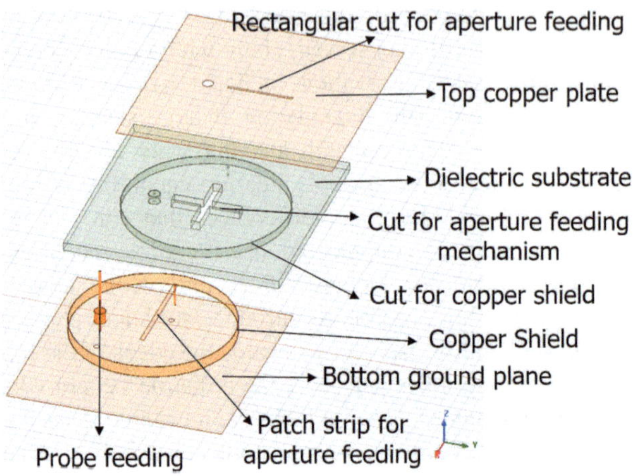


**Fig. 3.11** Stack diagram of feeding mechanisms of the proposed DRA element

polarized wave. Atop the dielectric substrate, a copper plate is affixed, featuring a strategically cut rectangular slot that serves as the aperture for wave excitation.

The dielectric substrate is composed of Rogers RO4350 material, characterized by a relative dielectric constant $\epsilon_r$ =3.7. Central to the substrate design is a cross-shaped cavity tailored to integrate the patch strip structure associated with the aperture feeding port. Aperture coupling with air cavity can be used to bind backward propagation and ground planes on top and bottom to prevent surface waves through the substrate medium. Surrounding the periphery of the two feeding vias, a circular copper shield is implemented. This shield functions as a barrier against potential wave leakage within the dielectric substrate, thereby preserving the integrity of the wave propagation and mitigating the interference. To block the undesirable radiation from the antenna's rear, a copper ground plane is positioned at the base of the dielectric substrate. The probe-feeding mechanism extends through the multi-layered structure and penetrates into the dielectric resonator. The aperture feeding mechanism is comprised of a feeding probe in conjunction with a rectangular copper strip, located at the cross-shaped incision in the dielectric substrate.

### *3.3.2 DRA Element Structure*

In the design process, the ANSYS HFSS Eigen Solver and Optimetrics Analyzer are used to optimize the structure dimensions. The final optimized DRA element structure for operating frequency at 3.5 GHz is illustrated in Fig. 3.12. It can be observed that a cylindrical dielectric resonator antenna is placed on top of the feeding mechanism. The dimensions of the resonator and the corresponding

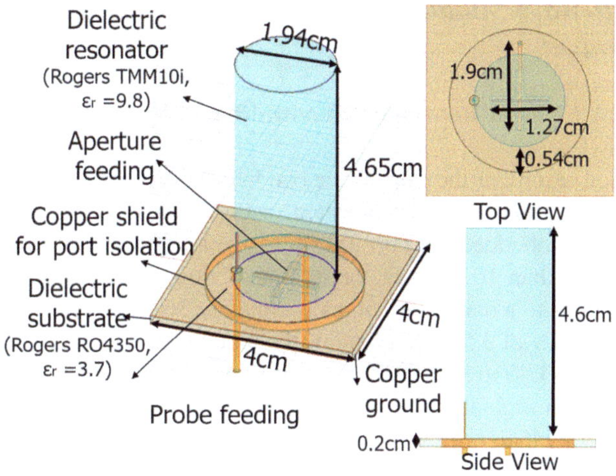


**Fig. 3.12** Proposed dual-polarized DRA element

resonant frequency ($f_r$) of Transverse Electric (TE) modes excited in the proposed resonator can be estimated via [20]

$$f_r = \frac{1}{2\pi\sqrt{\epsilon_r}}\sqrt{k_x^2 + k_y^2 + k_z^2}, \tag{3.3}$$

where $k_x$,$k_y$,$k_z$ are the mode wave numbers along the $x-$, $y-$, $z-$axis within the dielectric resonator.

One effective method to enhance the gain of a DRA is through the excitation of higher-order modes, which can be controlled by the ratio between the diameter and the height of the cylindrical resonator. While higher-order modes can increase gain, they may also affect the bandwidth and efficiency of the antenna. Higher-order modes can have narrower bandwidths and may be less efficient due to higher power losses. Furthermore, the substantial height of the resonator may impose practical manufacturing constraints. Therefore, a balance must be struck between gain, bandwidth, and efficiency when designing DRAs.

The proposed DRA structure uses Rogers TMM10i dielectric material ($\epsilon_r = 9.8$) to build the resonator. The 3rd-order height (where the first order is approximately $c/(2f\sqrt{\epsilon_r})$ $\approx$1.4 cm, after optimization, 1.55 cm is used) dielectric resonator is used in the design. The higher resonator also helps to reduce the diameter of the cylindrical structure to 1.94 cm (which is about $\lambda_0/4$), which leaves enough space for the circular copper shield to improve the isolation between neighboring elements.

## 3.3.3 *Full-Wave Simulation Results of the Proposed DRA Element*

### 3.3.3.1 DRA Element Impedance Bandwidth and Port Isolation

The simulation results of the impedance bandwidth and isolation between the two cross-polarized feeding ports are depicted in Fig. 3.13. Both the aperture feeding (port 1) and the probe feeding (port 2) exhibit a good effective bandwidth, with the return loss better than 10 dB over the frequency range of 3.31 to 3.67 GHz. This range corresponds to a bandwidth of approximately 360 MHz, or 10.2% relative to the center frequency of 3.5 GHz. Notably, the return loss is better than 20 dB within a bandwidth from 3.40 to 3.61 GHz, which spans about 210 MHz or 6% at 3.5 GHz. The observed high impedance matching across the bandwidth suggests a potential for high radiation efficiency of the proposed DRA structure.

Regarding the isolation between the two ports, the design achieves an impressive isolation level of 42.8 dB at 3.5 GHz, where the distance between the two ports is only 1.75 cm (or $0.2\lambda_0$), attributable to the effective cross-polarization technique employed. Moreover, the isolation remains better than 38.7 dB across a bandwidth from 3.40 to 3.61 GHz and better than 35 dB over a bandwidth from 3.31 to 3.67 GHz.

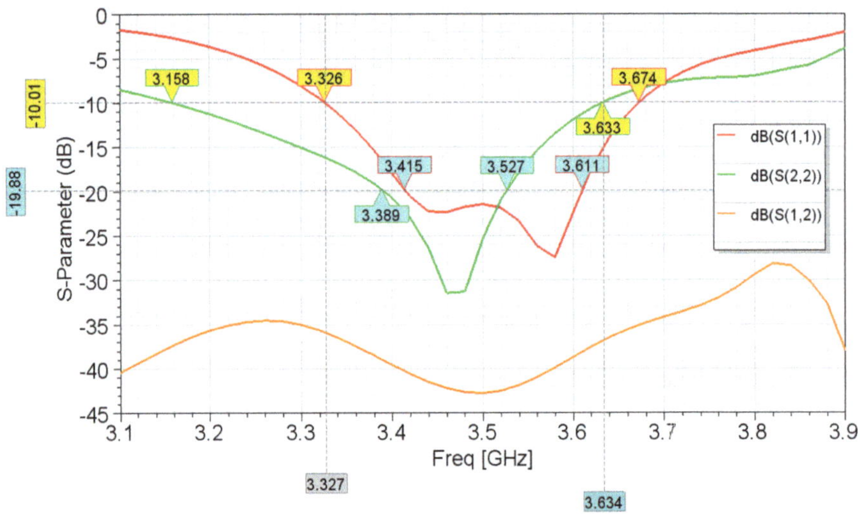


**Fig. 3.13** S-parameters of the proposed DRA element (simulation results)

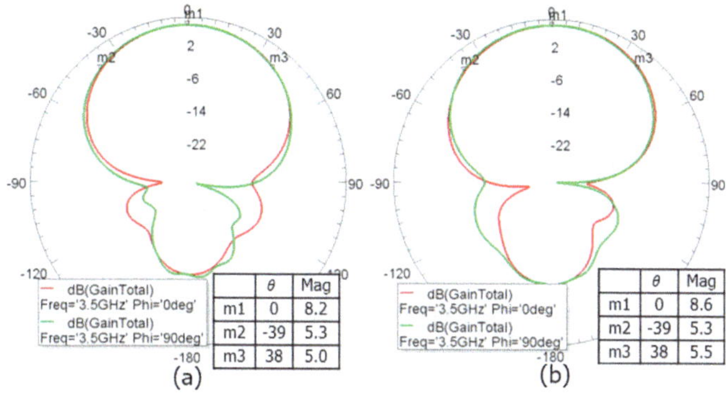


**Fig. 3.14** 2D-cut radiation patterns of the proposed DRA element at 3.5 GHz with (**a**) aperture feeding, (**b**) probe feeding (simulation results)

#### 3.3.3.2 DRA Element Radiation Patterns

Figure 3.14 illustrates the radiation patterns, expressed in terms of gain, for the proposed DRA element when excited via a single port, either the aperture feeding port or the probe feeding port. At the operating frequency of 3.5 GHz, the aperture feeding approach yields a realized gain of 8.2 dB, while the probe feeding method provides a slightly higher gain of 8.6 dB. The directivity associated with these two feeding mechanisms is measured at 8.6 dB and 8.7 dB, respectively, translating to high radiation efficiencies of 91.2% (or −0.4 dB) for the aperture feeding and 97.7% (or −0.1 dB) for the probe feeding.

Furthermore, as depicted in Fig. 3.14a,b, both feeding mechanisms produce comparable and symmetric radiation patterns along the $\phi = 0°$ and $\phi = 90°$ planes. The 3dB-beamwidth is observed to be roughly 77° in both cases. Additionally, a front-to-back radiation ratio of 15.3 dB is attained, which can be attributed to the presence of a ground plane at the base of the antenna structure.

### *3.3.4 Sample 1×8 DRA Array Radiation Patterns*

The antenna array is pivotal in focusing radiation power toward the intended direction while simultaneously reducing wave leakage in potentially interfering directions. As an example, we consider an 1×8 ULA using the proposed DRA element, as depicted in Fig. 3.15. To preclude the occurrence of grating lobes, which can impair beam resolution and elevate interference levels during the beamforming process, the center-to-center spacing between adjacent elements is maintained at 4 cm (approximately $0.47\lambda_0$ at the operating frequency of 3.5 GHz). In the preceding section, simulation results indicated that the two feeding mechanisms could offer comparably effective radiation characteristics. Therefore, in this section, all antenna

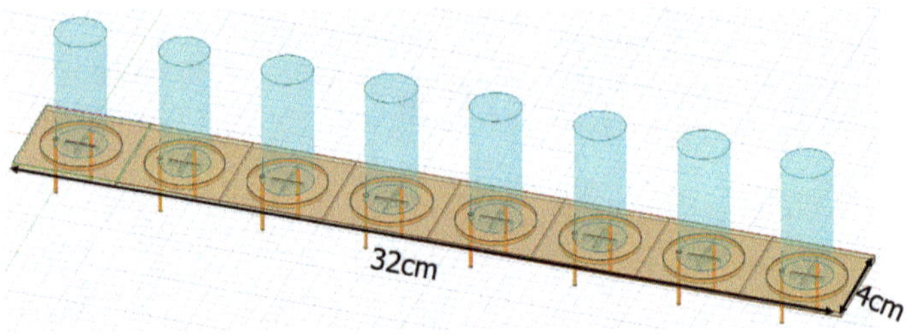


**Fig. 3.15** Sample 1×8 DRA array beamforming capability simulation setup

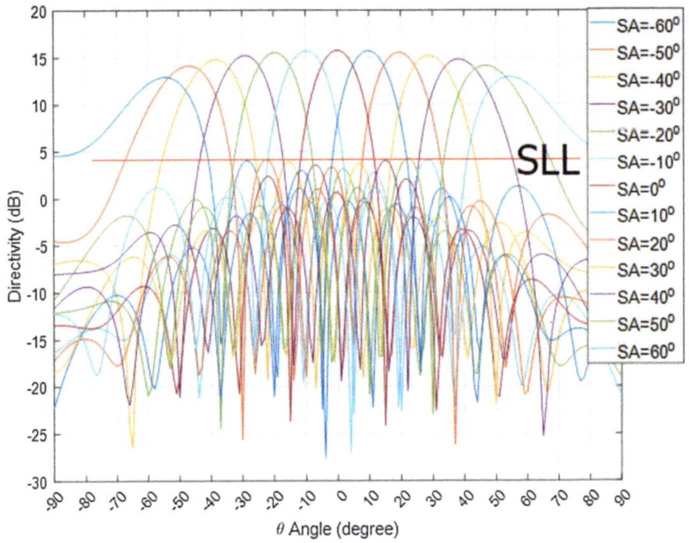


**Fig. 3.16** Sample 1×8 DRA array beamforming: 2D cuts along $\phi = 90°$ with steering angle (SA) from $-60°$ to $60°$ at 3.5 GHz (simulation results)

elements are excited with only the aperture feedings. The MDB vector is employed to maximize the directivity of the beam in the target steering angle $\theta_0$.

The sample 1×8 array is simulated over a steering angle range of $\theta_0 \in \{-60° : 10° : +60°\}$. Figure 3.16 depicts the 2D beamforming cuts along $\phi = 90°$ (narrow axis), while Table 3.4 details the radiation beam characteristics. At steering angle $\theta_0 = 0°$, an array realized gain of 15.2 dB is achieved, with a directivity of about 15.7 dB, corresponding to a radiation efficiency of 89.1% (or −0.5 dB). The radiation efficiency difference between the DRA element and the sample 1×8 array is only 2.1%.

The pronounced resonance characteristics of the DRA enable the confinement of EM fields within the resonator. This confinement allows for more focused radiation power, leading to better beam steerability and eliminating leakage waves that typically impair radiation efficiency. The array achieves a uniform beam pattern

**Table 3.4** Tx 1×8 DRA array beam steering HFSS simulation results

| Steering angle (deg) | Achieved angle (deg) | Directivity (dB) | 3 dB BW(deg) | 10 dB BW (deg) | SLL (dB) |
|---|---|---|---|---|---|
| −60 | −54 | 12.9 | 21.0 | 86.0 | −8.9 |
| −50 | −47 | 14.1 | 19.0 | 34.0 | −10.0 |
| −40 | −38 | 14.8 | 17.0 | 29.0 | −10.9 |
| −30 | −29 | 15.2 | 15.0 | 25.0 | −11.6 |
| −20 | −19 | 15.5 | 14.0 | 23.0 | −12.4 |
| −10 | −10 | 15.7 | 13.0 | 23.0 | −13.0 |
| 0 | 0 | 15.7 | 13.0 | 22.0 | −13.4 |
| 10 | 10 | 15.7 | 13.0 | 23.0 | −12.7 |
| 20 | 19 | 15.5 | 14.0 | 23.0 | −12.1 |
| 30 | 29 | 15.2 | 15.0 | 25.0 | −11.5 |
| 40 | 38 | 14.8 | 17.0 | 29.0 | −10.8 |
| 50 | 47 | 14.1 | 19.0 | 34.0 | −9.8 |
| 60 | 54 | 12.9 | 21.0 | 86.0 | −8.9 |

across the steering range from $-60^\circ$ to $+60^\circ$. Within this range, an average directivity of 14.9 dB is attained, with the 3 dB- and 10 dB-beamwidths averaging 16.2 and 35.5°, respectively. These simulation results demonstrate the sample $1 \times 8$ DRA array's capability to effectively focus radiation power across a 120° range. A less than 3 dB directivity degradation can be observed in this steering range. The normalized sidelobe level (relative to the beam peak) remains below $-8.9$ dB. When considering a narrower steering range of $-50^\circ$ to $+50^\circ$, the sidelobe level improves to below $-9.8$ dB. A beam directivity and 3 dB-beamwidth degradation of 1.6 dB and 6° can be observed at the edge of the steering angle $\pm 50^\circ$.

### 3.3.5 *Simulation Results of Full-Duplex Tx-to-Rx Ports Mutual Coupling*

The high isolation between the cross-polarized ports makes the proposed DRA element a good candidate for the monostatic FD antenna arrays. The co-location of Tx and Rx antennas diminishes the necessity for intricate synchronization across disparate stations, streamlining both design and operational complexities [29]. Additionally, the dual-functionality of a single array offers substantial space savings. In other scenarios where very high Tx-Rx isolation is required, the proposed DRA structures can also be implemented into bistatic arrays, as it allows for greater antenna spacing, which in turn further boosts isolation.

DRA element for concurrent Tx and Rx, with cross-polarization enabling the distinction between Tx/Rx ports, i.e., the aperture feeding ports (T1 to T8) and the probe feeding ports (RM1 to RM8) as illustrated in the left array of Fig. 3.17. The

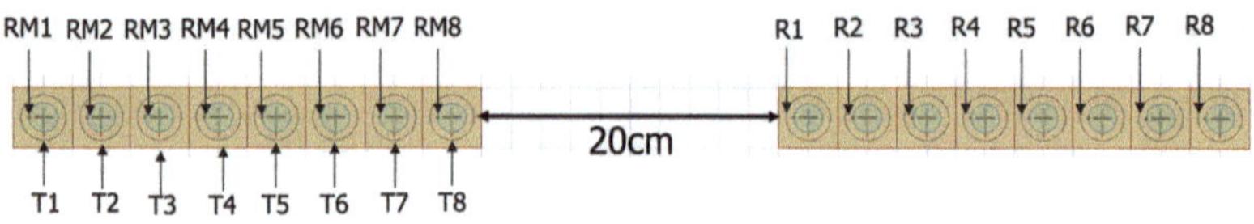


**Fig. 3.17** Full-duplex DRA arrays simulation setup

bistatic configuration utilizes two distinct arrays for transmitting and receiving in FD communications. The arrays are separated by 20 cm (approximately $2.5\lambda_0$ as shown in Fig. 3.17 for their 1×8 row sub-arrays). To maximize the cross-polarization within this setup, the Tx role is assigned to the aperture feeding ports (T1 to T8) of the left array, and the Rx role is designated to the probe feeding ports (R1 to R8) of the right array.

#### 3.3.5.1 Monostatic Array Configuration

The simulation outcomes for the Tx-to-Rx ports using a monostatic array configuration are presented in Fig. 3.18a. An average[2] MC of −53.8 dB can be achieved across 64 monostatic Tx-Rx port pairs. By leveraging a high cross-polarization ratio between the Tx and Rx ports, effectively confining EM waves, and employing a copper shield to suppress potential surface leakage and minimize spatial leakage waves, the measured MC values range from a best (minimum) of −71.9 dB to a maximum of −43.3 dB. Notably, the isolation level does not exhibit a strong correlation with the port separation distance. This considerable isolation level is crucial to guarantee the monostatic array performance, paving the way for a compact FD array design.

#### 3.3.5.2 Bistatic Array Configuration

The simulation results of the Tx-to-Rx ports with the bistatic array configuration are shown in Fig. 3.18b. The results show an average MC of −61.7 dB for 64 Tx-Rx port pairs, with the MC values ranging from −79.6 to −52.3 dB. A marked difference is observed in the MC values: for the more distant ports, such as T1 to T4 paired with R1 to R8, the average MC is −66.1 dB. In comparison, closer ports, like T5 to T8 combined with R1 to R8, exhibit a higher average MC, measured at −57.2 dB. Compared to the monostatic configuration, the separation between the Tx and Rx arrays yields an average isolation improvement of 7.9 dB and the worst Tx-Rx MC is improved by 9 dB.

[2] Geometric means are used to characterize the average mutual coupling, as the value represents a ratio of the received power to the transmitted power.

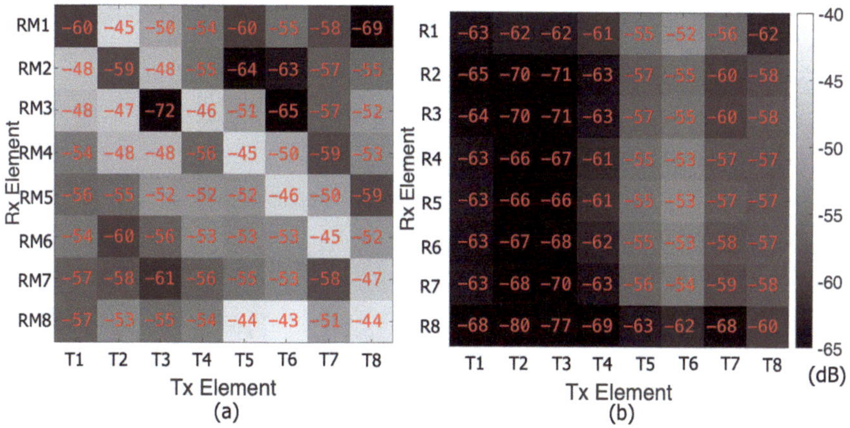


**Fig. 3.18** Tx-to-Rx port MC at 3.5 GHz: (**a**) monostatic, (**b**) bistatic (simulation results)

### *3.3.6 Proposed DRA Array Beam-Level Isolation in FD Communications*

Array beamforming plays a crucial role in further improving spatial isolation by generating directional radiation beams which diminish the power leakage in undesired directions and adjust the interactions between the Tx and Rx beams. Beam-level MC can be estimated via a summation of the MC magnitude modified by the feeding steering vectors across the ports/elements in the array and has the expression

$$S_b(\theta_{Tx}, \theta_{Rx}) = \sum_{j=1}^{8} \sum_{i=1}^{8} w_i(\theta_{Tx}) w_j(\theta_{Rx}) S_{ji}, \tag{3.4}$$

where the $S_{ji}$ is the complex S-parameter that represents the coupling between the Tx port $i$ and the Rx port $j$ in the array. $w_i(\theta_{Tx})$ and $w_j(\theta_{Rx})$ are the complex steering weights for the Tx port $i$ and Rx port $j$ at the downlink beamforming angle $\theta_{Tx}$ and uplink beamforming angle $\theta_{Rx}$ in the azimuth plane.

In the preceding section, we investigate the MC between Tx-Rx ports and the HFSS simulated S-parameter information is used as the $S_{ji}$ in Eq. (3.4) to investigate the Tx-Rx beam-level isolation with monostatic or bistatic array configurations. The Tx and Rx arrays are steering within a range of $-50°$ to $50°$, i.e., $\{\theta_{Tx}, \theta_{Rx}\} \in \{-50° : 10° : 50°\}$. Therefore, a total of $11 \times 11 = 121$ Tx-Rx beam pairs are simulated.

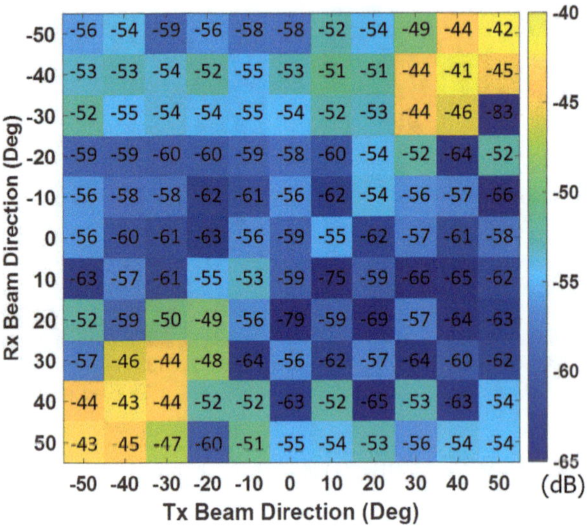


**Fig. 3.19** Tx-to-Rx beam-level isolation with monostatic array at 3.5 GHz (simulation results)

#### 3.3.6.1 Beam-Level Isolation with Monostatic Array Configuration

The Tx-Rx beam-level MC with monostatic array configuration is presented in Fig. 3.19. Across the 121 Tx-Rx beam pairs, an average beam-level MC of −56.0 dB can be achieved, which shows approximately 2.2 dB lower (better) MC compared with the Tx-Rx element level MC. The MC values span from −41.4 dB to −83.0 dB, suggesting that the maximum (worst-case) beam-level MC is constrained by the highest Tx-Rx port MC. Beam-level isolation deteriorates when the Tx and Rx beams are steered in significantly opposing directions (in the left bottom and right top corner of the graph). Due to the reason that the main source of the monostatic SI is from the near-field coupling and the beamforming is challenging in varying the interaction between the Tx/Rx beams, the monostatic array configuration makes it difficult to improve isolation with beamforming.

#### 3.3.6.2 Beam-Level Isolation Bistatic Array Configuration

Employing beamforming in a bistatic array configuration leads to significant isolation improvements, as depicted in Fig. 3.20. This technique achieves an average beam-level MC of −73.4 dB, with the best measured MC at −94.4 dB. These results represent an average isolation enhancement of 17.6 dB over the monostatic configuration and 11.9 dB over the element-level isolation in a bistatic setup. Furthermore, all 121 Tx-Rx beam pairs exhibit an MC lower (better) −60.4 dB. 81 Tx-Rx beam pairs (67%) achieve an isolation better than 70 dB. The direction of the Rx beams tends to have a greater influence on the beam-level isolation, that when $\theta_{Rx} \geq 40^{\circ}$ a degradation in the isolation performance can be observed.

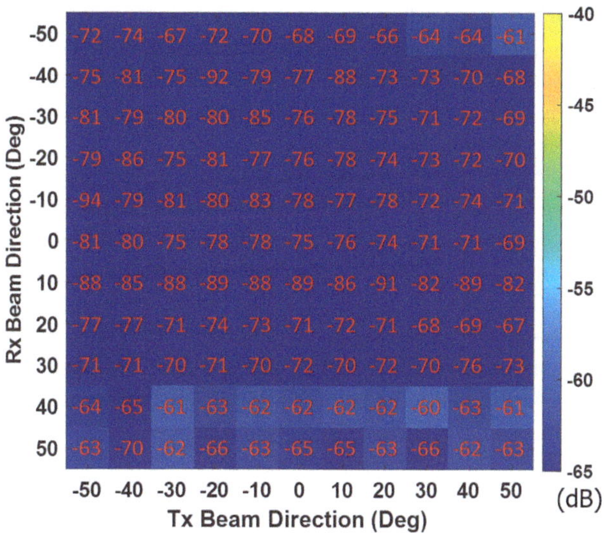


**Fig. 3.20** Tx-to-Rx Beam-level isolation with bistatic arrays at 3.5 GHz (simulation results)

## 3.4 Concluding Remarks

In this chapter, two antenna structures operating around 3.5 GHz were proposed: one employs a dual-layer metamaterial-based structure to jointly miniaturize the patch antenna size and improve inter-element isolation, and the other utilizes a DRA structure with dual-polarization feeding mechanisms. Full-wave simulation results of both proposed structures were presented to demonstrate good radiation characteristics in terms of element return loss, element gain, inter-element isolation, and array beamforming capabilities. The proposed DRA structure is shown to offer better performance in terms of enhanced impedance bandwidth, higher gain, and improved radiation efficiency, along with a considerable improvement in inter-port and inter-element isolation, which is achieved through a simple cylindrical resonator design. Additionally, its adaptability to various feeding mechanisms and notably high inter-element isolation make it a promising candidate for potential monostatic array configurations in FD communications.

However, implementing DRAs in large-scale arrays introduces unique fabrication challenges compared to conventional metallic or microstrip antenna arrays, including:

(1) Precision in Dielectric Material Properties: DRAs rely on specific permittivity values to achieve desired resonant modes. Even minor variations in the dielectric constant or loss tangent across array elements can adversely affect performance, resulting in inconsistencies in resonance and radiation characteristics.

(2) Complex and Nonstandard Geometries: DRA designs typically involve diverse shapes of dielectric resonators, such as cylindrical, rectangular, or hemispherical geometries. Fabricating these shapes with high dimensional accuracy is challenging and expensive, especially in large-scale production, and may require specialized manufacturing techniques.
(3) Bonding and Mounting: The choice of adhesive or bonding method is critical for preserving the antenna's dielectric environment. Inappropriate adhesives or mounting techniques, such as those that introduce additional layers of epoxy or solder, can lead to parasitic effects that adversely impact the antenna's performance. Therefore, in future work, we plan to investigate how DRA array designs can benefit from advanced fabrication technologies by exploring composite dielectric materials compatible with scalable manufacturing techniques, such as 3D printing and injection molding, to enable cost-effective, high-volume production.

On the other hand, the proposed metamaterial-based patch antenna structure offers the advantages of a low profile, design flexibility, and cost-effective fabrication. By leveraging modern PCB technologies, it can be easily scaled and integrated into an array. Therefore, a large-scale antenna array prototype based on the proposed metamaterial-based patch antenna structure will be developed and utilized for experimental evaluations and investigations in the following chapters.

## References

1. P. Xingdong, H. Wei, Y. Tianyang, L. Linsheng, Design and implementation of an active multibeam antenna system with 64 RF channels and 256 antenna elements for massive MIMO application in 5G wireless communications. China Commun. **11**(11), 16–23 (2014)
2. S. Malkowsky, J. Vieira, L. Liu, P. Harris, K. Nieman, N. Kundargi, I.C. Wong, F. Tufvesson, V. Öwall, O. Edfors, The world's first real-time testbed for massive MIMO: design, implementation, and validation. IEEE Access **5**, 9073–9088 (2017)
3. M. Di Renzo, A. Zappone, M. Debbah, M.S. Alouini, C. Yuen, J. De Rosny, S. Tretyakov, Smart radio environments empowered by reconfigurable intelligent surfaces: how it works, state of research, and the road ahead. IEEE J. Sel. Areas Commun. **38**(11), 2450–2525 (2020)
4. M. Jiang, Z.N. Chen, Y. Zhang, W. Hong, X. Xuan, Metamaterial-based thin planar lens antenna for spatial beamforming and multibeam massive MIMO. IEEE Trans. Antennas Propag. **65**(2), 464–472 (2016)
5. N. Shlezinger, G.C. Alexandropoulos, M.F. Imani, Y.C. Eldar, D.R. Smith, Dynamic metasurface antennas for 6G extreme massive MIMO communications. IEEE Wireless Commun. **28**(2), 106–113 (2021)
6. P. Mookiah, K.R. Dandekar, Metamaterial-substrate antenna array for MIMO communication system. IEEE Trans. Antennas Propag. **57**(10), 3283–3292 (2009)
7. R.O. Ouedraogo, E.J. Rothwell, A.R. Diaz, K. Fuchi, A. Temme, Miniaturization of patch antennas using a metamaterial-inspired technique. IEEE Trans. Antennas Propag. **60**(5), 2175–2182 (2012)
8. E. Korkontzila, D. Papafilippou, D. Chrissoulidis, Miniaturization of microstrip patch antenna for wireless applications by use of multilayered electromagnetic band gap substrate, in *2006 First European Conference on Antennas and Propagation* (IEEE, Piscataway, 2006), pp. 1–6

9. S. Ghosh, T.N. Tran, T. Le-Ngoc, Dual-layer EBG-based miniaturized multi-element antenna for MIMO systems. IEEE Trans. Antennas Propag. **62**(8), 3985–3997 (2014)
10. S. Ghosh, T.N. Tran, T. Le-Ngoc, A dual-layer EBG-based miniaturized patch multi-antenna structure, in *2011 IEEE International Symposium on Antennas and Propagation (APSURSI)* (IEEE, Piscataway, 2011), pp. 1828–1831
11. A. Askarian, J. Yao, Z. Lu, K. Wu, Surface-wave control technique for mutual coupling mitigation in array antenna. IEEE Microw. Wireless Compon. Lett. **32**(6), 623–626 (2022)
12. C.X. Mao, S.S. Gao, Y. Wang, J.T.S. Sumantyo, Compact broadband dual-sense circularly polarized microstrip antenna/array with enhanced isolation. IEEE Trans. Antennas Propag. **65**(12), 7073–7082 (2017)
13. K.L. Wu, C. Wei, X. Mei, Z.Y. Zhang, Array-antenna decoupling surface,. IEEE Trans. Antennas Propag. **65**(12), 6728–6738 (2017)
14. J. OuYang, F. Yang, Z. Wang, Reducing mutual coupling of closely spaced microstrip MIMO antennas for WLAN application. IEEE Antennas Wireless Propag. Lett. **10**, 310–313 (2011)
15. G. Zhai, Z.N. Chen, X. Qing, Enhanced isolation of a closely spaced four-element MIMO antenna system using metamaterial mushroom. IEEE Trans. Antennas Propag. **63**(8), 3362–3370 (2015)
16. Z. Iluz, R. Shavit, R. Bauer, Microstrip antenna phased array with electromagnetic bandgap substrate. IEEE Trans. Antennas Propag. **52**(6), 1446–1453 (2004)
17. H.S. Farahani, M. Veysi, M. Kamyab, A. Tadjalli, Mutual coupling reduction in patch antenna arrays using a UC-EBG superstrate. IEEE Antennas Wireless Propag. Lett. **9**, 57–59 (2010)
18. K. Payandehjoo, R. Abhari, Employing EBG structures in multiantenna systems for improving isolation and diversity gain. IEEE Antennas Wireless Propag. Lett. **8**, 1162–1165 (2009)
19. A. Petosa, A. Ittipiboon, Dielectric resonator antennas: a historical review and the current state of the art. IEEE Antennas Propag. Mag. **52**(5), 91–116 (2010)
20. A. Rashidian, L. Shafai, D.M. Klymyshyn, Compact wideband multimode dielectric resonator antennas fed with parallel standing strips. IEEE Trans. Antennas Propag. **60**(11), 5021–5031 (2012)
21. R.K. Mongia, A. Ittipiboon, Theoretical and experimental investigations on rectangular dielectric resonator antennas. IEEE Trans. Antennas Propag. **45**(9), 1348–1356 (1997)
22. M. Abedian, M. Khalily, P. Xiao, F. Wang, R. Tafazolli, A.A. Kishk, Mm-wave high isolated dual polarized dielectric resonator antenna for in-band full-duplex systems. IEEE Access **11**, 38218–38225 (2023)
23. Y.Z. Liang, F.C. Chen, W.F. Zeng, Q.X. Chu, Design of self decoupling dielectric resonator antenna with shared radiator. IEEE Trans. Antennas Propag. **71**(1), 1053–1058 (2022)
24. F. Yang, Y. Rahmat-Samii, Microstrip antennas integrated with electromagnetic band-gap (EBG) structures: a low mutual coupling design for array applications. IEEE Trans. Antennas Propag. **51**(10), 2936–2946 (2003)
25. Y. Pan, S. Zheng, B. Hu, Wideband and low-profile omnidirectional circularly polarized patch antenna. IEEE Trans. Antennas Propag. **62**(8), 4347–4351 (2014)
26. R. Li, G. DeJean, J. Laskar, M.M. Tentzeris, Investigation of circularly polarized loop antennas with a parasitic element for bandwidth enhancement. IEEE Trans. Antennas Propag. **53**(12), 3930–3939 (2005)
27. A. Petosa, *Dielectric Resonator Antenna Handbook* (Artech, Norwood, 2007)
28. S. Long, M. McAllister, L. Shen, The resonant cylindrical dielectric cavity antenna. IEEE Trans. Antennas Propag. **31**(3), 406–412 (1983)
29. J. Ha, M.A. Elmansouri, P.V. Prasannakumar, D.S. Filipovic, Monostatic co-polarized full-duplex antenna with left-or right-hand circular polarization. IEEE Trans. Antennas Propag. **65**(10), 5103–5111 (2017)

# Part I
# Half-Duplex Communications

# Chapter 4
# Hybrid Beamformer Design with Impaired Components for mMIMO

## 4.1 Introduction

As expressed in Chap. 2, the large antenna arrays in mMIMO systems bring two interesting challenges: (i) increased expense and energy consumption due to the use of large number of RF chains, and (ii) the use of significant amount of spectral resources and increased CSI overhead [1]. In order to overcome these problems, HBF has been proposed as a promising solution [1, 2], which is a two-stage precoder consisting of an analog RF beamforming stage and a digital baseband precoding stage, and with much-reduced system complexity/hardware cost, it can achieve the performance close to DBF which requires full instantaneous CSI. In HBF, large-dimensional processing is carried out by phase shifters (PSs) at the transceiver RF front-end, followed by low-dimensional processing at the baseband level. A few RF chains and digital-to-analog converters/analog-to-digital converters (DACs/ADCs) connect the RF and the baseband stages. Furthermore, at the RF-stage, two commonly used methods of processing input parameters are: (i) using fast time-varying CSI [3–9], and (ii) using slowly time varying instantaneous CSI [10–18]. The first strategy employs full CSI, whereas, in the second strategy, only the reduced-dimension effective instantaneous CSI seen from the BB-stage is required for the BB precoder design. In other words, the second strategy is capable of reducing both the hardware cost/complexity and the CSI overhead size.

Most of the HBF solutions have assumed the use of full-resolution PSs and DACs to achieve a satisfactory performance close to DBF structures. However, because of the increased complexity and costs, implementing $\infty$-bit PSs and DACs/ADCs is impractical. The recent studies show that there has been an increased interest in the use of small, power-efficient, and inexpensive devices for hardware efficient transceivers capable of beamforming. Therefore, the use of low-resolution PSs [19–27] and DACs/ADCs [28–33] allows a more viable HBF solution to be designed. Another important aspect in mMIMO is the number of the antenna elements as well as their arrangement at the transmitter side as the array configurations can be

 
T. Le-Ngoc et al., *Half-Duplex/Full-Duplex MU-mMIMO*, Wireless Networks,
https://doi.org/10.1007/978-3-032-09195-6_4

critical in the generation of narrower beams with reduced sidelobes, which can offer small interference to the unintended users. The existing research studies mostly use uniform linear array (ULA) at the BS for the HBF design [3–7, 10–14, 16, 17, 19–24, 28–32]. However, it is not applicable to deploy a large single-dimensional (1D) ULA at the BS because of: (i) spatial inefficiency, and (ii) restriction to illuminate both azimuth and elevation angles at the same time. The antenna elements can be arranged on a two-dimensional (2D) grid to overcome the aforementioned issues. In this chapter, we present the analysis of different 2D array structures (ULA, URA, UCA, and CCA) in mMIMO, where we compare the spatial, spectral and energy efficiencies by designing the HBF using low-resolution PSs and quantized DACs. The proposed HBF schemes include two cascaded stages: (i) the *RF-beamforming stage* is designed via the eigen-decomposition of mMIMO channel second-order correlation matrix, and (ii) the *baseband MU precoding stage* is constructed via the regularized zero-forcing (RZF) technique [34, 35] to mitigate the MU interference in the reduced-dimension effective MU-channel. The main contributions are summarized below:

- **Energy-Efficient HBF Design Using Low-Resolution PSs:** Most of the existing HBF designs (e.g., in [3–7, 10–17, 28–33]) assume the use of full-resolution PSs for the implementation of RF-stage. However, the components required for phase shift with high accuracy can be expensive [36]. Therefore, it is reasonable to use cost-effective PSs in HBF. One possible way to design the HBF with low-resolution PSs is to design the RF beamformer assuming infinite-resolution PSs and then, quantize each PSs value by a finite set [3]. This approach, however, is not suitable for systems with low-resolution phase constraint and requires full CSI. Furthermore, the HBF designs which are presented in [19–21, 23–26] are considered for SU-mMIMO and MU-MISO systems. In this chapter, we design the HBF using low-resolution PSs for MU-mMIMO by taking into account the few-bit phase constraint and formulate the optimization problem which is solved using orthogonal matching pursuit (OMP) [37, 38]. The algorithmic precoding solution takes the input as the optimal unconstrained (full-resolution) RF beamformer and approximates the constrained (quantized) RF beamformer by exploiting the dominant eigenvectors at the RF. We present the illustrative results of the proposed HBF scheme and show that it is possible to approach the performance limit of the unconstrained RF-stage with as low as 2 bit PSs (i.e., sum-rate degradation $\approx$ 2-3 bps/Hz).
- **Comparison of 2-D Antenna Array Structures:** In this chapter, we compare the spatial, spectral and energy efficiencies of different 2D antenna array structures using low-resolution PSs and DACs. The array structure used in [3–7, 10–14, 16, 17, 19–33] is either ULA or URA. However, in this work we compare four different 2D array structures using low-resolution PSs and DACs. The simulation results show that the sum-rate of both CCA and URA can approach the sum-rate of their DBF counterparts with much-reduced number of RF chains.
- **HBF Design Using Low-Resolution PSs and DACs:** As mentioned above, the HBF is designed using either only low-resolution PSs [19–27] or only low-

resolution DACs/ADCs [28–33]. To the best of our knowledge, the design of a MU-mMIMO HBF using both *low-resolution DACs* and *low-resolution PSs* has not been considered yet. Moreover, we also compare the performance of such a reduced-complexity HBF for different 2D array structures i.e., ULA, URA, UCA, and CCA. The simulation results give useful results about the combination of various low-resolution DACs and PSs for near optimal performance. The proposed HBF using 2-bit PSs and 5-bit DACs can achieve almost the same spectral efficiency while offering higher energy efficiency than the HBF using full-resolution PSs and DACs. Therefore, the designed HBF saves power consumption and cost while having a negligible impact on the performance.

- **Design of the Quantized Variable Gain RF Beamformer:** In this work, we design the HBF in which the RF beamforming stage is constructed using eigen beamforming based on user's angular location. Eigen beamforming can give sum-rate performance of HBF close to DBF [10–13], however, the RF beamformer results in having non-constant modulus entities. The quantization of such a variable-gain RF beamformer is challenging. Also, to the best of our knowledge, the quantization of analog precoder which is designed using eigen beamforming has not been considered yet. Therefore, we solve this problem by modifying our system model to have a constant modulus RF beamformer and introducing a transfer block at the baseband stage by formulating an optimization problem. The resulting RF beamformer requires a double-PS structure and follows the constant modulus constraint, i.e., the gain of each entity of the RF beamformer can vary between 0 and 2. The resulting HBF with constant-modulus gives the same sum-rate performance as the variable gain RF beamformer.

The rest of this chapter is organized as follows. In Sect. 4.2, we introduce the HBF design using full-resolution PSs and DACs. Section 4.3 presents the HBF solution when using quantized DACs/ADCs at BS. In Sect. 4.4, we discuss HBF design using low-resolution PSs followed by the algorithmic HBF solution for few-bit DACs and PSs in Sect. 4.5. The illustrative results are provided in Sect. 4.6 to compare the performance of different 2D array structures. Finally, the chapter is concluded in Sect. 4.7.

## 4.2 Hybrid Beamforming Design Using Full Resolution PSs and DACs

In this section, we discuss the HBF solution using full-resolution hardware components.

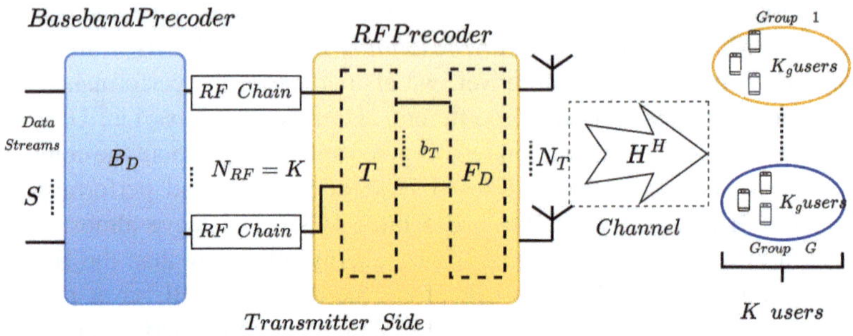


**Fig. 4.1** System model of the hybrid beamforming architecture

### 4.2.1 System Model

In the system model shown in Fig. 4.1, we assume that $N_T$ BS antennas at the transmitter are fed by $N_{RF}$ RF chains to serve $K$ single-antenna users. $b_T$ represents the number of orthogonal-beams spanning the degrees of freedom (DoF) provided by the channel, such that $K \leq N_{RF} \leq b_T \ll N_T$. $S$ denotes the independent data streams to be transmitted, which is equal to the number of simultaneously served users in this model (i.e., $S = K$). Furthermore, by selecting $N_{RF} = K$ RF chains, the hybrid precoding matrix $\mathbf{B} = \mathbf{F}_D\mathbf{T}\mathbf{B}_D$ where (i) RF-beamforming matrix $\mathbf{F}_D \in \mathbb{C}^{N\times b_T}$ is based on the channel second order statistics and reduces the CSI overhead and the number of RF chains (ii) transfer matrix $\mathbf{T} \in \mathbb{C}^{b_T\times N_{RF}}$ exploits the available DoF of the channel, and further reduces the number of RF chains without sacrificing the performance (iii) multi-user precoding matrix $\mathbf{B}_D \in \mathbb{C}^{N_{RF}\times S}$ is based on the reduced-dimension effective MU channel, and reduces the MU interference. The received signal at the $k^{\text{th}}$ user is expressed as:

$$y_k = \mathbf{h}_k^H\mathbf{B}\mathbf{s} + n_k. \tag{4.1}$$

$\mathbf{s} \in \mathbb{C}^{S\times 1}$ is the transmitted data signal satisfying $\mathbb{E}\big[||\mathbf{s}||_2^2\big] \leq \frac{P_T}{S}\mathbf{I}_s$, where $P_T$ is the transmit power at the BS and $n_k$ denotes the additive circular symmetric Gaussian noise such that $n_k \sim \mathcal{CN}(0, 1)$. Assuming no line of sight propagation, the channel $\mathbf{h}_k \sim \mathcal{CN}(0, \mathbf{R}_k)$, where $\mathbf{R}_k \in \mathbb{C}^{N\times N}$ is the covariance matrix.

The channel vector $\mathbf{h}_k \in \mathbb{C}^{N\times 1}$ for user $k$ can be expressed by using Karhunen-Loeve decomposition as:

$$\mathbf{h}_k = \mathbf{U}_k\mathbf{\Lambda}_k^{\frac{1}{2}}\mathbf{w}_k, \tag{4.2}$$

where $\mathbf{w}_k \in \mathbb{C}^{r\times \mathbf{1}} \sim \mathcal{CN}(0, \mathbf{I}_r)$ is the complex weight coefficient vector, $\mathbf{\Lambda}_k \in \mathbb{C}^{r\times r}$ is the diagonal matrix containing $r$ eigenvalues of $\mathbf{R}_k$ and $\mathbf{U_k} \in \mathbb{C}^{N\times r}$ is the matrix containing the eigenvectors corresponding to the non-zero eigenvalues.

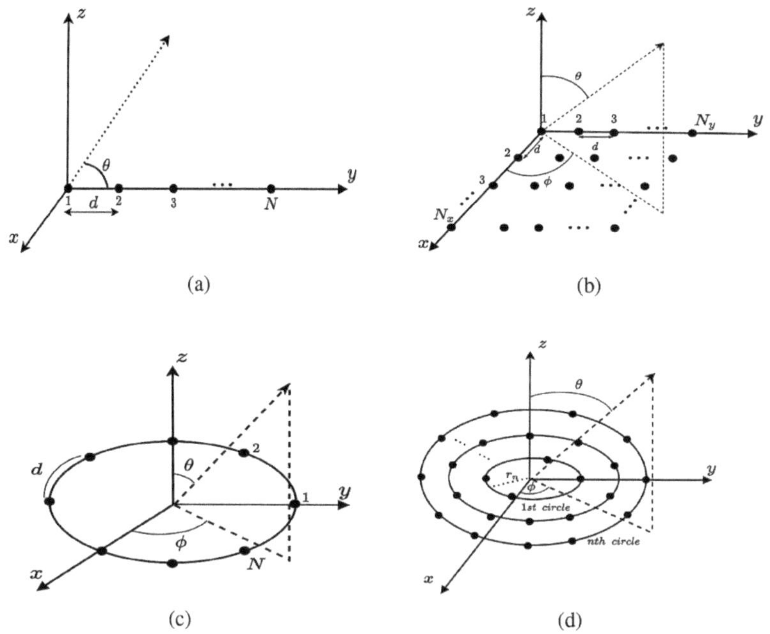


**Fig. 4.2** 2D array structures. (**a**) ULA. (**b**) URA. (**c**) UCA. (**d**) CCA

Assuming mutually independent channel vectors $\mathbf{h}_k$, the overall channel model can be expressed as

$$\mathbf{H} = [\mathbf{h}_1, \mathbf{h}_2, \ldots, \mathbf{h}_K]. \tag{4.3}$$

The correlation matrix $\mathbf{R}_k$ is obtained by using the one-ring model [39], where a user is located at some distance $r_d$ and the mean azimuth angle is $\theta$. For the 1D and 2D antenna structures as shown in Fig. 4.2, assuming a uniform distribution of power received from the antennas, the correlations between the channel coefficients of the antennas $1 \leq n, p \leq N$ are respectively given as:

$$\left[\mathbf{R}_{n,p}\right]_{1D} = \frac{1}{2\Delta} \int_{-\Delta+\theta}^{\Delta+\theta} e^{-j2\pi d(n-p)\sin(\Psi)}\, d\Psi, \tag{4.4}$$

$$\left[\mathbf{R}_{n,p}\right]_{2D} = \frac{1}{2\Delta} \int_{-\Delta}^{\Delta} e^{-j\frac{2\pi}{\lambda}\mathbf{k}(\Psi+\theta)(\mathbf{v}_n-\mathbf{v}_p)}\, d\Psi, \tag{4.5}$$

where $\lambda$ is the wavelength. $\Delta \approx \pm \arctan \frac{l}{r_d}$ is the angle spread around the mean azimuth angle for a user surrounded by a ring of scatters of radius $l$ , $\mathbf{k}(\Psi) = (\cos(\Psi), \sin(\Psi))^T$ is the wave vector for a planar wave impinging with angle-of-arrival (AoA) $\Psi$ and $\mathbf{v}_n, \mathbf{v}_p$ is $\mathbb{R}^2$. The correlation matrix $\mathbf{R}_{1D}$ for ULA is computed using (4.4) where $d$ is the distance between the $n^{\text{th}}$ and $p^{\text{th}}$ antenna. For UCA, URA, and CCA, (4.5) is used to calculate the correlation matrix $\mathbf{R}_{2D}$. Thus, the correlation matrix $\mathbf{R}$ depends on the antenna array configuration employed at the BS, and the corresponding rank of $\mathbf{R} = \mathbf{U}\boldsymbol{\Lambda}\mathbf{U}^H$ plays an important role in the sum-rate performance of the array structure.

In the eigen-beamforming, the hybrid precoder $\mathbf{B} \in \mathbb{C}^{N_T \times S}$ is constructed by concatenating the RF-beamforming $\mathbf{F}_D \in \mathbb{C}^{N_T \times b_T}$ and reduced-dimensional multi-user precoder (baseband-stage) $\mathbf{B}_D \in \mathbb{C}^{N_{RF} \times S}$ through an intermediate block called transfer block $\mathbf{T} \in \mathbb{C}^{b_T \times N_{RF}}$.

### 4.2.2 RF-Beamforming Stage

Suppose $K$ users are clustered into $G$ groups based on their AoD information, where each group contains $K_g$ number of users such that $K = \sum_{g=1}^{G} K_g$. For simplicity, we assume that the users in the same group have identical covariance matrix $\mathbf{R}_g$, where $g = 1, 2, \ldots, G$. The index $g_k = \sum_{g'=1}^{g-1} K_{g'} + k$ is used to denote the $k^{\text{th}}$ user in group $g$. Then, the channel vector for the user $k$ in group $g$ is $\mathbf{h}_{g_k} \sim \mathcal{CN}(0, \mathbf{R}_g)$, where $\mathbf{R}_g = \mathbf{U}_g\boldsymbol{\Lambda}_g\mathbf{U}_g^H$ is the covariance matrix of the group $g$. Based on the AoD of the users, the correlation matrix can be written as:

$$\mathbf{R}_g = [\mathbf{U}_g^+, \mathbf{U}_g^-][\boldsymbol{\Lambda}_g^+, \boldsymbol{\Lambda}_g^-][\mathbf{U}_g^+, \mathbf{U}_g^-]^H, \tag{4.6}$$

where $\mathbf{U}_g^+ \in \mathbb{C}^{N \times r_g^+}$ is the matrix of eigenvectors corresponding to the dominant eigenvalues for the user group $g$. Similarly, $\boldsymbol{\Lambda}_g^+ \in \mathbb{C}^{r_g^+ \times r_g^+}$ is the diagonal matrix of the dominant eigenvalues. As a result, the data streams per group is $S_g \in [0, \min\{K_g, r_g^+\}]$. The maximum number of independent data streams that can be transmitted is limited by the multiplexing gain $\min\{K_g, r_g^+\}$. The total data streams can be written as $S = \sum_{g=1}^{G} S_g$. By using Karhunen-Loeve decomposition, we have:

$$\mathbf{h}_{g_k} = \mathbf{U}_g\boldsymbol{\Lambda}_g^{\frac{1}{2}}\mathbf{w}_{g_k}. \tag{4.7}$$

The overall channel matrix $\mathbf{H}$ can be written as:

$$\mathbf{H} = [\mathbf{H}_1, \mathbf{H}_2, \ldots, H_G], \tag{4.8}$$

where $\mathbf{H}_g = [\mathbf{h}_{g_1}, \mathbf{h}_{g_2}, \ldots, h_{g_{K_g}}]$ is the channel matrix for group $g$. The performance of the hybrid precoder heavily depends on the choice of $b_T$ and $N_{RF}$. It must be seen that (**i**) $N_{RF} \geq S$ to ensure that the minimum number of RF chains used in the precoder is not less than the transmitted signals and (**ii**) $N_{RF} = \sum_{g=1}^{G} N_{RF_g}$. The RF-beamforming matrix $\mathbf{F}_D$ depends on the second-order statistics as it relies on the eigenvalues and eigenvectors extracted from the covariance matrix $\mathbf{R}_g$. If $\mathbf{B}_{A,g} = \mathbf{U}_g^+$ is the RF-beamforming matrix of group $g$ and by choosing $b_g = r_g^+$, the complete RF-beamforming matrix can be written as:

$$\mathbf{F}_D = [\mathbf{F}_{D,1}, \mathbf{F}_{D,2}, \ldots, \mathbf{F}_{D,G}]. \tag{4.9}$$

The overall reduced dimensional effective channel matrix is expressed as $\mathcal{H}^H = \mathbf{F}_D^H \mathbf{H}$. Using (4.8) and (4.9), $\mathcal{H}^H$ can be written as:

$$\mathcal{H}^H = \begin{bmatrix} \mathbf{H}_1^H \mathbf{F}_{D,1} & \mathbf{H}_1^H \mathbf{F}_{D,2} & \ldots & \mathbf{H}_1^H \mathbf{F}_{D,G} \\ \mathbf{H}_2^H \mathbf{F}_{D,1} & \mathbf{H}_2^H \mathbf{F}_{D,2} & \ldots & \mathbf{H}_2^H \mathbf{B}_{A,G} \\ \vdots & \vdots & \ddots & \vdots \\ \mathbf{H}_G^H \mathbf{F}_{D,1} & \mathbf{H}_G^H \mathbf{F}_{D,2} & \ldots & \mathbf{H}_G^H \mathbf{F}_{D,G} \end{bmatrix}, \tag{4.10}$$

where the diagonal matrices $\mathbf{H}_g^H \mathbf{F}_{D,g} \in \mathbb{C}^{K_g \times b_g}$ are the effective channel matrix for group $g$ and the off-diagonal matrices $\mathbf{H}_g^H \mathbf{F}_{D,q} \in \mathbb{C}^{K_g \times b_q}$ represent the effective interference channel matrix between groups $g$ and $q$, $\forall q \neq g$.

### 4.2.3 Transfer Block and Multiuser Precoder Design

The transfer block $\mathbf{T} \in \mathbb{C}^{b_T \times N_{RF}}$ is designed between the RF-beamforming matrix and the multi-user precoding matrix, reducing the number of RF chains $N_{RF}$ such that $K \leq N_{RF} < b_T$. $\mathbf{T}$ is placed at the RF precoder satisfying $\mathbb{E}\big[||\mathbf{s}||_2^2\big] = \text{tr}(\mathbf{B}_D^H \mathbf{T}^H \mathbf{T} \mathbf{B}_D) \leq P_T$, where $P_T$ is the transmit power. The transfer block $\mathbf{T}$ and the reduced dimensional multi-user precoder $\mathbf{B}_D$ is designed as:

$$\begin{aligned} \min_{\mathbf{T},\mathbf{B}_D} \quad & ||\hat{\mathbf{B}}_D - \mathbf{T}\mathbf{B}_D||_2^2, \\ \text{s.t.} \quad & \mathbf{T} \in \mathbb{T}_o, \\ & \text{tr}(\mathbf{B}_D^H \mathbf{T}^H \mathbf{T} \mathbf{B}_D) \leq P_T, \end{aligned} \tag{4.11}$$

where $\mathbb{T}_o$ represents the set of matrices of size $b_T \times N_{RF}$. $\hat{\mathbf{B}}_D = \gamma \mathbf{X} \mathcal{H}^H$ is the $N_{RF} \times K$ baseband precoder without using the transfer block $\mathbf{T}$. The optimization problem is formulated to design the reduced-size baseband precoder $\mathbf{B}_D$ giving the same performance as $\hat{\mathbf{B}}_D$. $\mathcal{H}^H$ is the reduced dimensional effective channel, $\mathbf{X} = [\mathcal{H}^H \mathcal{H} + \alpha b_T \mathbf{I}_{b_T}]^{-1}$, $\mathbf{I}_{b_T} \in \mathbb{C}^{b_T \times b_T}$, $\alpha$ is the regularization parameter and $\gamma$ is the normalization factor used to satisfy the power constraint, which can be written as:

$$\gamma = \sqrt{\frac{S}{tr\{\mathcal{H}^H \mathbf{X}^H \mathbf{F}_D^H \mathbf{F}_D \mathbf{X} \mathcal{H}\}}}. \tag{4.12}$$

Since (4.11) is non-convex, so $\mathbf{T}$ is dissolved into $\mathbf{T}_A \in \mathbb{C}^{b_T \times K}$ and $\mathbf{T}_B \in \mathbb{C}^{b_T \times k}$ by using $N_{RF} = K$ RF chains. Then, the optimal solution is given as [40]:

$$\begin{aligned}
\mathbf{T}_A(a,b) &= e^{j\left(\angle \hat{\mathbf{B}}_D(a,b) + \cos^{-1}\left(\frac{\hat{\mathbf{B}}_D(a,b)}{2\mu}\right)\right)}, \\
\mathbf{T}_B(a,b) &= e^{j\left(\angle \hat{\mathbf{B}}_D(a,b) - \cos^{-1}\left(\frac{\hat{\mathbf{B}}_D(a,b)}{2\mu}\right)\right)}, \\
\mathbf{B}_D &= (\mathbf{T}_A + \mathbf{T}_B)^{\dagger} \hat{\mathbf{B}}_D,
\end{aligned} \tag{4.13}$$

where $\mu = \frac{1}{2} \max_{a,b} |\hat{\mathbf{B}}_D(a,b)|$ is the half of the highest modulus element at $\hat{\mathbf{B}}_D$. Then, the joint-group processing (JGP) approach [10] is used for the SINR computation. Therefore, the SINR at the user $k$ in group $g$ can be computed as follows:

$$SINR_{g_k} = \frac{\frac{P_T}{S} |\mathbf{h}_{g_k}^H \mathbf{F}_D \hat{\mathbf{B}}_D \hat{\mathbf{B}}_D^H \mathbf{F}_D^H \mathbf{h}_{g_k}|^2}{1 + \frac{P_T}{S} \sum_{i \neq g_k} |\mathbf{h}_{g_k}^H \mathbf{F}_D \hat{\mathbf{B}}_D \hat{\mathbf{B}}_D^H \mathbf{F}_D^H \mathbf{h}_i|^2}. \tag{4.14}$$

Hence, the corresponding sum-rate performance can be computed as:

$$R = \sum_{K_g=1}^{G} \sum_{g_k=1}^{K_g} \mathbb{E}[\log_2(1 + SINR_{g_k})]. \tag{4.15}$$

## 4.3 Hybrid Beamforming Design Using Low-Resolution DACs

In this section, we discuss the HBF solution using low-resolution DACs/ADCs in mMIMO systems.

### *4.3.1 System Model*

In the system model shown in Fig. 4.3, we assume that $N_T$ BS antennas at the transmitter are fed by $N_{RF}$ RF chains to serve $K$ single-antenna users, where $K \leq N_{RF} \ll N_T$. The hybrid precoder $\mathbf{B} = \mathbf{F}_D \mathbf{B}_D$ is divided as the RF precoder $\mathbf{F}_D \in \mathbb{C}^{N_T \times N_{RF}}$ and the baseband precoder $\mathbf{B}_D \in \mathbb{C}^{N_{RF} \times K}$. The system model employs low resolution $q$-bit DACs between the RF and the baseband precoder to reduce the system complexity/cost. To model the precoded signal at the transmitter,

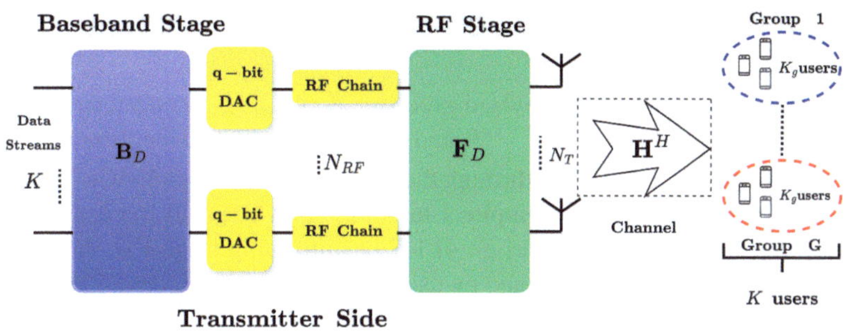


**Fig. 4.3** Hybrid beamforming architecture using few-bit DACs

**Table 4.1** Distortion factors $\mu$ in q-bit DACs ($q \leq 5$)

| q | 1 | 2 | 3 | 4 | 5 |
|---|---|---|---|---|---|
| $\mu$ | 0.6366 | 0.8825 | 0.96546 | 0.990503 | 0.997501 |

we consider the non-uniform quantizer, and adopt the additive quantization noise model (AQNM) as in [41, 42]. Then, the received signal at the $k^{\text{th}}$ user is expressed as:

$$\begin{aligned} y_k &= \mathbf{h}_k^H \mathbf{F}_D \mathbb{Q}\{\mathbf{B}_D \mathbf{s}\} + n_k, \\ &= \mathbf{h}_k^H \mathbf{F}_D \{\mu \mathbf{B}_D \mathbf{s} + n_q\} + n_k, \\ &= \underbrace{\mu \mathbf{h}_k^H \mathbf{F}_D \mathbf{B}_D \mathbf{s}}_{\text{Desired Signal}} + \underbrace{\mathbf{h}_k^H \mathbf{F}_D n_q}_{\text{Quantization Noise}} + \underbrace{n_k}_{\text{Noise}}, \end{aligned} \tag{4.16}$$

where $\mathbf{s} \in \mathbb{C}^K$ is the transmitted data signal satisfying the power constraint $P_T$ (i.e., $\mathbb{E}\{\|\mathbf{s}\|_2^2\} \leq P_T$). Here, $n_k$ denotes the additive circular symmetric Gaussian noise such that $n_k \sim \mathcal{CN}(0, 1)$, $\mathbb{Q}(.)$ is the quantizer function, $\mu$ is the distortion factor of $q$-bit DAC, and $n_q \sim \mathcal{CN}(0, \mathbf{R}_{n_q})$ is the additive Gaussian quantization noise which is uncorrelated with $\mathbf{s}$. For $q \leq 5$, the exact values of $\mu$ is given in Table 4.1. When quantization bit $q > 5$, then the distortion factor can be approximated as $\mu \approx 1 - \frac{\pi\sqrt{3}}{2} 2^{-2q}$ [42]. Furthermore, the covariance matrix of $n_q$ is given as:

$$\mathbf{R}_{n_q} = \mathbb{E}\left[n_q n_q^H\right] = \mu(1-\mu)\ \text{diag}(\mathbf{B}_D \mathbf{B}_D^H). \tag{4.17}$$

Assuming no line of sight propagation, the channel $\mathbf{h}_k \sim \mathcal{CN}(0, \mathbf{R}_k)$, where $\mathbf{R}_k \in \mathbb{C}^{N_T \times N_T}$ is the covariance matrix. The channel vector $\mathbf{h}_k \in \mathbb{C}^{N_T \times 1}$ for user $k$ can be expressed by using Karhunen-Loeve decomposition as given in (4.2). Similarly, the overall channel model can be expressed as given in (4.4).

### 4.3.2 RF and BB Stage Design with Low-Resolution DACs

In the eigen-beamforming, the hybrid precoder $\mathbf{B} \in \mathbb{C}^{N_T \times K}$ concatenates the RF-beamforming stage $\mathbf{F}_D \in \mathbb{C}^{N_T \times N_{RF}}$ and the reduced-dimensional baseband MU precoding stage $\mathbf{B}_D \in \mathbb{C}^{N_{RF} \times K}$ through the use of low-resolution DACs. The design of the RF beamforming stage employs full-resolution PSs and the details can be found in Sect. 4.2.2. After the design of RF beamforming stage $\mathbf{F}_D$, the baseband MU precoder $\mathbf{B}_D$ can be determined by using the joint group processing (JGP) [10]. Using regularized zero forcing, $\mathbf{B}_D$ can be defined as:

$$\mathbf{B}_D = \gamma \mathbf{X} \mathcal{H}^H, \tag{4.18}$$

$\mathcal{H}^H$ is the reduced-dimension effective channel as given in (4.10), $\mathbf{X} = [\mathcal{H}^{\mathcal{H}}\mathcal{H} + \alpha N_{RF}\mathbf{I}_{N_{RF}}]^{-1}$, $\mathbf{I}_{N_{RF}} \in \mathbb{C}^{N_{RF} \times N_{RF}}$, $\alpha$ is the regularization parameter and $\gamma$ is the normalization factor used to satisfy the power constraint, which can be written as:

$$\gamma = \sqrt{\frac{S}{\text{tr}\{\mathcal{H}^H \mathbf{X}^H \mathbf{F}_D^H \mathbf{F}_D \mathbf{X} \mathcal{H}\}}}. \tag{4.19}$$

Therefore, the signal to quantization, interference and noise ratio (SQINR) at the user $k$ in group $g$ can be computed as follows:

$$\text{SQINR}_{g_k} = \frac{\frac{P_T}{S} |\mathbf{h}_{g_k}^H \mathbf{F}_D \mathbf{X} \mathbf{F}_D^H \mathbf{h}_{g_k}|^2}{1 + \hat{a} + \hat{b}}, \tag{4.20}$$

where $\hat{a} = \frac{P_T}{S} \sum_{i \neq g_k} |\mathbf{h}_{g_k}^H \mathbf{F}_D \mathbf{X} \mathbf{F}_D^H \mathbf{h}_i|^2$ accounts for the multi-user interference and $\hat{b} = \frac{P_T}{S} |\mathbf{h}_{g_k}^H \mathbf{F}_D \mathbf{R}_{n_q} (\mathbf{F}_D^H \mathbf{h}_{g_k})^H|^2$ is the quantization noise. Hence, the corresponding sum-rate performance can be computed as:

$$\text{R} = \sum_{K_g=1}^{G} \sum_{g_k=1}^{K_g} \mathbb{E}[\log_2(1 + \text{SQINR}_{g_k})]. \tag{4.21}$$

## 4.4 Hybrid Beamforming Design Using Low-Resolution PSs

In this section, we present the HBF design using low-resolution PSs.

### 4.4.1 System Model

In this section, we introduce the system models of the proposed hybrid MU-mMIMO systems. We consider a MU-mMIMO system with three different HBF structures shown in Fig. 4.4. At first, the HBF architecture is depicted in Fig. 4.4a, which considers full-resolution hardware components i.e., $\infty$-bit DACs and PSs. Then, the HBF using $b$-bit PSs and $\infty$-bit DACs is shown in Fig. 4.4b, which offers an energy efficient structure due to less power consumption of finite-bit PSs. Finally, the least power consuming HBF model is presented in Fig. 4.4c, which uses $b$-bit PSs and $q$-bit DACs. The BS employs $N_T$ antenna elements at the transmitter, which

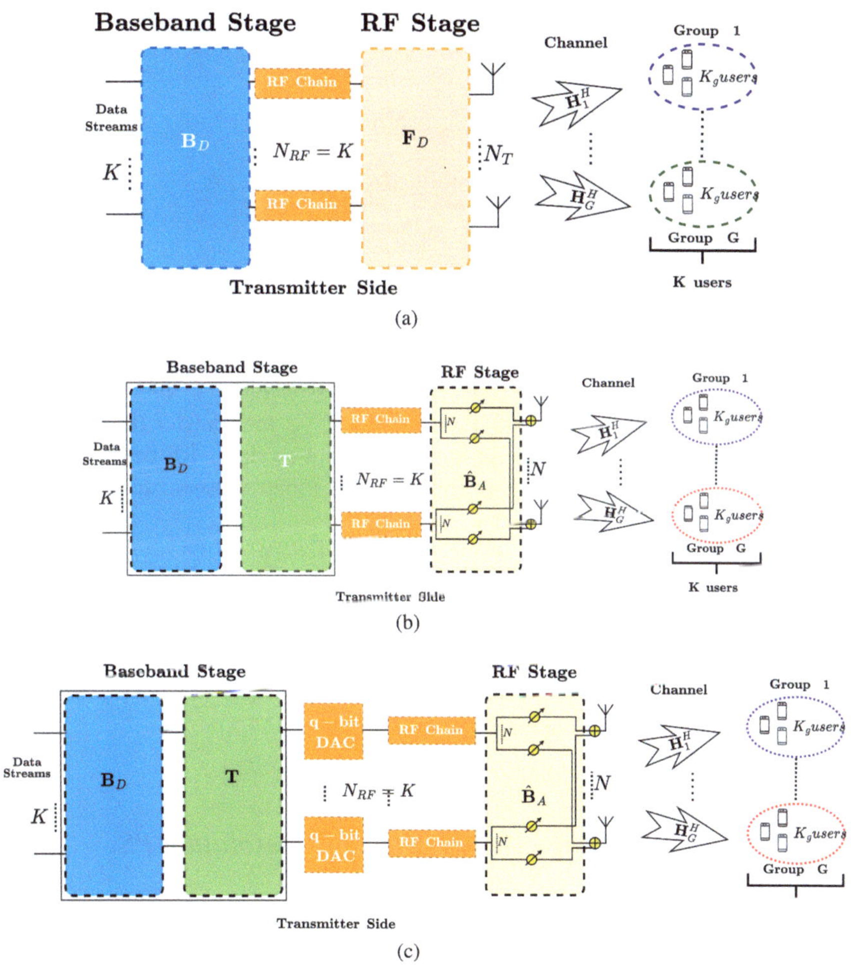


**Fig. 4.4** MU-mMIMO HBF architectures. (**a**) HBF with $\infty$-bits DACs and PSs. (**b**) HBF with $b$-bit PSs and $\infty$-bit DACs. (**c**) HBF with $q$-bit DACs and $b$-bit PSs

are fed by $N_{RF}$ RF chains to serve $K$ single-antenna users, where $K \leq N_{RF} \ll N_T$. As shown in Fig. 4.4a, the hybrid precoder $\mathbf{B} = \mathbf{F}_D\mathbf{B}_D$ is divided as the RF beamformer $\mathbf{F}_D \in \mathbb{C}^{N_T \times N_{RF}}$ and the baseband precoder $\mathbf{B}_D \in \mathbb{C}^{N_{RF} \times K}$. The design of HBF helps in reducing the number of RF chains from $N_T$ to $N_{RF}$. Then, the received signal at the $k^{\text{th}}$ user is expressed as

$$y_k = \mathbf{h}_k^H \mathbf{B}\mathbf{s} + n_k. \tag{4.22}$$

The system model shown in Fig. 1(c) employs low resolution $q$-bit DACs between the RF and the baseband precoders to reduce the system complexity/cost. To model the precoded signal at the transmitter, we consider the non-uniform quantizer, and adopt the additive quantization noise (AQN) model as in [41, 42]. Then, the received signal at the $k^{\text{th}}$ user is given as

$$\begin{aligned} y_k &= \mathbf{h}_k^H \mathbf{F}_D \mathbb{Q}\{\mathbf{T}\mathbf{B}_D\mathbf{s}\} + n_k, \\ &= \mathbf{h}_k^H \mathbf{F}_D \{\mu \mathbf{T}\mathbf{B}_D\mathbf{s} + n_q\} + n_k, \\ &= \underbrace{\mu \mathbf{h}_k^H \mathbf{F}_D \mathbf{T}\mathbf{B}_D\mathbf{s}}_{\text{Desired Signal}} + \underbrace{\mathbf{h}_k^H \mathbf{F}_D n_q}_{\text{Quantization Noise}} + \underbrace{n_k}_{\text{Noise}}, \end{aligned} \tag{4.23}$$

where $\mathbf{s} \in \mathbb{C}^K$ is the transmitted data signal satisfying the power constraint i.e., $\mathbb{E}\{\|\mathbf{s}\|_2^2\} \leq P_T$, where $P_T$ is the transmit power at the BS, $n_k$ denotes the additive circular symmetric Gaussian noise such that $n_k \sim \mathcal{CN}(0, 1)$, $\mu$ is the distortion factor of $q$-bit DAC, $n_q \sim \mathcal{CN}(0, \mathbf{R}_{n_q})$ is the additive Gaussian quantization noise which is uncorrelated with $\mathbf{s}$, and $\mathbf{T} \in \mathbb{C}^{N_{RF} \times N_{RF}}$ is the transfer block which is introduced at the baseband stage to have a constant-modulus RF beamformer $\hat{\mathbf{F}}_D \in \mathbb{C}^{N_T \times N_{RF}}$ such that $\hat{\mathbf{F}}_D\mathbf{T} = \mathbf{F}_D$. The covariance matrix of $n_q$ can be written as [33, 43]

$$\mathbf{R}_{n_q} = \mathbb{E}\left[\mathbf{n}_q\mathbf{n}_q^H\right] = \mu(1-\mu)\ \text{diag}(\mathbf{T}\mathbf{B}_D\mathbf{B}_D^H\mathbf{T}^H). \tag{4.24}$$

For $q = 1, 2, \ldots, 5$, the exact values of $\mu$ are 0.6366, 0.8825, 0.96546, 0.990503, and 0.997501, respectively, whereas for $q > 5$, the distortion factor $\mu$ can be approximated as $\mu \approx 1 - \frac{\pi\sqrt{3}}{2}2^{-2q}$ [42].

### 4.4.2 RF and BB Stage Design with Low-Resolution PSs

In the eigen beamforming, the hybrid precoder $\mathbf{B} \in \mathbb{C}^{N \times K}$ is constructed by concatenating the RF-beamforming stage $\mathbf{F}_D \in \mathbb{C}^{N_T \times N_{RF}}$ and reduced-dimensional MU baseband precoding stage $\mathbf{B}_D \in \mathbb{C}^{N_{RF} \times K}$. The HBF design based on eigen beamforming results in non-constant modulus entities at the RF stage. As discussed earlier, the quantization of this variable gain RF-stage is challenging. To solve this issue, we formulate an optimization problem and introduce a transfer block $\mathbf{T} \in \mathbb{C}^{N_{RF} \times N_{RF}}$ such that $\hat{\mathbf{F}}_D \mathbf{T} = \mathbf{F}_D$, where $\mathbf{F}_D \in \mathbb{C}^{N_T \times N_{RF}}$ is the variable gain RF beamformer, $\mathbf{T} \in \mathbb{C}^{N_{RF} \times N_{RF}}$ is the transfer block placed at the baseband stage and $\hat{\mathbf{F}}_D \in \mathbb{C}^{N_T \times N_{RF}}$ is the RF beamformer with constant-modulus entities. The design of the RF-stage together with the transfer block and the baseband stage is explained in the following sections.

#### 4.4.2.1 Variable-Gain RF Beamformer

We assume $K$ users are clustered into $G$ groups based on their angle-of-departure (AoD) information, where each group contains $K_g$ number of users such that $K = \sum_{g=1}^{G} K_g$. For simplicity, we also assume that the users in the same group have identical covariance matrix $\mathbf{R}_g$, where $g = 1, 2, \ldots, G$. The index $g_k = \sum_{g'=1}^{g-1} K_{g'} + k$ is used to denote the $k^{\text{th}}$ user in group $g$. Then, the channel vector for the user $k$ in group $g$ is $\mathbf{h}_{g_k} \sim \mathcal{CN}(0, \mathbf{R}_g)$, where $\mathbf{R}_g = \mathbf{U}_g \mathbf{\Lambda}_g \mathbf{U}_g^H$ is the covariance matrix of the group $g$. Based on the AoD of the users, the correlation matrix can be written as:

$$\mathbf{R}_g = [\mathbf{U}_g^+, \mathbf{U}_g^-][\mathbf{\Lambda}_g^+, \mathbf{\Lambda}_g^-][\mathbf{U}_g^+, \mathbf{U}_g^-]^H, \tag{4.25}$$

where $\mathbf{U}_g^+ \in \mathbb{C}^{N \times r_g^+}$ is the matrix of eigenvectors corresponding to the dominant eigenvalues for the user group $g$. Similarly, $\mathbf{\Lambda}_g^+ \in \mathbb{C}^{r_g^+ \times r_g^+}$ is the diagonal matrix of the dominant eigenvalues. As a result, the data streams per group is bounded by $S_g \in [0, \min\{K_g, r_g^+\}]$. Furthermore, the multiplexing gain limits the maximum number of independent data streams per group that can be transmitted, which is $\min\{K_g, r_g^+\}$. Thus, the total data streams can be written as $S = \sum_{g=1}^{G} S_g$. The selection of $N_{RF_g}$ is a design parameter and the performance of HBF heavily depends on its value such that: (i) $N_{RF_g} \geq S_g$ to ensure that the minimum number of RF chains used per group in HBF is not less than the transmitted signals per group, and (ii) the total number of RF chains is determined by the number of groups for $K$ users (i.e., $N_{RF} = \sum_{g=1}^{G} N_{RF_g}$). The RF beamforming matrix $\mathbf{B}_A$ depends on the second-order statistics as it relies on the eigenvalues and eigenvectors extracted from the covariance matrix $\mathbf{R}_g$. If $\mathbf{B}_{A_g} = \mathbf{U}_g^+$ is the RF beamforming matrix of group $g$, then the complete RF beamforming matrix can be written as:

$$\mathbf{F}_D = [\mathbf{F}_{D_1}, \mathbf{F}_{D_2}, \ldots, \mathbf{F}_{D_G}]. \tag{4.26}$$

#### 4.4.2.2 Design of Transfer Block and Constant-Gain RF Beamformer

From (4.26), we can see that $\mathbf{F}_D$ has non-constant modulus entities, which results in a variable-gain RF-stage. To the best of our knowledge, the design of constant-gain analog precoder in mMIMO, which is based on eigen beamforming has not been considered yet. For this challenging issue, we formulate an optimization problem and introduce a transfer block $\mathbf{T} \in \mathbb{C}^{N_{RF} \times N_{RF}}$ as shown in Fig. 1(b) such that $\mathbf{F}_D = \hat{\mathbf{F}}_D \mathbf{T}$, where $\mathbf{F}_D \in \mathbb{C}^{N_T \times N_{RF}}$ is the variable-gain RF beamformer as given in (4.26), $\mathbf{T} \in \mathbb{C}^{N_{RF} \times N_{RF}}$ is the transfer block placed at the baseband stage and $\hat{\mathbf{F}}_D \in \mathbb{C}^{N_T \times N_{RF}}$ is the RF beamformer with constant-gain. The transfer block $\mathbf{T}$ and the RF beamformer with constant-modulus $\hat{\mathbf{F}}_D$ is designed as

$$\begin{aligned} \underset{\mathbf{T}, \hat{\mathbf{F}}_D}{\operatorname{argmin}} \quad & \left\| \mathbf{F}_D - \hat{\mathbf{F}}_D \mathbf{T} \right\|_2^2, \\ \text{s.t.} \quad & \hat{\mathbf{F}}_D \in \mathbb{F}_{D_o}, \end{aligned} \tag{4.27}$$

where $\mathbb{F}_{D_o}$ represents the set of matrices of size $N_T \times N_{RF}$ satisfying the unit-modulus property for the constrained RF-beamformer $\hat{\mathbf{F}}_D$, which is constructed using double-PS structure as shown in Fig. 4.4b. The optimization problem is formulated to design the RF beamformer $\hat{\mathbf{F}}_D$ satisfying the modulus constraint and targeting the same performance as $\mathbf{F}_D$, where $\mathbf{F}_D$ is the RF beamformer whose entities does not satisfy the modulus constraint. Since (4.27) is non-convex because of the unit-modulus constraint, and as proven in [40, Lemma 1], we can dissolve $\hat{\mathbf{F}}_D$ into $\hat{\mathbf{F}}_{D_1} \in \mathbb{C}^{N_T \times N_{RF}}$ and $\hat{\mathbf{F}}_{D_2} \in \mathbb{C}^{N_T \times N_{RF}}$ based on double-PS structure, and can rewrite the equivalent optimization problem as:

$$\begin{aligned} \underset{\mathbf{T}, \hat{\mathbf{F}}_{D_1}, \hat{\mathbf{F}}_{D_2}}{\operatorname{argmin}} \quad & \left\| \mathbf{F}_D - (\hat{\mathbf{F}}_{D_1} + \hat{\mathbf{F}}_{D_2}) \mathbf{T} \right\|_2^2, \\ \text{s.t.} \quad & \hat{\mathbf{F}}_{D_1}, \hat{\mathbf{F}}_{D_2} \in \mathbb{F}_{D_o}, \\ & \hat{\mathbf{F}}_D = \hat{\mathbf{F}}_{D_1} + \hat{\mathbf{F}}_{D_2} \end{aligned} \tag{4.28}$$

Then, the optimal solution for the transfer block $\mathbf{T}$ and constant-modulus RF beamformer $\hat{\mathbf{F}}_D$ is given as:

$$\hat{\mathbf{F}}_{D_1}(a, b) = e^{j\left(\angle \mathbf{F}_D(a,b) + \cos^{-1}\left(\frac{\mathbf{F}_D(a,b)}{2\nu}\right)\right)}, \tag{4.29a}$$

$$\hat{\mathbf{F}}_{D_2}(a, b) = e^{j\left(\angle \mathbf{F}_D(a,b) - \cos^{-1}\left(\frac{\mathbf{F}_D(a,b)}{2\nu}\right)\right)}, \tag{4.29b}$$

$$\mathbf{T} = \left(\hat{\mathbf{F}}_{D_1} + \hat{\mathbf{F}}_{D_2}\right)^{\dagger} \mathbf{F}_D, \tag{4.29c}$$

where $\nu = \frac{1}{2} \max_{a,b} |\mathbf{F}_D(a, b)|$ is the half of the highest modulus element at $\mathbf{F}_D$. Each RF chain is connected to the corresponding antenna element through two

PSs, which are summed up together to formulate the RF beamforming gain. This summation allows the gain of each RF beamformer entity to vary between 0 and 2, i.e., by relaxing the unit-modulus constraint, the new constraint of the analog stage is $|\hat{\mathbf{F}}_D(a,b)| \leq 2 \; \forall \; a, b$. By implementing the double-PS structure, we can have PS-only RF beamformer without having any impact on the performance of eigen beamforming. In other words, without **T** and double-PS structure at RF stage, we would require variable-gain controllers together with PSs, which makes the implementation of HBF using low-resolution PSs challenging.

#### 4.4.2.3 RF Beamformer Quantization

By the introduction of the transfer block in the HBF architecture, each entity of the RF beamformer $\hat{\mathbf{F}}_D$, which is expressed in (4.29), can be converted to a modulus constraint within 0 and 2 (Fig. 1(c)). Thus, the analog RF beamformer $\hat{\mathbf{F}}_D$ can have only PSs for the RF processing. Since the phase of each entry of $\hat{\mathbf{F}}_D$ tends to be highly quantized as well as the use of full-resolution PSs is impractical because of the high cost and power consumption, therefore, we need to investigate the performance of our proposed HBF by using low-resolution PSs to be used in a more realistic scenario. For this purpose, we quantize the phase of each entry of $\hat{\mathbf{F}}_D$, i.e., phases of the $(2 \times N_{RF} N)$ entries of $\hat{\mathbf{F}}_D$ are quantized up to $b$ bits of precision. Each $(i, j)$th entry is quantized to its nearest neighbor based on the closest Euclidean distance. Thus, we can write as:

$$\hat{l} = \underset{l \in \mathbb{L}_o}{\operatorname{argmin}} \left| \vartheta - \frac{2\pi l}{L} \right|, \tag{4.30}$$

where $\vartheta$ is the unquantized phase of each entity of RF beamformer obtained from (4.29), $L = 2^b$ and $\mathbb{L}_o$ is the set of all possible quantized phase values i.e., $\mathbb{L}_o = \{0, \ldots, 2^{b-1}\}$. Then the quantized phase of each entry of $\hat{\mathbf{F}}_D$ can be written as:

$$\hat{\vartheta} = \frac{2\pi \hat{l}}{L}.$$

#### 4.4.2.4 RF Beamformer Design for Low-Resolution PSs

As discussed earlier, the straightforward approach for HBF using low-resolution PSs is to design the RF-stage based on full-resolution PSs first, and then quantizing the value of each PSs to a finite set. However, this approach yields a large sum-rate degradation and thus, it is not effective for very low-resolution PSs. To reduce the sum-rate degradation resulting from the constraint of low-resolution PSs, we design the RF beamformer by minimizing the Euclidean distance between the HBF using full-resolution PSs and the HBF using low-resolution PSs. The optimization problem can be formulated as:

$$\begin{aligned} \underset{\mathbf{T}^{(q)},\hat{\mathbf{F}}_D^{(q)}}{\operatorname{argmin}} \quad & \left\| \hat{\mathbf{F}}_D\mathbf{T} - \hat{\mathbf{F}}_D^{(q)}\mathbf{T}^{(q)} \right\|_F, \\ \text{s.t.} \quad & \hat{\mathbf{F}}_D^{(q)} \in \hat{\mathbb{F}}_D^{(q)}, \\ & \left\| \hat{\mathbf{F}}_D^{(q)}\mathbf{T}^{(q)} \right\|_F^2 = N_{RF}, \end{aligned} \tag{4.31}$$

where $\hat{\mathbb{F}}_D^{(q)}$ represents the set of matrices of size $N_T \times N_{RF}$ under the following constraints: (i) the gain of each entity of the matrix can vary between 0 and 2, and (ii) the phase of each matrix entity is quantized to enable the use of low-resolution PSs at the RF-stage, $\hat{\mathbf{F}}_D\mathbf{T}$ is the optimal unconstrained precoding matrix, $\hat{\mathbf{F}}_D^{(q)}$ and $\mathbf{T}^{(q)}$ are the RF beamformer and the transfer block designed for low-resolution PSs, respectively. The design problem can be defined as finding the projections of the optimal unconstrained precoder $\hat{\mathbf{F}}_D\mathbf{T}$ onto the set of $\hat{\mathbf{F}}_D^{(q)}$ and $\mathbf{T}$. Each entry of $\hat{\mathbf{F}}_D^{(q)}$ follows modulus constraint and can have a distinct phase value as depicted by the constraint $\hat{\mathbf{F}}_D^{(q)} \in \hat{\mathbb{F}}_D^{(q)}$. Furthermore, $\left\| \hat{\mathbf{F}}_D^{(q)}\mathbf{T}^{(q)} \right\|_F^2 = N_{RF}$ represents the normalized transmit power constraint of the RF beamformer. Due to the complex non-convex nature of the feasible set $\hat{\mathbb{F}}_D^{(q)}$, the problem of finding the projections is difficult. However, we can notice that there exits a connection between $\hat{\mathbf{F}}_D$ and $\hat{\mathbf{F}}_D^{(q)}$ by exploiting the structure of the channel $\mathbf{H}$. This can be further explained by the following *remarks*.

1. *Structure of unconstrained RF beamformer*: The design of unconstrained RF beamformer $\hat{\mathbf{F}}_D$ is based on Karhunen-Loeve decomposition of the covariance matrix $\mathbf{R}$. Thus, the eigenvectors (column vectors) for the corresponding dominant eigenvalues form an orthonormal basis.
2. *Relation between $\hat{\mathbf{F}}_D$ and $\hat{\mathbb{F}}_D^{(q)}$*: We notice that there exits an association between $\hat{\mathbf{F}}_D$ and $\hat{\mathbb{F}}_D^{(q)}$ as both are based on the dominant eigenvalues extracted from the covariance matrix. In fact, $\hat{\mathbb{F}}_D^{(q)}$ is the set of quantized matrices of $\mathbf{U}^+$, where $\mathbf{U}^+ = [\mathbf{U}_1^+, \mathbf{U}_2^+, \ldots, \mathbf{U}_G^+]$.
3. *Transfer block*: The problem of finding projection of $\hat{\mathbf{F}}_D$ onto $\hat{\mathbf{F}}_D^{(q)}$ alone is challenging. By using the transfer block $\mathbf{T}$, we can find the projection of $\hat{\mathbf{F}}_D\mathbf{T}$ onto the set of $\hat{\mathbf{F}}_D^{(q)}$ and $\mathbf{T}$.
4. *Design of $\hat{\mathbf{F}}_D^{(q)}$*: It must be seen that the feasible quantized RF beamformers in $\hat{\mathbb{F}}_D^{(q)}$ are of size $N_T \times N_{RF}$. Thus, each column of the constrained RF beamformer $\hat{\mathbf{F}}_D^{(q)}$ can be designed by applying $N_{RF}$ vectors, where each vector follows the constant-modulus and have distinct phase values only. Furthermore, by using the

transfer block $\mathbf{T}$, we can form the arbitrary linear combinations of $N_{RF}$ vectors and design the constrained RF beamformer by reducing the Euclidean distance i.e., $\left\|\hat{\mathbf{F}}_D\mathbf{T} - \hat{\mathbf{F}}_D^{(q)}\mathbf{T}^{(q)}\right\|_F$.

Based on remark 2, the optimal constrained RF beamformer can be found by replacing $\hat{\mathbb{F}}_A^{(q)}$ with $\hat{\mathbb{U}}^+$, where $\hat{\mathbb{U}}^+$ represents the set of matrices of quantized eigenvectors corresponding to the dominant eigenvalues, i.e., $\hat{\mathbf{U}}^+$. Subsequently, we can rewrite the optimization problem as follows:

$$\begin{aligned} \underset{\mathbf{T}^{(q)},\hat{\mathbf{F}}_D^{(q)}}{\operatorname{argmin}} \quad & \left\|\hat{\mathbf{F}}_D\mathbf{T} - \hat{\mathbf{F}}_D^{(q)}\mathbf{T}^{(q)}\right\|_F, \\ \text{s.t.} \quad & \hat{\mathbf{F}}_D^{(q)} \in \hat{\mathbb{U}}^+, \\ & \left\|\hat{\mathbf{F}}_D^{(q)}\mathbf{T}^{(q)}\right\|_F^2 = N_{RF}. \end{aligned} \tag{4.32}$$

We can append the constraint $\hat{\mathbf{F}}_D^{(q)}$ into the optimization objective and formulate the equivalent problem as [38]:

$$\begin{aligned} \underset{\mathbf{T}^{(q)}}{\operatorname{argmin}} \quad & \left\|\hat{\mathbf{F}}_D\mathbf{T} - \hat{\mathbf{U}}^+\mathbf{T}^{(q)}\right\|_F, \\ \text{s.t.} \quad & \left\|\operatorname{diag}(\mathbf{T}^{(q)}\mathbf{T}^{(q)*})\right\|_0 = N_{RF}, \end{aligned} \tag{4.33}$$

where the constraint $\left\|\operatorname{diag}(\mathbf{T}^{(q)}\mathbf{T}^{(q)*})\right\|_0 = N_{RF}$ states that the matrix $\mathbf{T}^{(q)}$ acts as an auxilliary variable and can only have a maximum of $N_{RF}$ non-zero rows. As a result, only $N_{RF}$ columns of $\hat{\mathbf{U}}^+$ are selected and thus, the transfer block $\mathbf{T}^{(q)}$ can be designed by $N_{RF}$ non-zero rows. The constrained RF beamformer is formulated by projecting the $N_{RF}$ columns of $\hat{\mathbf{U}}^+$ onto $\hat{\mathbf{F}}_D$. By employing the OMP technique proposed in [37, 38], we present an algorithmic solution to design the constrained RF beamformer for low-resolution PSs. The pseudo-code is given in Algorithm 4.1, which can be summarized as: (i) the algorithm starts by initializing the constrained precoder $\hat{\mathbf{F}}_{D_g}^{(q)}$ as an empty matrix, (ii) finds the vector along which the unconstrained precoder has the maximum projection, (ii) append the selected column vector to the empty matrix $\hat{\mathbf{F}}_{D_g}^{(q)}$, (iv) using least squares, find the transfer block $\mathbf{T}_g^{(q)}$, (v) remove the contribution of the selected vector, and (vi) finding the column along which the residual precoding matrix $\mathbf{F}_{res_g}$ has the largest projection. The algorithm continues until we find $N_{RF_g}$ precoding vectors. As a result, we get an $N_T \times N_{RF_g}$ RF preoding matrix and $N_{RF_g} \times N_{RF_g}$ transfer matrix. Finally, from Algorithm 4.1, step 11 ensures that the transmit power constraint is satisfied. The process is repeated for all $G$ user groups. Combining the RF and transfer matrix of all groups, we get a large

**Algorithm 4.1:** Constrained RF beamformer design via orthogonal matching pursuit

**Input**: Optimal unconstrained RF beamformer $\hat{\mathbf{F}}_D = [\hat{\mathbf{F}}_{D_1}, \hat{\mathbf{F}}_{D_2}, \ldots, \hat{\mathbf{F}}_{DG}]$.
**Output**: $\tilde{\mathbf{T}}^{(q)}, \hat{\mathbf{F}}_D^{(q)}$.
**for** $g = 1 : G$ **do**
  Initialize $\hat{\mathbf{F}}_{D_g}^{(q)} \leftarrow [\,]$.
  $\mathbf{F}_{res_g} \leftarrow \hat{\mathbf{F}}_{D_g}\mathbf{T}_g$.
  **for** $j \leq N_{RF_g}$ **do**
    $\Upsilon_g \leftarrow (\hat{\mathbf{U}}_g^{+})^* \mathbf{F}_{res_g}$.
    Find the index $l$ that maximizes $\Upsilon_g \Upsilon_g^*$.
    $\hat{\mathbf{F}}_{D_g}^{(q)} \leftarrow [\hat{\mathbf{F}}_{D_g}^{(q)} \mid \hat{\mathbf{U}}_g^{+(l)}]$.
    Compute $\mathbf{T}_g^{(q)}$ via least squares:
    $\mathbf{T}_g^{(q)} \leftarrow \left(\hat{\mathbf{F}}_{D_g}^{(q)*}\hat{\mathbf{F}}_{D_g}^{(q)}\right)^{-1}\hat{\mathbf{F}}_{D_g}^{(q)}\hat{\mathbf{F}}_{D_g}$.
    $\mathbf{F}_{res_g} \leftarrow \frac{\hat{\mathbf{F}}_{D_g}\mathbf{T}_g - \hat{\mathbf{F}}_{D_g}^{(q)}\mathbf{T}_g^{(q)}}{\left\|\hat{\mathbf{F}}_{D_g}\mathbf{T}_g - \hat{\mathbf{F}}_{D_g}^{(q)}\mathbf{T}_g^{(q)}\right\|_F}$.
  **end**
  $\tilde{\mathbf{T}}_g^{(q)} \leftarrow \frac{N_{RF_g}}{\left\|\hat{\mathbf{F}}_{D_g}^{(q)}\mathbf{T}_g^{(q)}\right\|_F}\mathbf{T}_g^{(q)}$.
**end**
$\tilde{\mathbf{T}}^{(q)} \leftarrow [\tilde{\mathbf{T}}_1^{(q)}, \tilde{\mathbf{T}}_2^{(q)}, \ldots, \tilde{\mathbf{T}}_G^{(q)}]$.
$\hat{\mathbf{F}}_D^{(q)} \leftarrow [\hat{\mathbf{F}}_{D_1}^{(q)}, \hat{\mathbf{F}}_{D_2}^{(q)}, \ldots, \hat{\mathbf{F}}_{DG}^{(q)}]$.

$N_T \times N_{RF}$ RF preoding matrix and $N_{RF} \times N_{RF}$ transfer matrix which minimizes $\left\|\hat{\mathbf{F}}_D\mathbf{T} - \hat{\mathbf{F}}_D^{(q)}\mathbf{T}^{(q)}\right\|_F^2$.

**Lemma 4.1** *If the Euclidean distance before normalization is* $\left\|\hat{\mathbf{F}}_{D_g}\mathbf{T}_g - \hat{\mathbf{F}}_{D_g}^{(q)}\mathbf{T}_g^{(q)}\right\|_F \leq \zeta$, *then the Euclidean distance after normalization is* $\left\|\hat{\mathbf{F}}_{D,g}\mathbf{T}_g - \hat{\mathbf{F}}_{D,g}^{(q)}\tilde{\mathbf{T}}_g^{(q)}\right\|_F \leq 2\zeta$.

***Proof*** See Appendix. □

#### 4.4.2.5 Baseband MU Precoder Design

After the design of RF beamformer $\hat{\mathbf{F}}_D^{(q)}$ for $b$-bit PSs, the baseband MU precoder $\mathbf{B}_D$ can be determined by using the joint group processing (JGP) [10]. The overall reduced dimensional effective channel matrix is expressed as $\mathcal{H} = \tilde{\mathbf{T}}^{(q)}\hat{\mathbf{B}}_A^{(q)H}\mathbf{H}$. Using (4.8) and (4.26), $\mathcal{H}^H$ can be written as:

$$\mathcal{H}^H = \begin{bmatrix} \mathbf{H}_1^H\hat{\mathbf{F}}_{D_1}^{(q)}\tilde{\mathbf{T}}_1^{(q)} & \mathbf{H}_1^H\hat{\mathbf{F}}_{D_2}^{(q)}\tilde{\mathbf{T}}_2^{(q)} & \dots & \mathbf{H}_1^H\hat{\mathbf{F}}_{D_G}^{(q)}\tilde{\mathbf{T}}_G^{(q)} \\ \mathbf{H}_2^H\hat{\mathbf{F}}_{D_1}^{(q)}\tilde{\mathbf{T}}_1^{(q)} & \mathbf{H}_2^H\hat{\mathbf{F}}_{D_2}^{(q)}\tilde{\mathbf{T}}_2^{(q)} & \dots & \mathbf{H}_2^H\hat{\mathbf{F}}_{D_G}^{(q)}\tilde{\mathbf{T}}_G^{(q)} \\ \vdots & \vdots & \ddots & \vdots \\ \mathbf{H}_G^H\hat{\mathbf{F}}_{D_1}^{(q)}\tilde{\mathbf{T}}_1^{(q)} & \mathbf{H}_G^H\hat{\mathbf{F}}_{D_2}^{(q)}\tilde{\mathbf{T}}_2^{(q)} & \dots & \mathbf{H}_G^H\hat{\mathbf{F}}_{D_G}^{(q)}\tilde{\mathbf{T}}_G^{(q)} \end{bmatrix}, \tag{4.34}$$

where the diagonal matrices $\mathbf{H}_g^H\hat{\mathbf{F}}_{D_g}^{(q)}\tilde{\mathbf{T}}_g \in \mathbb{C}^{K_g \times N_{RF_g}}$ are the effective channel matrix for group $g$ and the off-diagonal matrices $\mathbf{H}_g^H\hat{\mathbf{F}}_{D_{\hat{g}}}^{(q)}\tilde{\mathbf{T}}_{\hat{g}} \in \mathbb{C}^{K_g \times N_{RF_{\hat{g}}}}$ represent the effective interference channel matrix between groups $g$ and $\hat{g}$, $\forall \hat{g} \neq g$. By applying the well-known RZF technique [34, 35], $\mathbf{B}_D$ can be defined as:

$$\mathbf{B}_D = \gamma \mathbf{X}\mathcal{H}^H, \tag{4.35}$$

where $\mathcal{H}^H$ is the reduced-dimension effective channel as given in (4.34), $\mathbf{X} = [\mathcal{H}^H\mathcal{H} + \alpha N_{RF}\mathbf{I}_{N_{RF}}]^{-1}$, $\mathbf{I}_{N_{RF}} \in \mathbb{C}^{N_{RF} \times N_{RF}}$, $\alpha$ is the regularization parameter and $\gamma$ is the normalization factor used to satisfy the power constraint, which can be written as:

$$\gamma = \sqrt{\frac{S}{\mathrm{tr}\{\mathcal{H}^H\mathbf{X}^H\tilde{\mathbf{T}}^{(q)H}\hat{\mathbf{F}}_D^{(q)H}\hat{\mathbf{F}}_D^{(q)}\tilde{\mathbf{T}}^{(q)}\mathbf{X}\mathcal{H}\}}}. \tag{4.36}$$

Therefore, the signal-to-interference-plus-noise ratio (SINR) at the user $k$ in group $g$ can be computed as follows:

$$\mathrm{SINR}_{g_k} = \frac{\frac{P_T}{S}|\mathbf{h}_{g_k}^H\hat{\mathbf{F}}_D^{(q)}\tilde{\mathbf{T}}^{(q)}\mathbf{X}\tilde{\mathbf{T}}^{(q)H}\hat{\mathbf{F}}_D^{(q)H}\mathbf{h}_{g_k}|^2}{1 + \frac{P_T}{S}\sum_{i \neq g_k}|\mathbf{h}_{g_k}^H\hat{\mathbf{F}}_D^{(q)}\tilde{\mathbf{T}}^{(q)}\mathbf{X}\tilde{\mathbf{T}}^{(q)H}\hat{\mathbf{F}}_D^{(q)H}\mathbf{h}_i|^2}, \tag{4.37}$$

where $\frac{P_T}{S}\sum_{i \neq g_k}|\mathbf{h}_{g_k}^H\hat{\mathbf{F}}_D^{(q)}\tilde{\mathbf{T}}^{(q)}\mathbf{X}\tilde{\mathbf{T}}^{(q)H}\hat{\mathbf{F}}_D^{(q)H}\mathbf{h}_i|^2$ accounts for MU interference. Hence, the corresponding sum-rate performance can be computed as:

$$R_{\mathrm{sum}} = \sum_{g=1}^{G}\sum_{g_k=1}^{K_g}\mathbb{E}[\log_2(1 + \mathrm{SINR}_{g_k})]. \tag{4.38}$$

## 4.5 Hybrid Beamforming Design Using Low-Resolution PSs and DACs

In this section, we design the HBF using low-resolution hardware components (i.e., $q$-bit DACs and $b$-bit PSs). The HBF structure is shown in Fig. 4.4c. Based on the

AoD information of users, we design the RF beamformer and the baseband MU precoder along with a transfer block such that quantization error is minimized.

### 4.5.1 RF Beamformer and Transfer Block Design

Following the same user distribution as discussed in Sect. 4.4.2.1, $K$ users are clustered into $G$ groups. Based on the AoD of the users, the correlation matrix of group $g$ is given in (4.25), which consists of the dominant eigenvectors $\mathbf{U}_g^+ \in \mathbb{C}^{N\times r_g^+}$ corresponding to the dominant eigenvalues $\mathbf{\Lambda}_g^+ \in \mathbb{C}^{r_g^+\times r_g^+}$. Moreover, $\min\{K_g, r_g^+\}$ limits the maximum number of the independent data streams that can be transmitted. By using Karhunen-Loeve decomposition, we can write the user channel vector $\mathbf{h}_{g_k}$ and the overall channel matrix $\mathbf{H}$ as given in (4.7) and (4.8), respectively. If $\mathbf{F}_{D_g} = \mathbf{U}_g^+$ is the RF beamforming matrix of group $g$ having non-constant modulus, then the complete RF beamforming matrix can be written as given in (4.26). As discussed in Sect. 4.4.2.2, the variable-gain RF beamforming matrix $\mathbf{F}_{D_g}$ can be converted to a constant-gain beamforming matrix $\hat{\mathbf{F}}_{D_g}$ by using a transfer block $\mathbf{T}_g$ at the baseband. Using this approach, we can rewrite (4.26) as:

$$\mathbf{F}_D = [\hat{\mathbf{F}}_{D_1}\mathbf{T}_1, \hat{\mathbf{F}}_{D_2}\mathbf{T}_2, \ldots, \hat{\mathbf{F}}_{D_G}\mathbf{T}_G]. \tag{4.39}$$

$\hat{\mathbf{F}}_{D_g}$ and $\mathbf{T}_g$ can be found using (4.29), respectively. However, since the RF beamformer employs $b$-bit PSs for the phase quantization, we can rewrite (4.39) as:

$$\mathbf{F}_D^{(q)} = [\hat{\mathbf{F}}_{D_1}^{(q)}\mathbf{T}_1^{(q)}, \hat{\mathbf{F}}_{D_2}^{(q)}\mathbf{T}_2^{(q)}, \ldots, \hat{\mathbf{F}}_{D_G}^{(q)}\mathbf{T}_G^{(q)}]. \tag{4.40}$$

By using the solution given in Algorithm 4.1, we can find $\hat{\mathbf{F}}_{D_g}^{(q)}$ and $\mathbf{T}_g^{(q)}$ for HBF using $b$-bit PSs at the RF-stage. The overall RF beamforming matrix and the transfer matrix for low-resolution PSs can be given as:

$$\hat{\mathbf{F}}_D^{(q)} = [\hat{\mathbf{F}}_{D_1}^{(q)}, \hat{\mathbf{F}}_{D_2}^{(q)}, \ldots, \hat{\mathbf{F}}_{D_G}^{(q)}]. \tag{4.41}$$

$$\mathbf{T}^{(q)} = [\mathbf{T}_1^{(q)}, \mathbf{T}_2^{(q)} \ldots, \mathbf{T}_G^{(q)}]. \tag{4.42}$$

### 4.5.2 Baseband MU Precoder Design With Finite-Resolution DACs

In this section, the design of the baseband MU precoder $\hat{\mathbf{B}}_D^{(q)}$ using $q$-bit DACs is presented. After designing RF beamformer $\hat{\mathbf{F}}_D^{(q)}$ and the transfer block $\mathbf{T}^{(q)}$

for $b$-bit PSs as given in (4.41) and (4.42), respectively, the quantized baseband precoder $\hat{\mathbf{B}}_D^{(q)}$ can be designed using RZF as given in (4.35). The use of $q$-bit DACs introduces the quantization noise. Therefore, the signal to quantization, interference and noise ratio (SQINR) at the user $k$ in group $g$ can be computed as given in (4.43), where $\frac{P_T}{S}\sum_{i\neq g_k}|\mathbf{h}_{g_k}^H\hat{\mathbf{F}}_D^{(q)}\tilde{\mathbf{T}}^{(q)}\mathbf{X}\tilde{\mathbf{T}}^{(q)H}\hat{\mathbf{F}}_D^{(q)H}\mathbf{h}_i|^2$ accounts for the interference experienced by users in group $g$ from the users in group $\hat{g} \neq g$ and $\frac{P_T}{S}|\mathbf{h}_{g_k}^H\hat{\mathbf{F}}_D^{(q)}\tilde{\mathbf{T}}^{(q)}\mathbf{R}_{n_q}(\tilde{\mathbf{T}}^{(q)H}\hat{\mathbf{F}}_D^{(q)H}\mathbf{h}_{g_k})^H|^2$ is the quantization noise when using $q$-bit DACs. Hence, the corresponding sum-rate performance of HBF with $b$-bit PSs and $q$-bit DACs can be computed as:

$$\text{SQINR}_{g_k} = \frac{\frac{P_T}{S}|\mathbf{h}_{g_k}^H\hat{\mathbf{F}}_D^{(q)}\tilde{\mathbf{T}}^{(q)}\mathbf{X}\tilde{\mathbf{T}}^{(q)H}\hat{\mathbf{F}}_D^{(q)H}\mathbf{h}_{g_k}|^2}{\begin{array}{l}1+\frac{P}{S}\sum_{i\neq g_k}|\mathbf{h}_{g_k}^H\hat{\mathbf{F}}_D^{(q)}\tilde{\mathbf{T}}^{(q)}\mathbf{X}\tilde{\mathbf{T}}^{(q)H}\hat{\mathbf{F}}_D^{(q)H}\mathbf{h}_i|^2\\ +\frac{P_T}{S}|\mathbf{h}_{g_k}^H\hat{\mathbf{F}}_D^{(q)}\tilde{\mathbf{T}}^{(q)}\mathbf{R}_{n_q}(\tilde{\mathbf{T}}^{(q)H}\hat{\mathbf{F}}_D^{(q)H}\mathbf{h}_{g_k})^H|^2\end{array}}, \tag{4.43}$$

$$R_{\text{sum}} = \sum_{g=1}^{G}\sum_{g_k=1}^{K_g}\mathbb{E}[\log_2(1+\text{SQINR}_{g_k})]. \tag{4.44}$$

## 4.6 Simulation Results

In this section, the Monte-Carlo simulation results are presented based on the hybrid precoding for various array structures. For the presented results, the BS is equipped with 100 isotropic antennas, arranged in $1 \times 100$ ULA and $10 \times 10$ URA configurations. In UCA, the antennas are equally distributed on a circle of radius $\lambda\nabla$ where $\nabla = \frac{1}{2\sqrt{(1-\cos(\frac{2\pi}{N_T}))^2+\sin(\frac{2\pi}{N_T})^2}}$, whereas the antennas are distributed on multiple circular rings in CCA. The radius of $n$th ring is $\lambda R_n$, where $R_n = nL$ [44]. Then, the corresponding number of antennas on $n$th ring is $N_n = \frac{2\pi R_n}{d}$, where $d$ is the antenna spacing which is selected as $\frac{\lambda}{2}$ for all array structures. $L$ is the inter-ring spacing which is selected as $0.55\lambda$. Using this configuration of CCA, 100 antennas can be arranged in just 5 rings. The signal-to-noise-ratio (SNR) is defined as $\frac{P}{\sigma^2 K}$ and the regularization parameter $\alpha = \frac{\sigma^2}{P}$ [35]. In the following analysis, the array structures performance is compared based on: (i) *spatial efficiency*, (ii) *sum-rate versus users AoD*, (iii) *sum-rate versus number of the users.*

### 4.6.1 *Spatial Efficiency*

To arrange 100 antenna elements with an inter-element spacing $d$, It can be seen that the $1 \times 100$ ULA requires $50\lambda$ space in one direction. On the other hand, $10 \times 10$ URA requires just $5\lambda$ space in horizontal and vertical directions, which is equivalent to a space of $25\lambda^2$. Similarly, UCA with a radius of about $7.96\lambda$ (D = 15.92 $\lambda$) requires an array size of approximately $199\lambda^2$. It can be seen that URA is more spatially efficient structure compared to both ULA and UCA. In CCA, 100 antennas can be arranged in 5 rings as discussed before. The maximum ring radius in CCA is about $2.75\lambda$, which is more than *three times* smaller in size than UCA having the same number of antennas. In other words, this is equivalent to a space of approximately $24\lambda^2$. To better understand the spatial efficiency, consider the signal frequency is 6 GHz (sub-6 GHz mMIMO frequency). Then, the array sizes of ULA, URA, UCA, and CCA are respectively 2.5 m, 0.0625 $m^2$, 0.49 $m^2$, and 0.0592 $m^2$. Thus, CCA offers the highest spatial efficiency, whereas ULA is the least spatially efficient structure.

### 4.6.2 *Performance Comparison with Full-Resolution Components*

In this section, we first compare the performance of different 2D array structures using full-resolution hardware components.

#### 4.6.2.1 Sum-Rate Versus User AoD

In this section, the sum-rate of the array structures is compared based on the AoD information of the users around the BS. As discussed, the hybrid precoder is designed using the RF-beamforming stage and the optimized reduced-size baseband stage. The RF-beamforming stage depends on the dominant eigenvalues extracted from the covariance matrix **R** [see (4.9)]. Based on the different geometrical configuration of ULA, URA, UCA, and CCA, the rank of the covariance matrix can vary [see (4.4), (4.5)], playing an important role in the sum-rate performance of the array structures. For $N_T = 100$, the rank $r$ is 21 and 17 for UCA and ULA respectively and 13 for CCA and URA. Also, some of the non-zero eigenvalues are too small that can be ignored. So, the dominant eigenvalues $r_g^+$ of UCA, ULA, CCA, and URA become 10, 9, 4, and 3, respectively. In Table 4.2, a comparison of numbers of the dominant eigenvalues of different array structures is presented. With the increased number of antennas, both ULA and UCA can have a higher value of $r_g^+$ and can support an increased number of independent data streams. On the other hand, URA and CCA can have a small increase in $r_g^+$ because of the much-reduced array size. As a result, the correlation between the antennas is increased, resulting

**Table 4.2** Number of dominant eigenvalues of ULA, URA, UCA, and CCA versus number of antennas

| Number of antennas ($\mathbf{N}_T$) | URA | CCA | ULA | UCA |
|---|---|---|---|---|
| **64** | 3 | 4 | 6 | 8 |
| **128** | 4 | 6 | 10 | 13 |
| **256** | 9 | 8 | 19 | 26 |
| **512** | 14 | 11 | 39 | 49 |
| **1024** | 20 | 14 | 71 | 89 |

in small increment in dominant eigenvalues in URA and CCA. In Fig. 4.5, the sum-rate is calculated for multiple user groups, when $N_T = 100$. We consider three user groups ($G = 3$) are located around the BS at mean azimuth locations $\theta_1 = 30°$, $\theta_2 = 90°$, $\theta_3 = 150°$ with $\Delta = 15°$. The sum-rate performance of hybrid precoding for the array structures is compared with their DBF counterparts using RZF. Figure 4.5a shows the result of the hybrid precoding without using the transfer block, when $N_{RF_g} = K_g = 3$, $N_{RF} = K = 9$ and $b_g = 3$. As expected, DBF provides the highest sum-rate. For the hybrid beamforming sum-rates, URA and CCA give the highest sum-rate whereas ULA gives the lowest sum-rate. On the other hand, the sum-rate of UCA is comparable to the sum-rate of URA and CCA. By using the transfer block in the hybrid precoder, when, $N_{RF_g} = K_g = 3$, $N_{RF} = K = 9$ and $b_g = 5$, it can be seen in Fig. 4.5b that the sum-rate of the array structures is increased significantly. The sum-rate of URA and CCA approaches to the sum-rate of their DBF equivalents. Though ULA still provides the lowest sum-rate among the array structures, it shows a sum-rate increase of about 20 bps/Hz after using the transfer block. Thus, we can conclude that for the same number of RF chains, the use of transfer block greatly increases the sum-rate performance of the array structures. Furthermore, both URA and CCA provides the best performance among the array structures by converging to their DBF sum-rate. Secondly, the sum-rate is compared versus users AoD. For a fixed SNR = 10 dB, the sum-rate using hybrid precoding with a transfer block is compared for the mean azimuth user locations between $\theta = 0°$ to $\theta = 180°$ as shown in Fig. 4.6. URA and CCA give the best sum-rate at all azimuth user locations, when $G = 1$, $N_{RF_g} = K_g = 3$, $N_{RF} = K = 3$, $b_g = b = 4$, $\Delta = \pm 15°$. The sum-rate of ULA is the lowest, whereas UCA gives a reasonably high sum-rate. Furthermore, the sum-rate variation of UCA and CCA is $\ll 1$ bps/Hz, whereas the variation is about 2 bps/Hz for URA. On the other hand, the sum-rate of ULA varies about 5 bps/Hz. Thus, it can be concluded that for a fixed number of users, the hybrid precoding sum-rate can vary w.r.t the *users AoD* depending on the array structure. Both CCA and URA can provide a very high sum-rate at all azimuth user locations.

#### 4.6.2.2 Sum-Rate Versus Number of Users

Finally, a comparison of the sum-rate versus the number of users is provided now. In Fig. 4.7, the sum-rate of the array structures using the hybrid precoding is compared

**Fig. 4.5** Sum-rate comparison of ULA, URA, CCA, and UCA, $N_T = 100$, $G \triangleq 3$, $N_{RF_g} = K_g = 3$, $N_{RF} = K = 9$, $\theta_1 = 30°$, $\theta_2 = 90°$, $\theta_3 = 150°$, $\Delta = \pm 15°$: (**a**) $b_g = 3$ without transfer block (**b**) $b_g = 5$ with transfer block

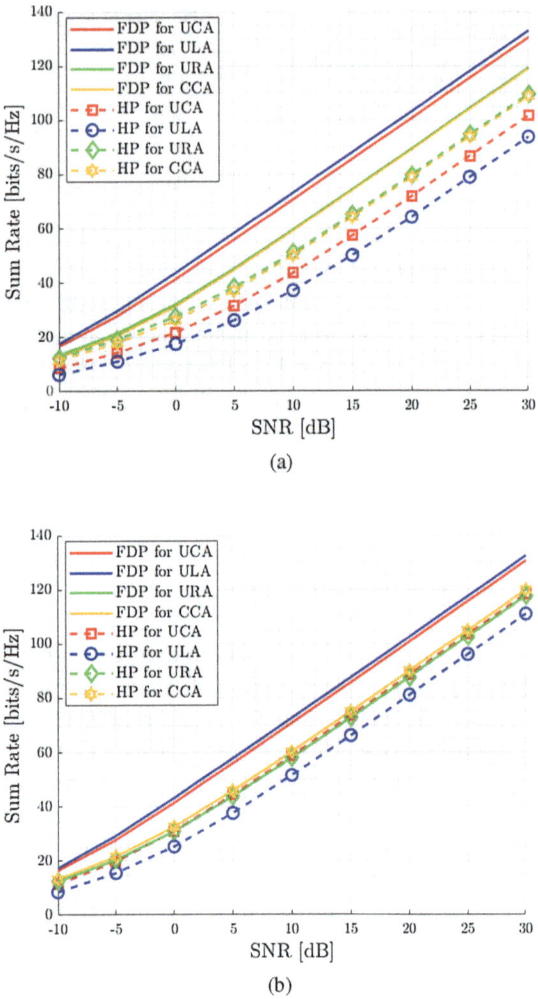


for different number of users. The result is shown for a single user group ($G = 1$) at fixed SNR = 10 dB, when $N_{RF_g} = K_g = 3$, $N_{RF} = K = 3$, $b_g = b = 4$, $\theta = 60°$, $\Delta = \pm 15°$. Based on the dominant eigenvalue $r_g^+$ as discussed previously, the users are gradually increased from 1 to 3, and the corresponding sum-rate of the array structure is plotted. The DBF sum-rates are compared with the sum-rate of the hybrid precoding with the transfer block. ULA provides the lowest sum-rate performance among the array structures. Also, a large performance gap exists between the DBF and hybrid precoding results of ULA. UCA provides a reasonably high sum-rate versus $K$ and it can approach its DBF sum-rate only when $K = 1$. On the other hand, the sum-rate of URA and CCA using the hybrid precoding can approach their respective DBF sum-rates at all $K$ values as shown in Fig. 4.7.

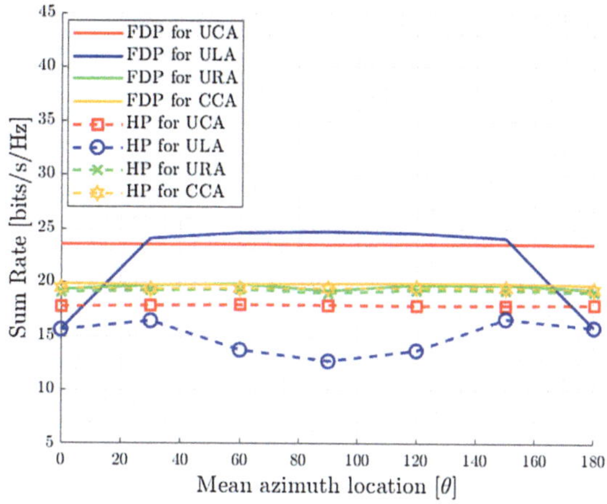


**Fig. 4.6** Sum-rate comparison of ULA, URA, CCA, and UCA versus mean azimuth user location, $N_T = 100$, $G = 1$, $N_{RF} = K = 3$, $b = 4$, $\Delta = \pm 15°$, SNR = 10 dB

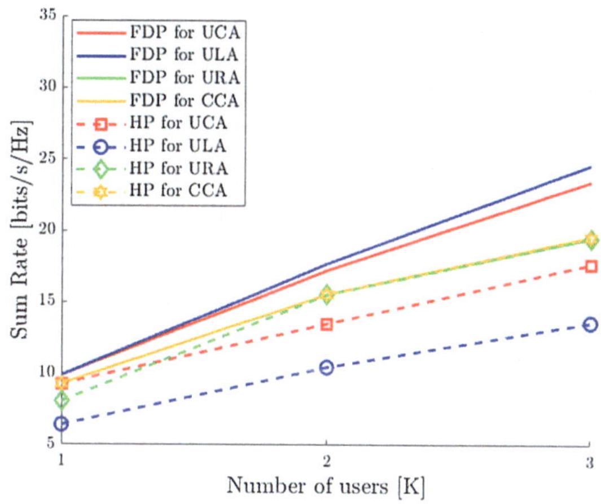


**Fig. 4.7** Sum-rate comparison of ULA, URA, CCA, and UCA versus number of users, $N_T = 100$, $G = 1$, $N_{RF} = K = 3$, $b = 4$, $\theta = 60°$, $\Delta = \pm 15°$, SNR = 10 dB

Thus, both URA and CCA can provide the highest sum-rate versus $K$ among the array structures. The overall performance comparison of the array structures is summarized in Table 4.3.

### *4.6.3 Performance Comparison with Low-Resolution DACs*

In the following analysis, 2D array structures are compared based on: (i) spectral efficiency using $q$-bit DACs, (ii) energy efficiency using $q$-bit DACs.

**Table 4.3** Performance comparison of ULA, UCA, URA and CCA

| | ULA | UCA | URA | CCA |
|---|---|---|---|---|
| Spatial efficiency | Low | Low | High | High |
| Sum-rate versus mean azimuth location $\boldsymbol{\theta} \in [0°,180°]$ | Low sum-rate. The sum-rate varies about 5 bps/Hz for fixed $K$. (see Fig. 4.6) | High sum-rate. The sum-rate remains same for fixed $K$ (see Fig. 4.6) | High sum-rate. The sum-rate varies about 2 bps/Hz for a fixed $K$ (see Fig. 4.6) | High sum-rate. The sum-rate remains same for fixed $K$ (see Fig. 4.6) |
| Sum-rate versus number of users | Low sum-rate (see Fig. 4.7) | High sum-rate (see Fig. 4.7) | High sum-rate (see Fig. 4.7) | High sum-rate (see Fig. 4.7) |

#### 4.6.3.1 Spectral Efficiency Using $q$-Bit DACs

In this section, we analyze the spectral efficiency of ULA, URA, UCA, and CCA by computing the sum-rate of the HBF using few-bit DACs. At first, we compare the sum-rate of the HBF using $q$-bit DACs for ULA, URA, UCA, and CCA.

In Fig. 4.8, the sum-rate is calculated for multiple user groups, when $N_T = 100$. We consider three user groups ($G = 3$) are located around the BS at mean elevation angle $\theta = 73°$, mean azimuth locations $(\phi_1, \phi_2, \phi_3) = (30°, 90°, 150°)$ with $\delta_a = 15°$ and $\delta_e = 12.5°$. The sum-rate of HBF for the array structures is compared with their DBF counterparts using RZF. Figure 4.8 shows the sum-rate of HBF of different 2D array structures by using different quantization levels of DAC. It can be seen that 1-bit quantization gives the lowest sum-rate for all array structures, though having the advantage of least DAC power consumption. The sum-rate increases with an increase in the number of DAC quantization bits. Moreover, CCA gives the highest sum-rate when compared to ULA, URA, and UCA for both full and low-resolution DACs.

#### 4.6.3.2 Energy Efficiency Using $q$-Bit DACs

In this section, the energy efficiency of HBF using $q$-bit DACs is compared for different 2D array structures. The energy efficiency is defined as the ratio of the sum-rate $R_{sum}$ and the total power consumption $P_{total}$ [45], i.e.,

$$\epsilon = \frac{R_{sum}}{P_{total}} = \frac{R_{sum}}{\mathrm{tr}(\mathbf{BB}^*) + N_{RF} P_{RF} 2^{2q}}, \tag{4.45}$$

where $\mathrm{tr}(\mathbf{BB}^*)$ represents the total transmission power, $P_{RF}$ is the power consumption per RF chain, $N_T$ is the total number of the antennas, $N_{RF}$ is the number of the RF chains and $q$ is the different number of the bits used for DAC quantization.

The use of few-bit DACs in HBF significantly reduces the power consumption, which increases the overall energy efficiency. However, this increase in energy

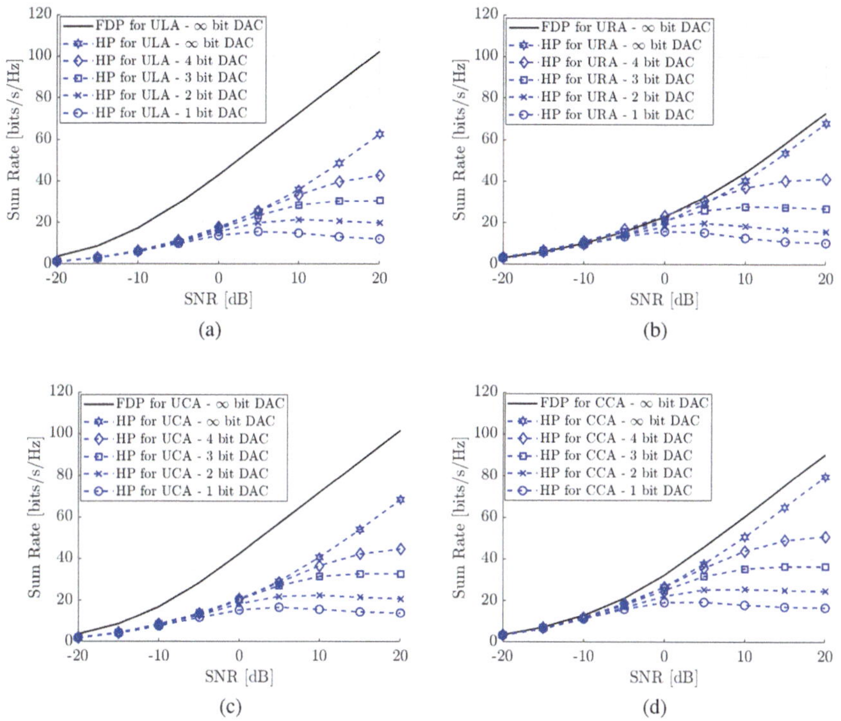


**Fig. 4.8** Sum-rate versus SNR comparison using few-bit DACs, $N_T = 100$, $G = 3$, $N_{RF_g} = K_g = 3$, $N_{RF} = K = 9$, $\theta = 73°$, $(\phi_1, \phi_2, \phi_3) = (30°, 90°, 150°)$, $(\delta_a, \delta_e) = (\pm 15°, \pm 12.5°)$ (**a**) ULA. (**b**) URA. (**c**) UCA. (**d**) CCA

efficiency comes at the expense of slight decrease in the spectral efficiency. Figure 4.9 compares the energy efficiency of HBF using ULA, URA, UCA, and CCA with the energy efficiency of DBF (here we are taking only DBF of UCA for comparison as it provides the highest DBF sum-rate among 2D array structures) for different $q$ bit DACs, where $q = 1, 2, 3, \ldots, 10$. Figure 4.9a shows the energy efficiency at very low SNR at -20 dB, whereas Fig. 4.9b shows the energy efficiency at high SNR at 20 dB. CCA and URA, both gives the highest energy efficiency at very low SNR, and this highest value can be achieved even by using 1-bit DAC as shown in Fig. 4.9a. At high SNR, CCA still provides the highest energy efficiency among the compared array structures. The highest efficiency value is achieved by using $(q = 5)$-bit DAC. In comparison, DBF offers very low energy efficiency due to the use of large number of power-hungry RF chains $(N_{RF} = N_T)$.

Finally, we examine the energy efficiency of 2D array structures versus $q$ at different values of SNR. Figure 4.10 compares the performance of ULA, URA, UCA, and CCA in terms of energy efficiency. For all compared array structures, we can achieve very high energy efficiency by using $(q \leq 5)$-bit DACs in HBF. From Fig. 4.10, it can be seen that: (i) 1-bit DACs can contribute the highest

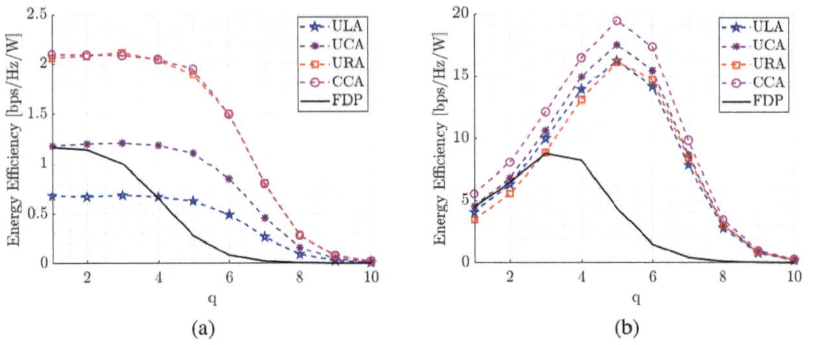


**Fig. 4.9** Energy efficiency versus $q$ for different array structures, $N_T = 100$, $G = 3$, $N_{RF_g} = K_g = 3$, $N_{RF} = K = 9$, $\delta_a = \pm 15°$, $\delta_e = \pm 12.5°$, $\theta = 73°$, $(\phi_1, \phi_2, \phi_3) = (30°, 90°, 150°)$, (**a**) SNR = −20 dB. (**b**) 20 dB

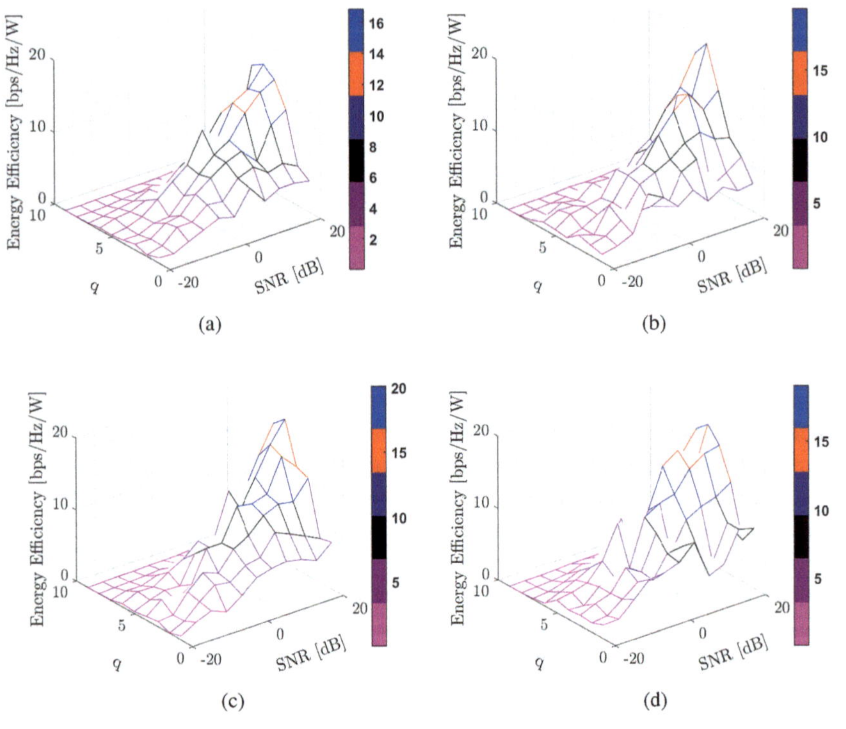


**Fig. 4.10** Energy efficiency versus $q$ and SNR, $N_T = 100$, $G = 3$, $N_{RF_g} = K_g = 3$, $N_{RF} = K = 9$, $\theta = 73°$, $(\phi_1, \phi_2, \phi_3) = (30°, 90°, 150°)$, $\delta_a = \pm 15°$, $\delta_e = \pm 12.5°$ (**a**) ULA. (**b**) URA. (**c**) UCA. (**d**) CCA

**Table 4.4** Performance comparison of ULA, UCA, URA, and CCA with few-bit DACs

| | ULA | UCA | URA | CCA |
|---|---|---|---|---|
| Sum-rate using few-bit DACs (see Fig. 4.8) | Low sum-rate | Low sum-rate | High sum-rate | High sum-rate |
| Energy efficiency using few-bit DACs (see Figs. 4.9 and 4.10) | Low efficiency | Low efficiency | High efficiency | High efficiency |

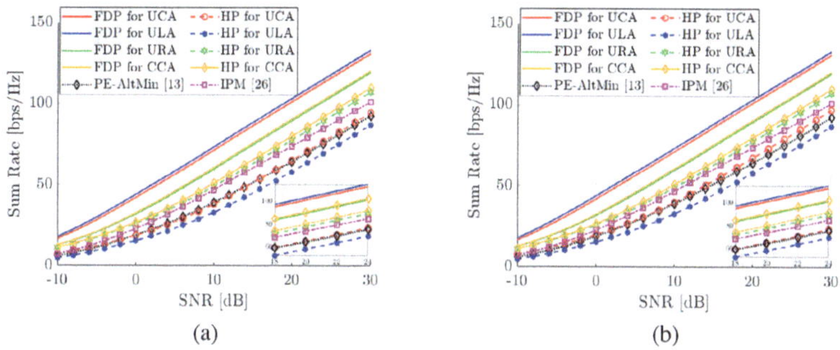


**Fig. 4.11** Sum-rate comparison of ULA, UCA, URA and CCA using full-resolution PSs and DACs: (**a**) variable-gain RF beamformer. (**b**) constant-modulus RF beamformer (with transfer block)

energy efficiency only at very low SNR (ii) at very high SNR, only (3–5)-bit DACs are sufficient to provide the highest energy efficiency along with very good spectral efficiency. The overall performance comparison of the array structures is summarized in Table 4.4.

### *4.6.4 Performance Comparison with Low-Resolution PSs and DACs*

In this section, we compare the performance of different 2D array structures using low-resolution PSs and DACs. Since the RF beamformer can be designed for constant and variable-gain entities, therefore, we first compare the performance of the HBF for the two cases: (i) HBF with variable-gain RF beamformer, and (ii) HBF with transfer block and constant-gain RF beamformer. To compare the performance of these two HBF schemes, we assume full-resolution hardware components (i.e., DACs and PSs).

In Fig. 4.11a, the sum-rate of proposed HBF scheme with variable-gain RF beamformer is compared to DBF as well as different HBF techniques (i.e., phase extraction alternative minimization (PE-AltMin) method in [9] and iterative phase matching (IPM) HBF technique in [22]) for different array structures, whereas in Fig. 4.11b, the sum-rate of HBF with transfer block and constant-gain RF beamformer is presented. It can be seen that both HBF schemes (i.e., two-stage HBF as in Fig. 4.11a and three-stage HBF as in Fig. 4.11b) yield similar performance for ULA, URA, UCA, and CCA. Furthermore, DBF achieves higher sum-rate at the expense of large number of RF chains i.e., $N_{RF} = N_T = 100$, whereas for both proposed HBF schemes, we can approach the sum-rate of DBF with relatively small number of RF chains, i.e., $N_{RF} = 9$. Also, both CCA and URA can give high sum-rate when compared to ULA and UCA as well as the HBF solutions presented in [9] and [22]. In the following analysis, we provide an extensive comparison of spectral and energy efficiencies of the array structures for the case of: (i) HBF with $b$-bit PSs and $\infty$-bit DACs, and (ii) HBF with $b$-bit PSs and $q$-bit DACs.

#### 4.6.4.1 $b$-Bit PSs AND $\infty$-Bit DACs

In this section, we analyze the performance of ULA, URA, UCA, and CCA by evaluating the proposed HBF schemes when using few-bit PSs only. Figure 4.12 depicts the sum-rate of different array structures for the following cases: (i) DBF, (ii) HBF using $\infty$-bit PSs, (iii) HBF using 1-bit PSs (phase quantization of (ii)), (iv) HBF using 1-bit PSs (OMP), (v) HBF using PE-AltMin [9], and (vi) HBF using IPM [22]. By using the same simulation parameters as mentioned earlier, it can be seen that the proposed OMP based HBF design for 1-bit PSs can provide higher sum-rates than the HBF designed with simple phase quantization as well as different HBF solutions (i.e., [9, 22]) for URA and CCA array structures. Also, the sum-rate degradation is reduced to around (3–4) [bps/Hz] at low SNR, whereas at high SNR, the degradation is $\approx$(8–10) [bps/Hz] due to high quantization error. However, by using only 2-bit PSs, we can approach the sum-rate of $\infty$-bit PSs as shown in Fig. 4.13. In this case, the sum-rate degradation is reduced to $\approx$(2–3) [bps/Hz] at all SNR values. From Figs. 4.12 and 4.13, we can also see that URA and CCA can give higher sum-rate than ULA and UCA for both 1-bit and 2-bit PSs.

Figure 4.14 plot the sum-rate versus $b$, where $b$ is the number of the bits of the PSs, and compares the spectral efficiency of ULA, URA, UCA, and CCA. The results are presented at SNR of 0, 10 and 20 dB. Figure 4.14 indicates that the proposed HBF scheme with 2-bit PSs can approach the performance of HBF with $\infty$-bit PSs. For a given SNR, the sum-rate performance gap between the DBF and HBF with $\infty$-bit PSs in the case of URA and CCA is much smaller than in the case of ULA and UCA. Furthermore, for all array structures, the sum-rate gap increases with increasing SNR. Similarly, for the same SNR values, we investigate the sum-rate versus $N_{RF_g}$ of DBF, HBF's with $\infty$-bit and 1-bit PS in different array structures as plotted in Fig. 4.15. The results show that both URA and CCA can give high sum-rate with a slight increase in $N_{RF_g}$. The sum-rate increases by approximately 20% for ULA and UCA, and almost 10% for URA and CCA by increasing $N_{RF_g}$ from 3

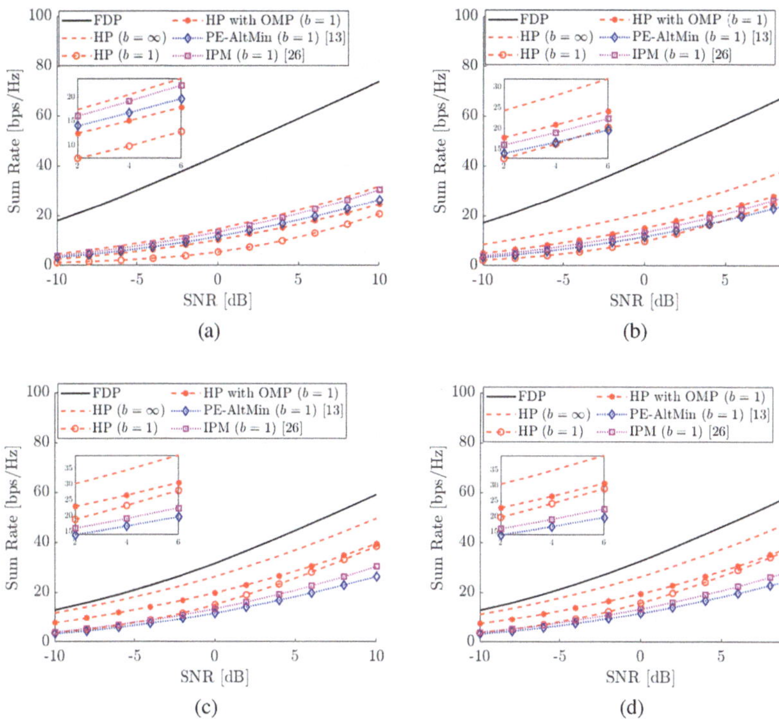


**Fig. 4.12** Sum-rate comparison with 1-bit PSs: (**a**) ULA. (**b**) UCA. (**c**) URA. (**d**) CCA

to 4. Also, in Fig. 4.16, we plot the sum-rate versus $N_{RF_g}$ of DBF, HBF's with $\infty$-bit and 2-bit PS. It can be seen that by increasing $N_{RF_g}$ from 3 to 4, the proposed HBF scheme using either URA or CCA can approach the performance of DBF with $\infty$-bit PSs, and offers a slightly lower sum-rate with 2-bit PSs. On the other hand, the propsed HBF scheme using ULA and UCA requires more RF chains to approach the performance of the DBF.

The use of low-resolution PSs with the proposed HBF scheme not only offers high spectral efficiency but can also provide high energy efficiency compared to DBF and HBF using full-resolution PSs. The energy efficiency is defined as the ratio of the sum-rate $R_{sum}$ and the total power consumption $P_{total}$, i.e.,

$$\epsilon = \frac{R_{sum}}{P_{total}} = \frac{R_{sum}}{P_T + N_{RF}P_{RF} + N_{PS}P_{PS_b}}, \tag{4.46}$$

where $P_T$ represents the total transmission power, $N_{RF}$ is the total number of RF chains, $P_{RF}$ is the power consumption of each RF chain, and $P_{PS_b}$ is the power consumption of $b$-bit PSs. As in [40], we use $P_T = 1$ W and $P_{RF} = 250$ mW. The use of transfer block yields a constant-gain RF beamformer but also doubles the number of PSs per each RF chain due to double-PS structure (as discussed in

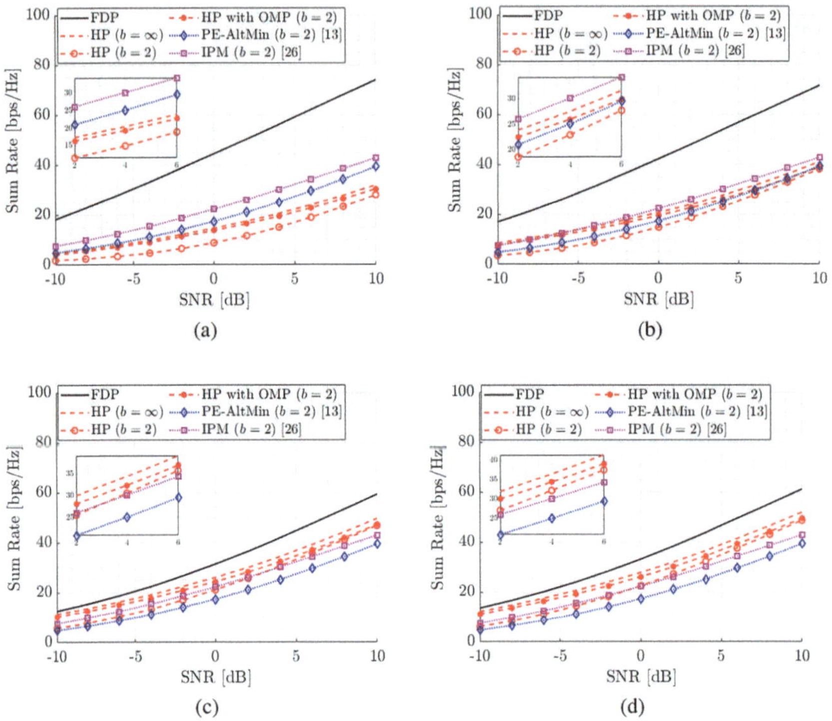


**Fig. 4.13** Sum-rate comparison with 2-bit PSs: (**a**) ULA. (**b**) UCA. (**c**) URA. (**d**) CCA

Sect. 4.4.2.2). Thus, $N_{PS} = 2 \times N_{RF} \times N$. We use 5 mW and 10 mW as the power consumption values for 1-bit and 2-bit PSs i.e., ($P_{PS_1} = 5$ mW, $P_{PS_2} = 10$ mW) [46, 47]. Figure 4.17 analyzes the energy efficiency of ULA, UCA, URA, and CCA versus transmit power, which varies from -30 dB to 30 dB and compares $\epsilon$ at SNR = 15 dB for four cases which are: (i) DBF, (ii) HBF using $\infty$-bit PSs, (iii) HBF using 1-bit PSs, and (iv) HBF using 2-bit PSs. For the full-resolution PSs, we assume 5-bit PSs be a good choice to replicate the performance of $\infty$-bit PSs.

Figure 4.17 shows that both 1-bit and 2-bit PSs can offer high energy efficiency compared to DBF and HBF using $\infty$-bit PSs. However, the energy efficiency decreases sharply beyond 0 dB for all array structures because of high transmit power. Compared to ULA and UCA, both URA and CCA can offer higher energy efficiency when using only 1 or 2-bit PSs at the RF-stage.

#### 4.6.4.2 *b*-Bit PSs AND *q*-Bit DACs

In this section, we examine the spectral effiency of the proposed HBF scheme with low-resolution DACs and PSs, i.e., the combined effect of $q$-bit DACs and $b$-bit PSs

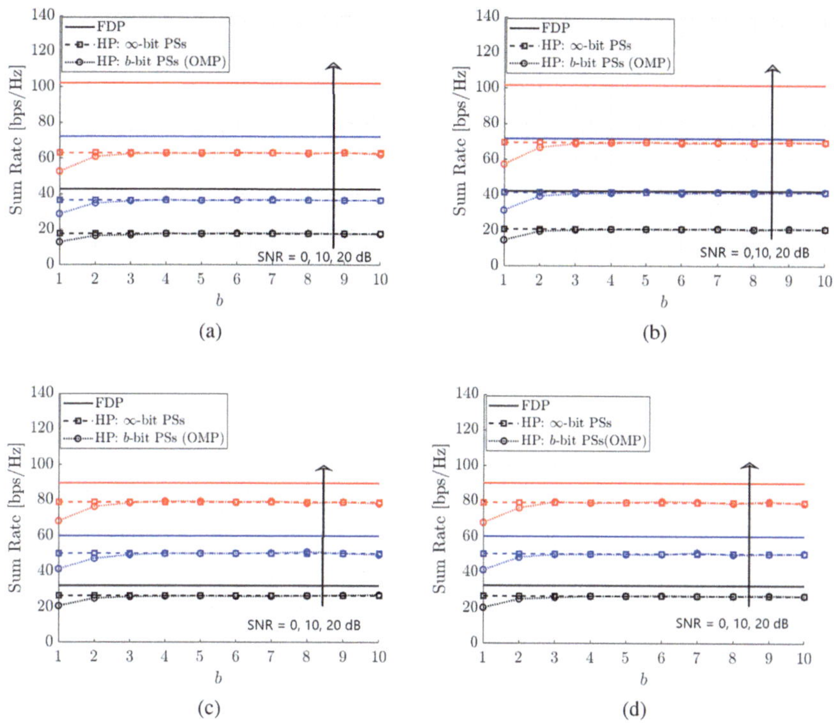


**Fig. 4.14** Sum-rate vs $b$ at SNR $= (0, 10, 20)$ dB: (**a**) ULA. (**b**) UCA. (**c**) URA. (**d**) CCA

on the sum-rate performance for ULA, URA, UCA, and CCA. To better understand the effect of low-resolution PSs and DACs on the sum-rate degradation $R_{deg}$, we first plot $R_{deg}$ due to each individual hardware component (i.e., DACs or PSs). Figure 4.18 shows various $R_{deg}$ plots versus $b$ for different values of $q$. It can be seen that 1-bit DAC can give large sum-rate degradation ( $\approx$ 65%). The large quantization noise introduced by the use of only 1-bit DAC impede the performance improvement even by increasing PSs resolution (i.e., $b$). However, by increasing DAC resolution, we can see that $R_{deg}$ decreases significantly, e.g., $R_{deg}$ reduces to $\approx$ 25% with 5-bit DACs and 1-bit PSs. In other words, we can approach the performance of HBF with $\infty$-bit resolution with just 5-bit DACs, offering significant advantages in power consumption and cost.

Figure 4.19 plots $R_{deg}$ versus $q$ for different PSs resolution. Similar to the case of 1-bit DACs, HBF using 1-bit PSs can give high $R_{deg}$ ( $\approx$ 70%). However, this degradation can be decreased significantly by increasing the resolution of DACs. Thus, HBF with 1-bit PSs can provide higher sum-rate than HBF with 1-bit DACs. Also, for $b \geq 2$, we can see that $R_{deg}$ is almost the same for all resolution levels i.e, $b = 2, 3, 4, 5$, which indicates that the performance of HBF depends more on DAC resolution than on PS resolution. Furthermore, this analysis gives us the least

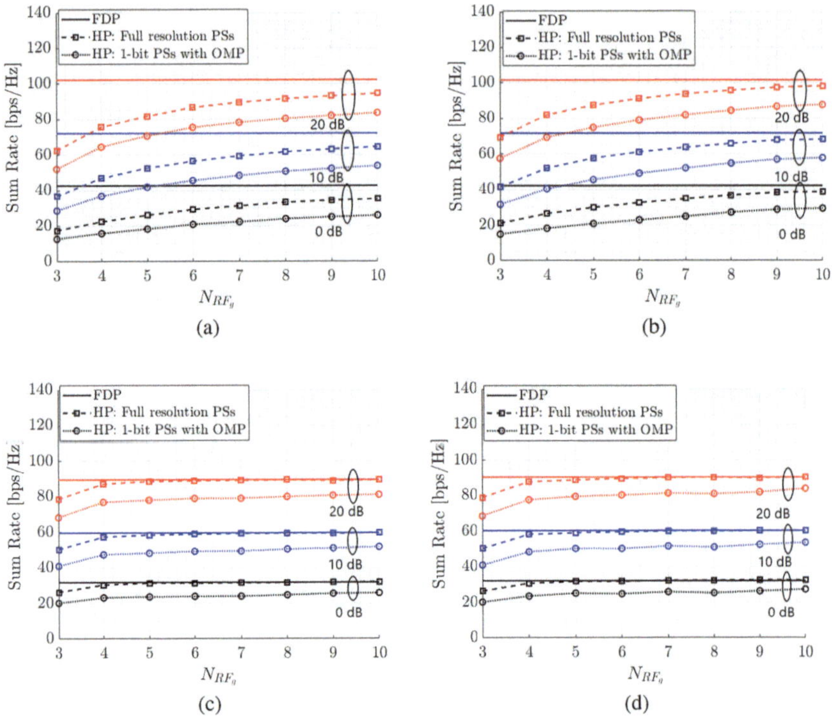


**Fig. 4.15** Sum-rate vs $N_{RF_g}$ with 1-bit PSs at SNR $=$ (0, 10, 20) dB: (**a**) ULA. (**b**) UCA. (**c**) URA. (**d**) CCA

numbers of bits of DACs and PSs required to achieve a satisfactory performance close to DBF and HBF using $\infty$-bit components.

Figure 4.20 plots the sum-rate versus $q$ and $b$ for ULA, UCA, URA, and CCA at SNR $= 10$ dB and $b$ and $q$ varying from 1 to 10. The sum-rate performance gap between the DBF and proposed HBF using ULA and UCA is larger than the case using URA and CCA. Moreover, the analysis of the results obtained in Figs. 4.18 and 4.19 gives some following useful notes: (i) the use of low-resolution DACs significantly degrades the sum-rate performance as compared to the use of low-resolution PSs (e.g., $R_{(q=1,b=10)} \ll R_{(q=10,b=1)}$), (ii) single-layered array structures (i.e., ULA and UCA) give low sum-rate, whereas, the multi-layered array structures (i.e., URA and CCA) can approach the sum-rate of DBF, and (iii) the optimal sum-rate can be obtained by using different combinations of $q$ and $b$ and thus, offers the choice to use which combination of low-resolution hardware components for increased sum-rate. This can also be better understood by the equal-sum-rate contour plots shown in Fig. 4.21. It can be observed that we can get reasonably high sum-rate by using the proposed HBF with ($q \geq 5, b \geq 2$), which is, in particular, close to that of DBF for both CCA and URA. Finally, Table 4.5 gives

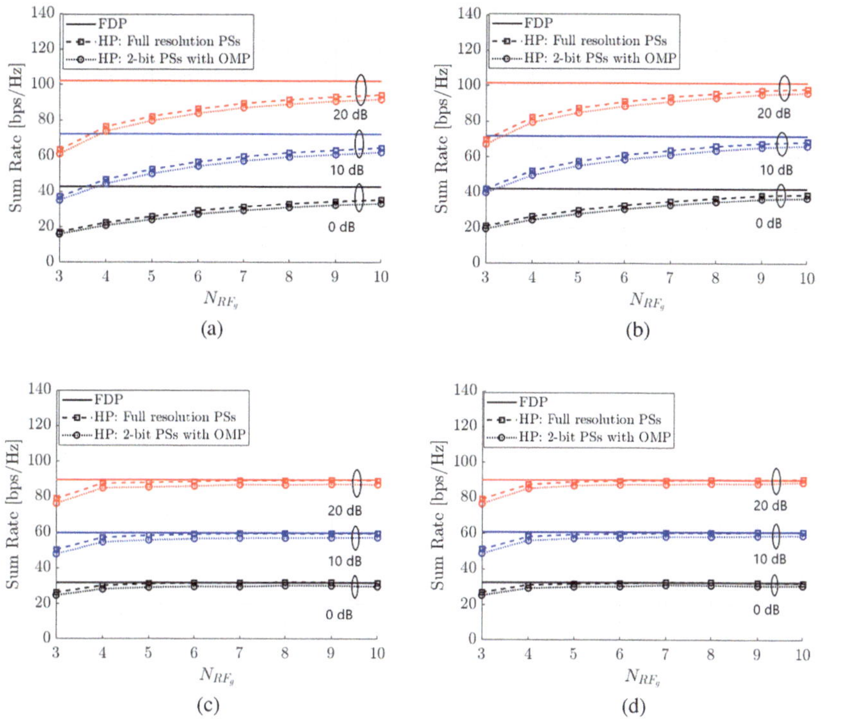


**Fig. 4.16** Sum-rate vs $N_{RF_g}$ with 2-bit PSs at SNR $= (0, 10, 20)$ dB: (**a**) ULA. (**b**) UCA. (**c**) URA. (**d**) CCA

the comprehensive comparison of spatial, spectral and energy efficiencies of ULA, UCA, URA, and CCA when using few-bit DACs and PSs.

## 4.7 Concluding Remarks

In this chapter, we have presented the MU-mMIMO hybrid beamforming (HBF) design for low-resolution PSs and/or DACs, and investigated its performance using different 2D antenna array structures. In the HBF design, the RF precoder has been developed via the slow time-varying AoD information, whereas the reduced-size effective CSI is utilized at the baseband precoder. Considering the hardware complexity, we have considered the following cases in the HBF design: (i) low-resolution PSs only, and (ii) low-resolution PSs and DACs. An optimization problem has been formulated to provide an RF beamformer with uniform gain entities by introducing a transfer block at the baseband stage. Furthermore, we have proposed an algorithmic solution using OMP to reduce the Euclidean distance between the HBF with full-resolution PSs and the HBF with low-resolution PSs. Based on OMP, we have first designed HBF using $b$-bit PSs and $\infty$-bit DACs, then constructed the

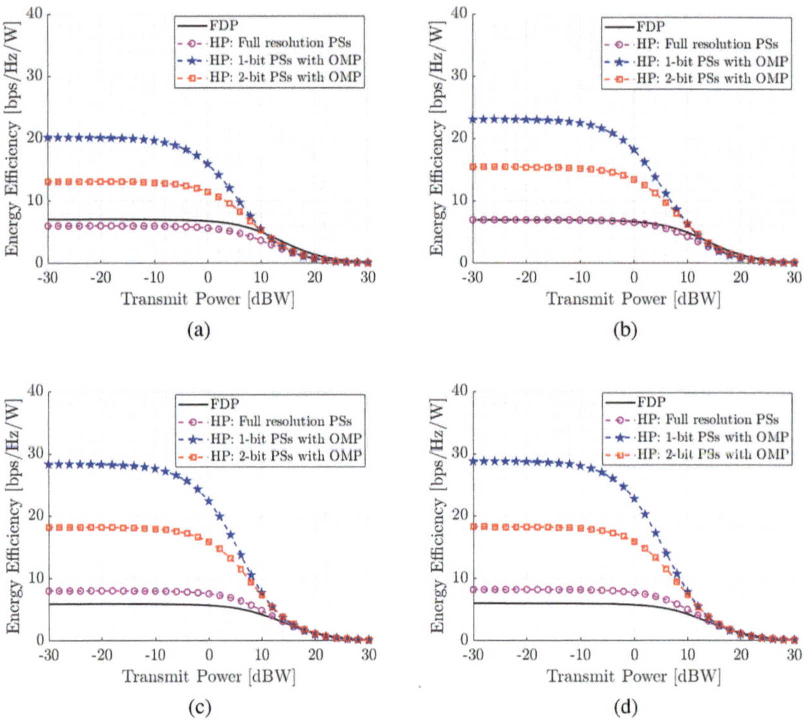


**Fig. 4.17** Energy efficiency comparison versus transmit power at SNR = 15 dB: (**a**) ULA. (**b**) UCA. (**c**) URA. (**d**) CCA

HBF using $b$-bit PSs and $q$-bit DACs to significantly reduce hardware complexity and costs in MU-mMIMO HBF systems. The spectral and energy efficiencies of ULA, UCA, URA, and CCA have been compared using Monte Carlo simulations. Illustrative results indicate that the proposed HBF using both URA and CCA can provide high sum-rates, which are comparable to their DBF counterparts. Also, the use of low-resolution PSs and DACs can provide high energy efficiency. Finally, we have presented different combinations of PS and DAC quantization levels to achieve near-optimal sum-rate. It has been shown that HBF using only 2-bit PSs and 5-bit DACs can provide high performance close to DBF.

## Appendix: Proof of Lemma 4.1

Let the normalization factor $\frac{N_{RF_g}}{\left\|\hat{\mathbf{F}}_{D_g}^{(q)}\mathbf{T}_g^{(q)}\right\|_F}$ is equal to $\frac{1}{\rho}$. Then, we can write:

$$\left\|\hat{\mathbf{F}}_{D_g}^{(q)}\mathbf{T}_g^{(q)}\right\|_F = \rho N_{RF_g} = \rho \left\|\hat{\mathbf{F}}_{D_g}\mathbf{T}_g\right\|_F . \tag{4.47}$$

Using norm inequality, we have:

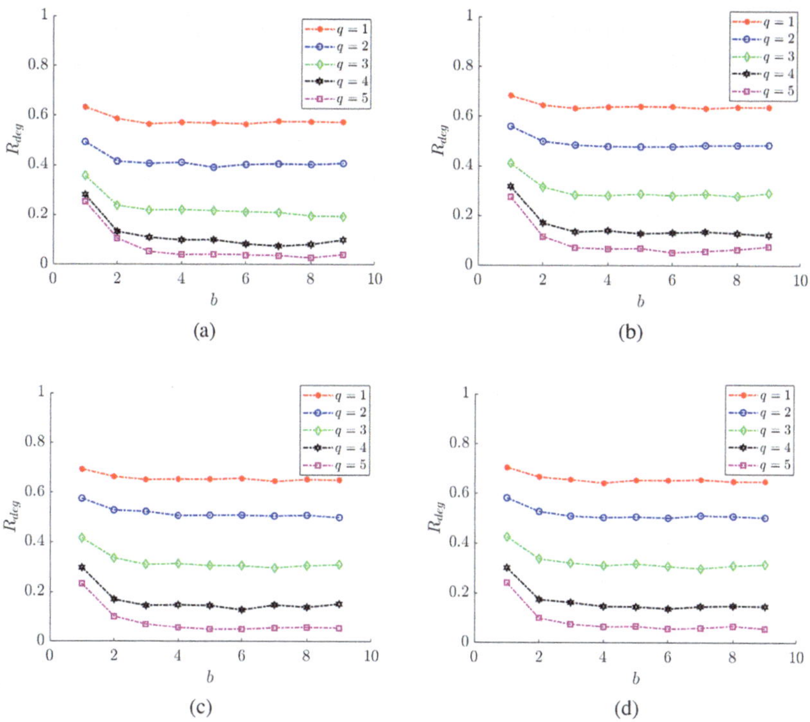


**Fig. 4.18** Sum-rate degradation versus $b$ for different DAC resolutions at SNR = 10 dB: (**a**) ULA. (**b**) UCA. (**c**) URA. (**d**) CCA

$$\left\|\hat{\mathbf{F}}_{D_g}\mathbf{T}_g - \hat{\mathbf{F}}_{D_g}^{(q)}\mathbf{T}_g^{(q)}\right\|_F \geq \left|\left\|\hat{\mathbf{F}}_{D_g}\mathbf{T}_g\right\|_F - \left\|\hat{\mathbf{F}}_{D_g}^{(q)}\mathbf{T}_g^{(q)}\right\|_F\right|,$$
$$= |1-\rho|\left\|\hat{\mathbf{F}}_{D_g}\mathbf{T}_g\right\|_F, \tag{4.48}$$

which is equivalent to $\left\|\hat{\mathbf{F}}_{D_g}\mathbf{T}_g\right\|_F \leq \frac{1}{|\rho-1|}\zeta$. When $\rho \neq 1$, we have $\left\|\hat{\mathbf{F}}_{D_g}\mathbf{T}_g - \hat{\mathbf{F}}_{D_g}^{(q)}\mathbf{T}_g^{(q)}\right\|_F \neq 0$. Then,

$$\begin{aligned}
\left\|\hat{\mathbf{F}}_{D_g}\mathbf{T}_g - \hat{\mathbf{F}}_{D_g}^{(q)}\tilde{\mathbf{T}}_g^{(q)}\right\|_F &= \left\|\hat{\mathbf{F}}_{D_g}\mathbf{T}_g - \hat{\mathbf{F}}_{D_g}^{(q)}\mathbf{T}_g^{(q)} + \left(1-\frac{1}{\rho}\hat{\mathbf{F}}_{D_g}^{(q)}\mathbf{T}_g^{(q)}\right)\right\|_F, \\
&\leq \left\|\hat{\mathbf{F}}_{D_g}\mathbf{T}_g - \hat{\mathbf{F}}_{D_g}^{(q)}\mathbf{T}_g^{(q)}\right\|_F + \left|1-\frac{1}{\rho}\right| \left\|\hat{\mathbf{F}}_{D_g}^{(q)}\tilde{\mathbf{T}}_g^{(q)}\right\|_F, \\
&\leq \zeta + |\rho-1|\left\|\hat{\mathbf{F}}_{D_g}\mathbf{T}_g\right\|_F
\end{aligned}$$

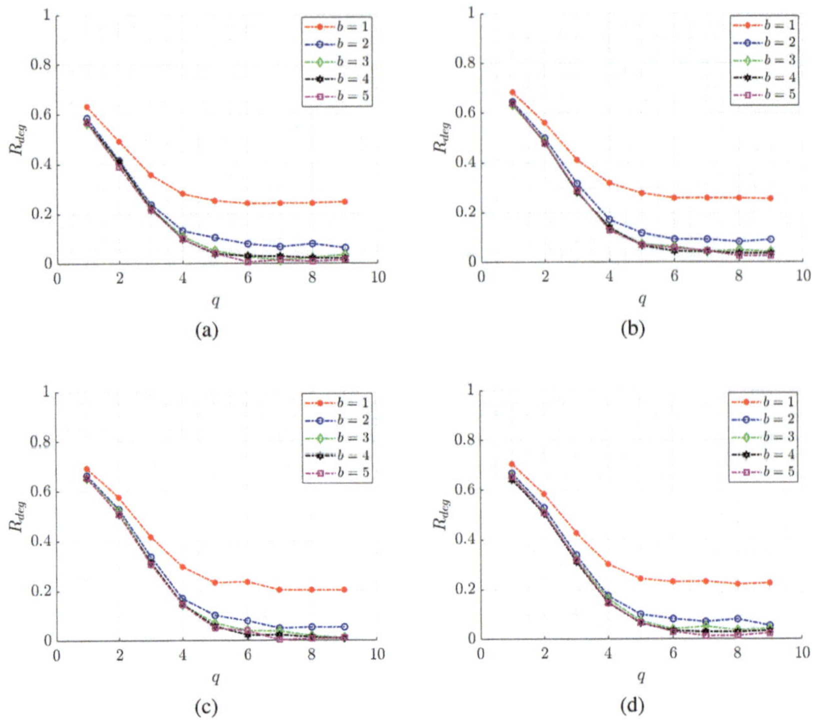


**Fig. 4.19** Sum-rate degradation versus $q$ for different PS resolutions at SNR = 10 dB: (**a**) ULA. (**b**) UCA. (**c**) URA. (**d**) CCA

$$
\begin{aligned}
&\leq \zeta + \frac{|\rho - 1|}{|\rho - 1|}\zeta, \\
&= 2\zeta.
\end{aligned}
\tag{4.49}
$$

From (4.49), we can write as:

$$
\left\| \hat{\mathbf{F}}_{D_g} \mathbf{T}_g - \hat{\mathbf{F}}_{D_g}^{(q)} \tilde{\mathbf{T}}_g^{(q)} \right\|_F = 2\zeta. \tag{4.50}
$$

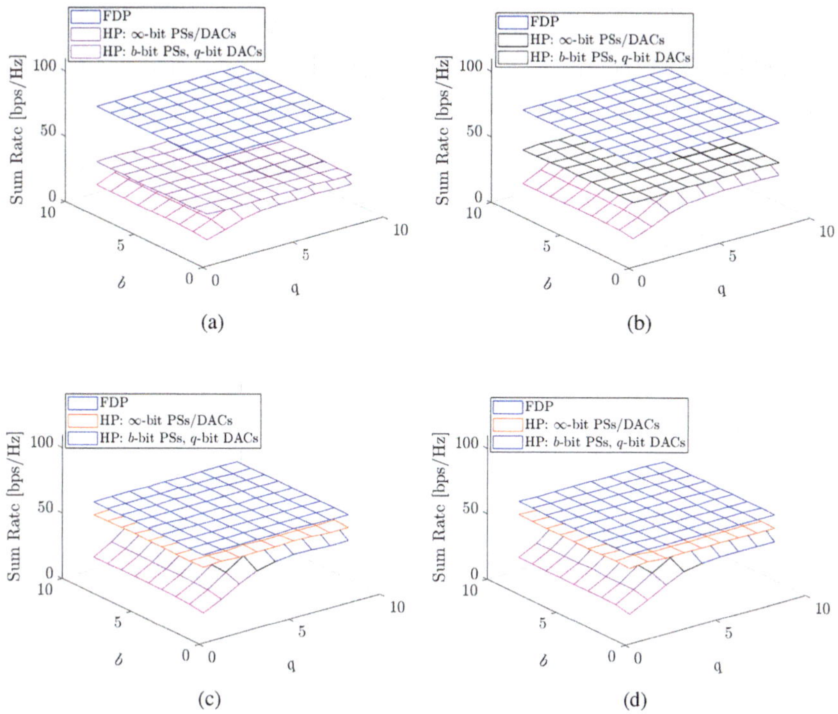


**Fig. 4.20** Sum-rate versus $q$ and $b$ at SNR = 10 dB: (**a**) ULA. (**b**) UCA. (**c**) URA. (**d**) CCA

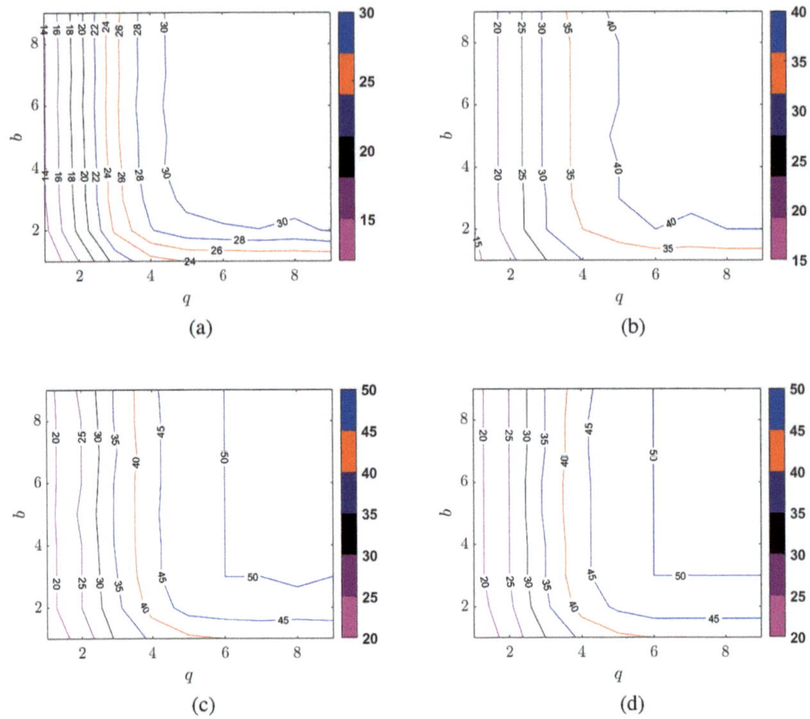


**Fig. 4.21** Equal-sum-rate contour plots at SNR = 10 dB: (**a**) ULA. (**b**) UCA. (**c**) URA. (**d**) CCA

**Table 4.5** Performance comparison of ULA, UCA, URA and CCA using low-resolution PSs and DACs

| | ULA | UCA | URA | CCA |
|---|---|---|---|---|
| Spatial efficiency | Low | Low | High | High |
| Sum-rate using $b$-bit PSs (see Figs. 4.12 and 4.16) | Low | High | High | High |
| Energy efficiency using $b$-bit PSs (see Fig. 4.17) | Low | High | High | High |
| Sum-rate using $q$-bit DACs and $b$-bit PSs (see Figs. 4.18 and 4.19) | Low | Low | High | High |

# References

1. A.F. Molisch, V.V. Ratnam, S. Han, Z. Li, S.L.H. Nguyen, L. Li, K. Haneda, Hybrid beamforming for massive MIMO: a survey. IEEE Commun. Mag. **55**(9), 134–141 (2017)
2. E. Ali, M. Ismail, R. Nordin, N.F. Abdulah, Beamforming techniques for massive MIMO systems in 5G: overview, classification, and trends for future research. Front. Inf. Technol. Electron. Eng. **18**, 753–772 (2017)
3. L. Liang, W. Xu, X. Dong, Low-complexity hybrid precoding in massive multiuser MIMO systems. IEEE Wireless Commun. Lett. **3**(6), 653–656 (2014)

4. T.E. Bogale, L.B. Le, A. Haghighat, L. Vandendorpe, On the number of RF chains and phase shifters, and scheduling design with hybrid analog–digital beamforming. IEEE Trans. Wireless Commun. **15**(5), 3311–3326 (2016)
5. W. Ni, X. Dong, W. Lu, Near-optimal hybrid processing for massive MIMO systems via matrix decomposition. IEEE Trans. Signal Process. **65**(15), 3922–3933 (2017)
6. S. He, C. Qi, Y. Wu, Y. Huang, Energy-efficient transceiver design for hybrid sub-array architecture MIMO systems. IEEE Access **4**, 9895–9905 (2016)
7. W. Ni, X. Dong, Hybrid block diagonalization for massive multiuser MIMO systems. IEEE Trans. Commun. **64**(1), 201–211 (2016)
8. A. Li, C. Masouros, Hybrid analog-digital Millimeter-Wave MU-MIMO transmission with virtual path selection. IEEE Commun. Lett. **21**(2), 438–441 (2017)
9. X. Yu, J.C. Shen, J. Zhang, K.B. Letaief, Alternating minimization algorithms for hybrid precoding in millimeter wave MIMO systems. IEEE J. Sel. Topics Signal Process. **10**(3), 485–500 (2016)
10. A. Adhikary, J. Nam, J.Y. Ahn, G. Caire, Joint spatial division and multiplexing–the large-scale array regime. IEEE Trans. Inf. Theory **59**(10), 6441–6463 (2013)
11. J. Nam, A. Adhikary, J. Ahn, G. Caire, Joint spatial division and multiplexing: opportunistic beamforming, user grouping and simplified downlink scheduling. IEEE J. Sel. Topics Signal Process. **8**(5), 876–890 (2014)
12. D. Kim, G. Lee, Y. Sung, Two-stage beamformer design for massive MIMO downlink by trace quotient formulation. IEEE Trans. Commun. **63**(6), 2200–2211 (2015)
13. S. Park, A. Alkhateeb, R.W. Heath, Dynamic subarrays for hybrid precoding in wideband mmWave MIMO systems. IEEE Trans. Wireless Commun. **16**(5), 2907–2920 (2017)
14. Y. Jeon, C. Song, S. Lee, S. Maeng, J. Jung, I. Lee, New beamforming designs for joint spatial division and multiplexing in large-scale MISO multi-user systems. IEEE Trans. Wireless Commun. **16**(5), 3029–3041 (2017)
15. A. Koc, A. Masmoudi, T. Le-Ngoc, Hybrid beamforming for uniform circular arrays in multi-user massive MIMO systems, in *Proceedings of the 32nd IEEE 32nd Canadian Conference on Electrical and Computer Engineering (CCECE), Edmonton, Canada, May. 5–8* (2019)
16. R. Mai, T. Le-Ngoc, Nonlinear hybrid precoding for coordinated multi-cell massive MIMO systems. IEEE Trans. Veh. Technol. **68**(3), 2459–2471 (2019)
17. S. Zarei, W. Gerstacker, R. Schober, Low-complexity hybrid Linear/Tomlinson-Harashima precoding for downlink large-scale MU-MIMO systems, in *IEEE Globecom Workshops (GC Wkshps), Washington, DC USA, Dec. 4–8* (2016)
18. C. Pradhan, A. Li, L. Zhuo, Y. Li, B. Vucetic, Hybrid-precoding for mmWave multi-user communications in the presence of beam-misalignment. IEEE Trans. Wireless Commun. **19**(9), 6083–6099 (2020)
19. F. Sohrabi, W. Yu, Hybrid beamforming with finite-resolution phase shifters for large scale MIMO systems, in *Proceedings of IEEE 16th International Workshop on Signal Processing Advances in Wireless Communications (SPAWC), Stockholm, Sweden, Jun. 27–July 1* (2015)
20. F. Sohrabi, W. Yu, Hybrid digital and analog beamforming design for large-scale antenna arrays. IEEE J. Sel. Topics Signal Process. **10**(3), 501–513 (2016)
21. S. Payami, M. Ghoraishi, M. Dianati, Hybrid beamforming for large antenna arrays with phase shifter selection. IEEE Trans. Wireless Commun. **15**(11), 7258–7271 (2016)
22. Z. Wang, M. Li, Q. Liu, A.L. Swindlehurst, Hybrid precoder and combiner design with low-resolution phase shifters in mmWave MIMO systems. IEEE J. Sel. Topics Signal Process. **12**(2), 256–269 (2018)
23. H. Li, Q. Liu, Z. Wang, M. Li, Transmit antenna selection and analog beamforming with low-resolution phase shifters in mmWave MISO systems. IEEE Commun. Lett. **22**(9), 1878–1881 (2018)
24. H. Li, Q. Liu, Z. Wang, M. Li, Joint antenna selection and analog precoder design with low-resolution phase shifters. IEEE Trans. Veh. Technol. **68**(1), 967–971 (2018)
25. H. Li, M. Li, Q. Liu, Hybrid beamforming with dynamic subarrays and low-resolution PSs for mmWave MU-MISO systems. IEEE Trans. Commun. **68**(1), 602–614 (2019)

26. H. Seleem, A.I. Sulyman, A. Alsanie, Hybrid precoding-beamforming design with Hadamard RF codebook for mmWave large-scale MIMO systems. IEEE Access **5**, 6813–6823 (2017)
27. H. Li, M. Li, Q. Liu, A.L. Swindlehurst, Dynamic hybrid beamforming with low-resolution PSs for wideband mmWave MIMO-OFDM systems. IEEE J. Sel. Areas Commun. **38**(9), 2168–2181 (2020)
28. C. Wang, C. Wen, S. Jin, S. Tsai, Finite-alphabet precoding for massive MU-MIMO with low-resolution DACs. IEEE Trans. Wireless Commun. **17**(7), 4706–4720 (2018)
29. S. Jacobsson, G. Durisi, M. Coldrey, T. Goldstein, C. Studer, Quantized precoding for massive MU-MIMO. IEEE Trans. Commun. **65**, 4670–4684 (2017)
30. L.N. Ribeiro, S. Schwarz, M. Rupp, A.L.F. de Almeida, Energy efficiency of mmWave massive MIMO precoding with low-resolution DACs. IEEE J. Sel. Topics Signal Process. **12**(2), 298–312 (2018)
31. J. Mo, A. Alkhateeb, S. Abu-Surra, R.W. Heath, Hybrid architectures with few-bit ADC receivers: achievable rates and energy-rate tradeoffs. IEEE Trans. Wireless Commun. **16**(4), 2274–2287 (2017)
32. K. Roth, H. Pirzadeh, A.L. Swindlehurst, J.A. Nossek, A comparison of hybrid beamforming and digital beamforming with low-resolution ADCs for multiple users and imperfect CSI. IEEE J. Sel. Topics Signal Process. **12**(3), 484–498 (2018)
33. A. Koc, T. Le-Ngoc, Hybrid millimeter-wave massive MIMO systems with low CSI overhead and Few-Bit DACs/ADCs, in *Proceedings of IEEE 92th Vehicular Technology Conference (VTC-Fall), Victoria, Canada, Nov. 18–Dec. 16* (2020)
34. J. Hoydis, S. ten Brink, M. Debbah, Massive MIMO in the UL/DL of cellular networks: how many antennas do we need? IEEE J. Sel. Areas Commun. **31**(2), 160–171 (2013)
35. C.B. Peel, B.M. Hochwald, A.L. Swindlehurst, A vector-perturbation technique for near-capacity multiantenna multiuser communication-part I: channel inversion and regularization. IEEE Trans. Commun. **53**(1), 195–202 (2005)
36. J.D. Krieger, C. Yeang, G.W. Wornell, Dense delta-sigma phased arrays. IEEE Trans. Antennas Propag. **61**(4), 1825–1837 (2013)
37. J.A. Tropp, A.C. Gilbert, Signal recovery from random measurements via orthogonal matching pursuit. IEEE Trans. Inf. Theory **53**(12), 4655–4666 (2007)
38. O. El Ayach, S. Rajagopal, S. Abu-Surra, Z. Pi, R.W. Heath, Spatially sparse precoding in millimeter wave MIMO systems. IEEE Trans. Wireless Commun. **13**(3), 1499–1513 (2014)
39. Da-Shan Shiu, G.J. Foschini, M.J. Gans, J.M. Kahn, Fading correlation and its effect on the capacity of multielement antenna systems. IEEE Trans. Commun. **48**(3), 502–513 (2000)
40. A. Koc, A. Masmoudi, T. Le-Ngoc, 3D angular-based hybrid precoding and user grouping for uniform rectangular arrays in massive MU-MIMO systems. IEEE Access **8**, 84689–84712 (2020)
41. Y. Xiong, Achievable rates for massive MIMO relaying systems with variable-bit ADCs/DACs. IEEE Commun. Lett. **24**(5), 991–994 (2020)
42. C. Kong, C. Zhong, S. Jin, S. Yang, H. Lin, Z. Zhang, Full-duplex massive MIMO relaying systems with low-resolution ADCs. IEEE Trans. Wireless Commun. **16**(8), 5033–5047 (2017)
43. W.B. Abbas, F. Gomez-Cuba, M. Zorzi, Millimeter wave receiver efficiency: a comprehensive comparison of beamforming schemes with low resolution ADCs. IEEE Trans. Wireless Commun. **16**(12), 8131–8146 (2017)
44. R.L. Haupt, Optimized element spacing for low sidelobe concentric ring arrays. IEEE Trans. Antennas Propag. **56**(1), 266–268 (2008)
45. E. Vlachos, A. Kaushik, J. Thompson, Energy efficient transmitter with low resolution DACs for massive MIMO with partially connected hybrid architecture, in *2018 IEEE 87th Vehicular Technology Conference (VTC Spring)* (2018), pp. 1–5
46. J.C. Chen, Energy-efficient analog combiner design using low-resolution phase shifters and antenna selection for mmWave D2D communications. IEEE Trans. Veh. Technol. **69**(11), 13979–13984 (2020)
47. R. Méndez-Rial, C. Rusu, N. González-Prelcic, A. Alkhateeb, R.W. Heath, Hybrid MIMO architectures for millimeter wave communications: phase shifters or switches? IEEE Access **4**, 247–267 (2016)

# Chapter 5
# Non-planar Antenna Array Structures for mMIMO

## 5.1 Introduction

As discussed in Chap. 4, the antenna elements at the BS can be configured in a number of ways, resulting in narrower beams to reduce MU interference. A comparison of different 2D array structures in MU-mMIMO is presented in [1, 2], showing higher spectral and energy efficiencies of uniform rectangular array (URA) and concentric circular array (CCA) than uniform linear array (ULA) and uniform circular array (UCA). Given the massive wireless connectivity requirements of IoT, mMTC, and future aerial applications involving mobile drones, UAVs, and satellites, existing array structures lack the ability to transmit signals in arbitrary locations in 3D space. As a result, in future wireless applications, the 3D configuration of antenna elements will be critical. The design of a uniform spherical array (USA) is presented in [3], which shows significant advantages in mMIMO. Antenna elements can also be placed in a 3D cylindrical array to increase SE as shown in [4]. The majority of existing research on HBF solutions in mmWave MU-mMIMO systems is focused on ULA or URA (e.g., [5–8]). Therefore, analyzing the performance of various 3D array topologies remains an unaddressed research problem.

In this chapter, we compare the spectral and energy efficiencies of three different 3D array structures in mmWave MU-mMIMO systems, i.e., cylindrical array (CA), hemi-spherical array (HSA), and spherical array (SA) and analyze the computational complexity as well as robustness of the proposed HBF scheme. The main contributions of this chapter are outlined as follows:

- **SVD-Based HBF Design:** We consider the design of HBF for mmWave MU-mMIMO systems, which involves two stages: (i) RF beamforming based on SVD of channel matrix, and baseband MU precoding based on the instantaneous effective baseband channel to mitigate MU-interference by RZF technique.
- **Comparison of 3D Antenna Array Structures:** We compare the spectral and energy efficiencies of different 3D antenna array structures in mmWave



T. Le-Ngoc et al., *Half-Duplex/Full-Duplex MU-mMIMO*, Wireless Networks,
https://doi.org/10.1007/978-3-032-09195-6_5

MU-mMIMO systems. Particularly, we consider three different 3D array configurations, namely, SA, HSA, and CA, and compare their performance versus URA (2D array configuration). We show that in comparison to 2D arrays, 3D array structures can provide high spectral and energy efficiencies as well as non-varying achievable rate independent from the user angular location.

- **Robustness of HBF to CSI Errors:** We compare the effect of channel estimation error on the sum-rate performance for different array structures and show that the proposed HBF scheme is not sensitive to the accuracy of CSI.

The rest of this chapter is organized as follows. In Sect. 5.2, we discuss the system and channel model for mmWave MU-mMIMO systems. Section 5.3 presents the SVD-based HBF solution to maximize the total achievable rate. The illustrative results are provided in Sect. 5.4 to compare the performance of different 3D array structures. Finally, the chapter is concluded in Sect. 5.5.

## 5.2 System and Channel Model

In this section, we introduce the system and channel models of the proposed hybrid mmWave MU-mMIMO systems.

### 5.2.1 System Model

We consider the downlink of a mmWave MU-mMIMO system with HBF structure, as depicted in Fig. 5.1. The BS employs $N$ antenna elements at the transmitter, which are fed by $N_{RF}$ RF chains simultaneously communicating with $K$ single-antenna users. At the BS, the hybrid precoder $\mathbf{B} = \mathbf{F}_D\mathbf{B}_D$ consists of the digital baseband precoder $\mathbf{B}_D \in \mathbb{C}^{N_{RF}\times K}$ and the analog RF beamformer $\mathbf{F}_D \in \mathbb{C}^{N_T\times N_{RF}}$

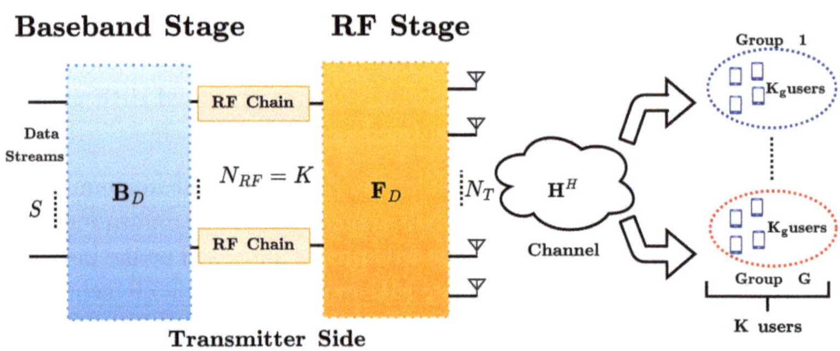


**Fig. 5.1** System model for the proposed mmWave MU-mMIMO hybrid precoding

that is implemented using phase shifters and thus, imposes a constant-modulus contraint., i.e., $|\mathbf{F}_D(i,j)| = \frac{1}{\sqrt{N_T}}$ $\forall i, j$. The design of HBF reduces the number of RF chains from $N_T$ to $N_{RF}$, while satisfying $K \leq N_{RF} \ll N_T$. Assuming the number of data streams $S$ equal to the number of the users $K$., i.e., $S = K$, the precoded signal at the transmitter is given by:

$$\mathbf{s} = \mathbf{F}_D \mathbf{B}_D \mathbf{d}, \tag{5.1}$$

where $\mathbf{d} \in \mathbb{C}^K$ is the data signal with $\mathbb{E}\{\mathbf{d}\mathbf{d}^H\} = \mathbf{I}_K$. The transmitted signal satisfies the power constraint., i.e., $\mathbb{E}\{\|\mathbf{s}\|_2^2\} \leq P_T$, where $P_T$ is the transmit power at the BS. Then, the received signal at $k^{th}$ user is given by:

$$\begin{aligned} y_k &= \mathbf{h}_k^H \mathbf{B}\mathbf{s} + n_k, \\ &= \underbrace{\mathbf{h}_k^H \mathbf{F}_D \mathbf{b}_{d_k} d_k}_{\text{Desired Signal}} + \underbrace{\sum_{\hat{k} \neq k}^{K} \mathbf{h}_k^H \mathbf{F}_D \mathbf{b}_{d_{\hat{k}}} d_{\hat{k}}}_{\text{IUI}} + \underbrace{n_k}_{\text{Noise}}, \end{aligned} \tag{5.2}$$

where $\mathbf{h}_k \in \mathbb{C}^{N_T}$ and $\mathbf{s}_k$ are the narrowband flat-fading channel vector and data signal for $k^{th}$ user, respectively, $\mathbf{b}_{d_k}$ is the $k^{th}$ column of $\mathbf{B}_D$, and $n_k$ denotes the additive circular symmetric Gaussian noise such that $n_k \sim \mathcal{CN}(0, \sigma^2)$.

By using (5.1) and (5.2), the instantaneous signal-to-interference-plus-noise ratio (SINR) expression at $k^{th}$ user is derived as follows:

$$\text{SINR}_k = \frac{|\mathbf{h}_k^H \mathbf{F}_D \mathbf{b}_{d_k}|^2}{\sum_{\hat{k} \neq k}^{K} |\mathbf{h}_k^H \mathbf{F}_D \mathbf{b}_{d_{\hat{k}}}|^2 + \sigma^2}. \tag{5.3}$$

By using the instantaneous SINR, the ergodic sum-rate capacity $R_{\text{sum}}$ for the MU-mMIMO systems is given by:

$$R_{\text{sum}} = \sum_{k=1}^{K} \mathbb{E}[\log_2(1 + \text{SINR}_k)]. \tag{5.4}$$

The design of $\mathbf{F}_D, \mathbf{B}_D$ aims at maximizing the SE of mmWave MU-mMIMO systems. Then, the optimization problem can be formulated as follows:

$$\max_{\mathbf{F}_D, \mathbf{B}_D} \quad R_{\text{sum}}\big(\mathbf{F}_D, \mathbf{B}_D\big) \tag{5.5a}$$

$$\text{s.t.} \quad |\mathbf{F}_D(i,j)| = \frac{1}{\sqrt{N_T}}, \tag{5.5b}$$

$$\mathbb{E}\{\|\mathbf{s}\|_2^2\} \leq P_T. \tag{5.5c}$$

### 5.2.2 Channel Model

We consider a narrowband clustered channel model, based on the extended Saleh-Valenzuela model to characterize the sparse scattering of the mmWave communication channel [5]. We assume the channel matrix $\mathbf{H} \in \mathbb{C}^{N_T \times K}$ be the sum of $N_c$ scattering clusters, where each cluster contributes $N_P$ propagation paths. Then, the narrowband channel vector $\mathbf{h}_k$ for $k^{th}$ user can be written as:

$$\mathbf{h}_k = \sqrt{\frac{N_T}{N_c L}} \sum_{m=0}^{N_C-1} \sum_{l=0}^{L-1} \alpha_k^{m,l} \mathbf{a}_k(\phi_k^{m,l}, \theta_k^{m,l}), \tag{5.6}$$

where $\alpha_k^{m,l}$ corresponds to the complex gain of the $lth$ multipath ray in the $mth$ cluster, $\mathbf{a}_k(\phi_k^{m,l}, \theta_k^{m,l})$ is the transmit array response vector, where $\phi_k^{m,l} \in [\phi_k - \delta_e, \phi_k + \delta_e]$ represents the mean elevation angle of the corresponding path with angle spread $\delta_e$ and $\theta_k^{m,l} \in [\theta_k - \delta_a, \theta_k + \delta_a]$ represents the azimuth angle of departure (AoD) with azimuth angle spread $\delta_a$. We assume complex weights $\alpha^{m,l}$ are i.i.d $\mathcal{CN}(0, \sigma_{\alpha,m}^2)$, where $\sigma_{\alpha,m}^2$ is the average power of the $mth$ cluster such that $\sum_{m=0}^{N_C-1} \sigma_{\alpha,m}^2 = \sqrt{\frac{N_T}{N_C L}}$. The array steering vector is given as [3, 4, 8]:

$$\mathbf{a}(\phi, \theta) = \frac{1}{\sqrt{N_T}} [a_1(\phi, \theta), a_2(\phi, \theta), \ldots, a_N(\phi, \theta)]^T, \tag{5.7}$$

where $a_n(\phi, \theta) = e^{j\frac{2\pi}{\lambda}\mathbf{r}_n^T \mathbf{u}(\phi,\theta)}$, $N_T$ is the number of the antenna elements at BS, $\mathbf{r}_n = [x_n, y_n, z_n]^T$ is the position of the $n^{th}$ antenna element and $\mathbf{u}(\phi, \theta) = [\cos\theta \sin\phi, \sin\theta \sin\phi, \cos\theta]^T$. Different from the conventional array structures, we assume the antenna elements arranged in 3D space namely, CA, HSA, and SA, and compare their performance with that of 2D array i.e., URA, as shown in Fig. 5.2.

## 5.3 Hybrid Beamforming Design

In this section, we present the design of proposed hybrid precoder $\mathbf{B} \in \mathbb{C}^{N_T \times K}$, which maximizes the spectral and energy efficiencies of mmWave MU-mMIMO systems. The HBF is constructed by concatenating the RF beamformer $\mathbf{F}_D \in \mathbb{C}^{N_T \times N_{RF}}$ and reduced-dimensional MU baseband precoder $\mathbf{B}_D \in \mathbb{C}^{N_{RF} \times K}$. The HBF solutions presented in literature (e.g., [5–7]) can achieve a high sum-rate performance but depends on finding the optimal fully digital precoder $\mathbf{F}_{opt}$, which increases the computational complexity. Eigen beamforming (EBF), on the other hand, can give high performance with reduced complexity. However, the RF beamformer in this case constitutes variable-gain entities and thus, does not meet the constraint given in (5.5b). To design the constant-modulus RF beamformer

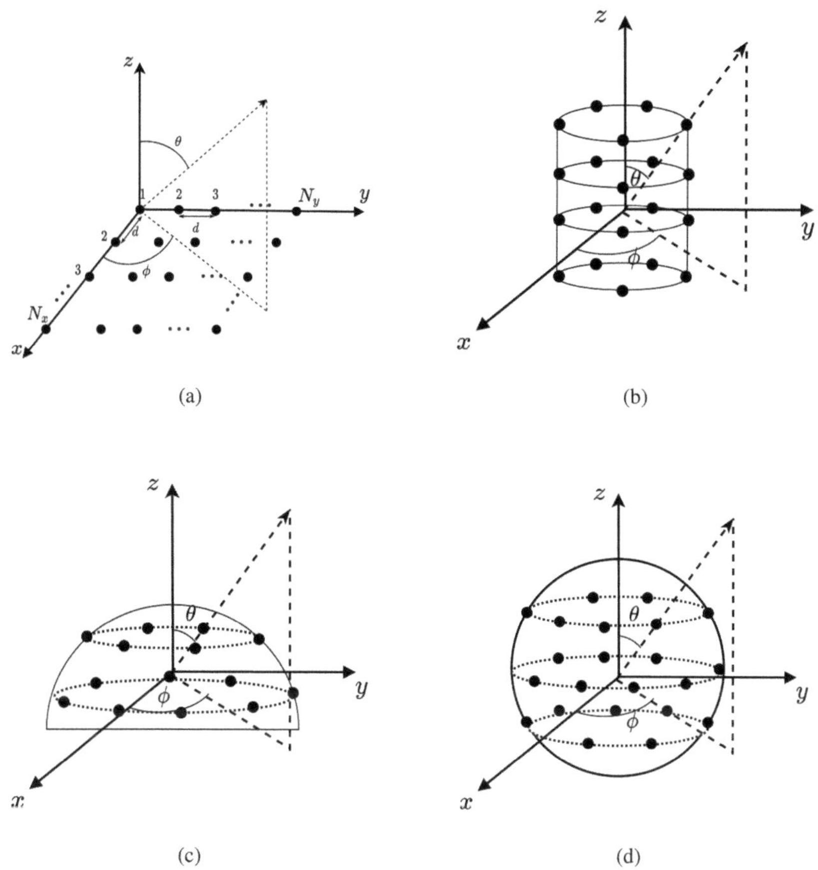


**Fig. 5.2** mmWave MU-mMIMO array structures: (**a**) URA. (**b**) CA. (**c**) HSA. (**d**) SA

while using EBF is a challenging problem. The simple solution is to formulate the correlation matrix $\mathbf{R}$ based on the array response vector $\mathbf{a}_k$ [9]. Then, the correlation matrix for $k^{th}$ user can be written as:

$$\mathbf{R}_k = \mathbf{a}_k \mathbf{a}_k^T. \tag{5.8}$$

By using Karhunen-Loeve decomposition, we let $\mathbf{R}_k = \mathbf{U}_k \boldsymbol{\Lambda}_k \mathbf{U}_k^H$. Then, the RF beamformer can be designed using the dominant eigenvalues., i.e., $\mathbf{U}_k^*$. However, this approach gives low performance. In [10], a three-stage HBF is designed using EBF but it requires the use of doubling phase shifter structure. In the proposed HBF design, we introduce a low-complexity solution, which uses ordered SVD of the mmWave MU-mMIMO channel and achieves high performance without requiring $\mathbf{F}_{opt}$ [11].

**Algorithm 5.1:** SVD based RF beamformer design

**Input**: $[\theta_k^{(g)}, \phi_k^{(g)}]$ for $k = 1, \ldots, K$.
**Output**: $\mathbf{F}_D$.
**for** $g = 1 : G$ **do**
  Compute channel matrix $\mathbf{h}_k^{(g)}$ via (5.6).
  Using (5.9), form the unitary matrices $\mathbf{Q}^{(g)}$ and $\mathbf{V}^{(g)}$.
  **for** $n = 1 : N_T$ **do**
    Calculate the phase of each element of $\mathbf{V}(:, n)^{(g)}$, i.e., $\psi(:, n)^{(g)}$.
    Construct the phase vector $\gamma_n^{(g)}$.
  **end**
  Formulate $\mathbf{\Gamma}^{(g)}$ via (5.10).
  Compute $\mathbf{W}^{(g)}$ using $N_{RF}^{(g)}$ via (5.11).
**end**
$\mathbf{F}_D \leftarrow [\mathbf{W}^{(1)}, \mathbf{W}^{(2)}, \ldots, \mathbf{W}^{(G)}]$.

### 5.3.1 RF Beamformer Design

We assume $K$ users are clustered into $G$ groups based on their AoD information, where each group contains $K_g$ number of users such that $K = \sum_{g=1}^{G} K_g$. The index $g_k = \sum_{g'=1}^{g-1} K_{g'} + k$ is used to denote the $k^{th}$ user in group $g$. Then, the channel matrix $\mathbf{H}^{(g)}$ for $gth$ group can be written as:

$$\mathbf{H}^{(g)} = \mathbf{Q}^{(g)}\mathbf{\Sigma}^{(g)}\mathbf{V}^{H(g)}, \tag{5.9}$$

where $\mathbf{Q}^{(g)}$ and $\mathbf{V}^{(g)}$ are $K_g \times K_g$ and $N_T \times N_T$ unitary matrices, respectively. $\mathbf{\Sigma}^{(g)}$ is a $K_g \times N_T$ rectangular diagonal matrix with non-negative real numbers on the diagonal, and the elements are arranged in decreasing order. Let $\mathbf{\Gamma}^{(g)}$ be the matrix containing $N_T$ column vectors. Then, we can write as:

$$\mathbf{\Gamma}^{(g)} = [\upsilon_1^{(g)}, \upsilon_2^{(g)}, \ldots, \upsilon_{N_T}^{(g)}], \tag{5.10}$$

where $\upsilon_n^{(g)} = [e^{j\psi_{1,n}^{(g)}}, e^{j\psi_{2,n}^{(g)}}, \ldots, e^{j\psi_{N,n}^{(g)}}]^T \in \mathbb{C}^N$ is the phase vector of group $g$, and $\psi_{m,n}^{(g)}$ is the phase angle of the $(m, n) - th$ entry of $\mathbf{V}$. Comparing (5.9) and (5.10), we can see that each entry of $\upsilon_n^{(g)}$ and $\mathbf{V}_g(n)$ have the same phase angle., i.e., $\psi$. Let $\mathbf{W}^{(g)}$ be the set containing $N_{RF}^{(g)}$ columns where each $nth$ column represents the $nth$ largest singular value of channel matrix $\mathbf{H}^{(g)}$. We can write $\mathbf{W}^{(g)}$ as:

$$\mathbf{W}^{(g)} = [\upsilon_1^{(g)}, \upsilon_2^{(g)}, \ldots, \upsilon_{N_{RF}}^{(g)}] \tag{5.11}$$

The pseudo-code for RF beamformer design is given in Algorithm 5.1.

### 5.3.2 Baseband MU Precoder Design

After the design of RF beamformer $\mathbf{F}_D$, the baseband MU precoder $\mathbf{B}_D$ can be determined by using the joint group processing (JGP) [8]. The overall reduced dimensional effective channel matrix is expressed as $\mathcal{H}^H = \mathbf{F}_D^H\mathbf{H}$. The effective channel $\mathcal{H}$ can be written as:

$$\mathcal{H} = \begin{bmatrix} \mathbf{H}^{H(1)}\mathbf{F}_D^{(1)} & \mathbf{H}^{H(1)}\mathbf{F}_D^{(2)} & \dots & \mathbf{H}^{H(1)}\mathbf{F}_D^{(G)} \\ \mathbf{H}^{H(2)}\mathbf{F}_D^{(1)} & \mathbf{H}^{H(2)}\mathbf{F}_D^{(2)} & \dots & \mathbf{H}^{H(2)}\mathbf{F}_D^{(G)} \\ \vdots & \vdots & \ddots & \vdots \\ \mathbf{H}^{H(G)}\mathbf{F}_D^{(1)} & \mathbf{H}^{H(G)}\mathbf{F}_D^{(2)} & \dots & \mathbf{H}^{H(G)}\mathbf{F}_D^{(G)} \end{bmatrix}, \tag{5.12}$$

where the diagonal matrices $\mathbf{H}^{H(g)}\mathbf{F}_D^{(g)} \in \mathbb{C}^{K(g)\times N_{RF(g)}}$ are the effective channel matrix for group $g$ and the off-diagonal matrices $\mathbf{H}^{H(g)}\mathbf{F}_D^{(\hat{g})} \in \mathbb{C}^{K(g)\times N_{RF\hat{g}}}$ represent the effective interference channel matrix between groups $g$ and $\hat{g}, \forall \hat{g} \neq g$. By applying the well-known RZF technique, $\mathbf{B}_D$ is defined as:

$$\mathbf{B}_D = \gamma\mathbf{T}\mathcal{H}, \tag{5.13}$$

where $\mathcal{H}$ is the reduced-dimension effective channel as provided in (5.12), $\mathbf{T} = [\mathcal{H}\mathcal{H}^H + \alpha N_{RF}\mathbf{I}_{N_{RF}}]^{-1}$, $\mathbf{I}_{N_{RF}} \in \mathbb{C}^{N_{RF}\times N_{RF}}$, $\alpha$ is the regularization parameter and $\gamma$ is the normalization factor used to satisfy the power constraint, which can be written as:

$$\gamma = \sqrt{\frac{S}{\mathrm{tr}\{\mathcal{H}^H\mathbf{T}^H\mathbf{F}_D^H\mathbf{F}_D\mathbf{T}\mathcal{H}\}}}. \tag{5.14}$$

## 5.4 Illustrative Results

In this section, the Monte-Carlo simulation results are presented based on the proposed hybrid precoding for various array structures. For the presented results, we assume the BS is equipped with different array configurations as depicted in Fig. 5.2. For URA(CA), we arrange the antenna elements in 20 rows(rings), where each row (ring) has 10 elements with a spacing of $d_s = \frac{\lambda}{2}$ [4, 8]. For HSA and SA, the antenna elements are arranged having uniform angle spacing [3]. The signal-to-noise-ratio (SNR) is defined as $\frac{P_T}{\sigma^2 K}$ and the regularization parameter $\alpha = \frac{\sigma^2}{P_T}$ [10].

### 5.4.1 Spectral Efficiency (SE) Comparison

In this section, we compare the spectral efficiency of different 3D array structures in mmWave MU-mMIMO systems both with 2D array structure (i.e., URA) and HBF solutions presented in [5, 7, 9]. In Fig. 5.3, we compare the sum-rate performance of different array structures, when $N = 200$. We consider $K = 12$ users are equally distributed into $G = 3$ groups (i.e., $K_g = 4$, $\forall g$), which are located around the BS at mean elevation angle $\phi = 73°$, mean azimuth locations $(\theta_1, \theta_2, \theta_3) = (30°, 90°, 150°)$ with $\delta_a = 15°$ and $\delta_e = 12.5°$. Each user having minimum number of RF chains (i.e., $N_{RF_g} = K_g = 4$). For SNR varying from -10 dB to 30 dB, we plot the sum-rate of URA, CA, HSA and SA in Fig. 5.3. FDBF achieves higher sum-rate but it requires the utilization of 200 RF chains. On the other hand, by using the proposed HBF, the sum-rate capacity of FDBF can be closely achieved with HBF by only using 12 RF chains., i.e., a 94% reduction of RF chains. Also, the use of 3D array structure can provide a higher sum-rate when compared to URA (e.g., the sum-rate increases by approximately 10 bps/Hz via CA, HSA or SA) as well as the HBF schemes in [5, 7, 9]. In Fig. 5.4, we investigate the variation of sum-rate for different

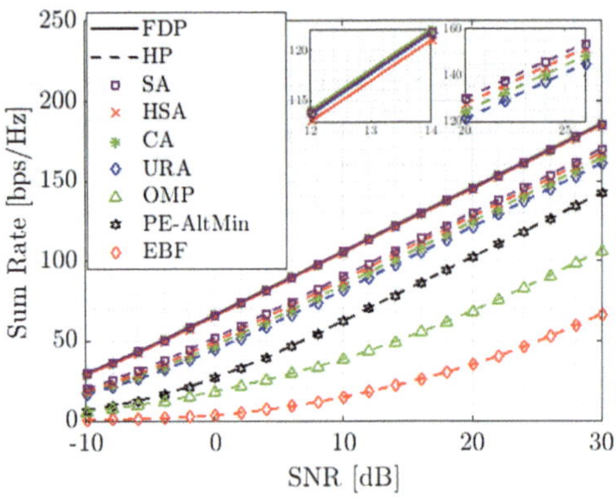


**Fig. 5.3** Sum-rate versus SNR comparison for URA, CA, HSA and SA

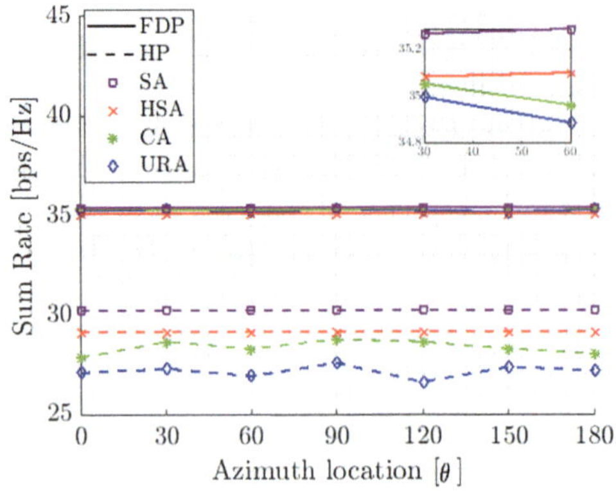


**Fig. 5.4** Sum-rate versus azimuth location $\theta$, $\phi = 73°$, SNR = 10 dB

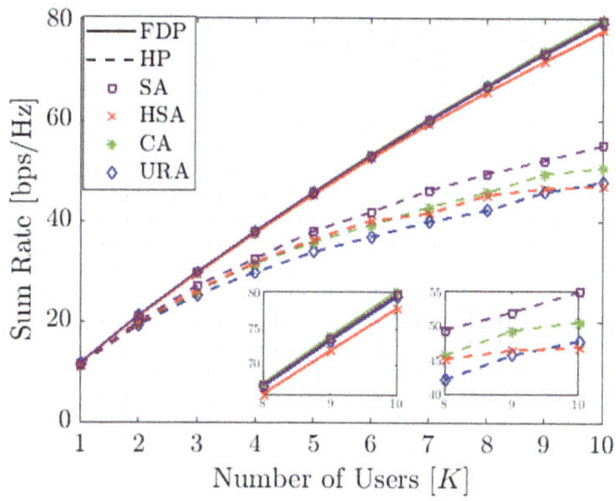


**Fig. 5.5** Sum-rate versus number of users $K$, $N = 200$, $G = 1$, $N_{RF} = K$, $\phi = 73°$, $\theta = 90°$, $(\delta_a, \delta_e) = (\pm 15°, \pm 12.5°)$, SNR = 20 dB

array structures when the user changes its angular location. We consider single user group., i.e., $G = 1$ having $N_{RF_g} = K_g = 4$. Figure 5.4 shows the sum-rate versus the mean azimuth location of a user group $\theta$, where $\theta$ varies between [0°:30°:180°] and $\phi = 73°$ at SNR = 10 dB. It can be seen that both HSA and SA can provide a non-varying/uniform sum-rate irrespective of the angular location $\theta$ because of a more symmetrical configuration, whereas the sum-rate can vary slightly for URA and CA. Also, both SA and HSA provide a higher sum-rate.

Figure 5.5 plots the sum-rate versus the number of users $K$ for a single user group ($G = 1$) located at $\theta = 90°$ and evaluated at SNR = 20 dB. 3D array structures can provide a higher sum-rate, where in particular SA can offer the highest sum-rate among all the array structures. Considering the future massive wireless connectivity requirements of IoT and for different aerial applications (e.g., in UAVs), different 3D array structures can be considered, which can provide a high sum-rate while supporting more number of users/devices.

## 5.4.2 *Energy Efficiency Comparison*

In this section, we compare the energy efficiency of the proposed HBF in mmWave MU-mMIMO systems using different 3D array structures. The energy efficiency $\epsilon$ is defined as the ratio of the sum-rate $R_{sum}$ and the total power consumption $P_{total}$, i.e.,

$$\epsilon = \frac{R_{sum}}{P_{total}} = \frac{R_{sum}}{P_T + N_{RF}P_{RF} + N_{PS}P_{PS}}, \tag{5.15}$$

where $P_T$ represents the total transmission power, $N_{RF}$ is the total number of RF chains, $P_{RF}$ and $P_{PS}$ are the power consumptions of each RF chain and PS, respectively. $N_{PS} = N_{RF} \times N_T$ is the total number of PSs used at the RF front end. As in [8], we use $P_T = 1$ W, $P_{RF} = 250$ mW and $P_{PS} = 1$ mW. In Fig. 5.6, we plot

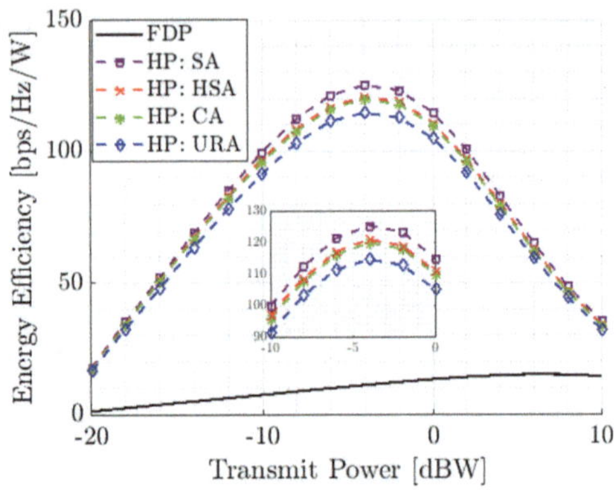


**Fig. 5.6** Energy efficiency versus transmit power $P_T$

energy efficiency versus transmit power for single user group (i.e., $G = 1$) located at $\phi = 73°$, $\theta = 30°$. We analyze the performance of different 3D array structures at SNR = 10 dB. FDBF provides very low performance due to use of large number of RF chains. On the other hand, the proposed HBF can provide a higher energy efficiency as it uses only a few number of RF chains (in this case, $N_{RF} = K = 4$). Furthermore, 3D array structures can give a more energy-efficient HBF design in mmWave MU-mMIMO systems when compared to 2D array structures.

### *5.4.3 Complexity Analysis*

The design of FDBF requires $O(N^3 + KN^2)$ operations, where $O(KN^2)$ is required for computation of $\mathbf{HH}^H$ and matrix inversion requires $O(N^3)$. For the proposed HBF, to obtain RF beamformer $\mathbf{F}_D$ as given in Algorithm 5.1, the complexity is $O(GN)$, where $N_T$ is the total number of antennas and $G$ is the total number of user groups. Then, the design of baseband precoder (as given in (5.13)) $\mathbf{B}_D$ requires $O(N_{RF}^3 + KN_{RF}^2)$ operations, which makes the computational complexity of the proposed HBF solution as $O(GN_T + N_{RF}^3 + KN_{RF}^2)$. For the HBF solution in [5] and [7], the computational complexity of $\mathbf{F}_{opt}$ requires $O(K^2N_T)$ operations. Thus, the overall complexity becomes $O\big(K^2N_T + N_T^2KN_{RF}\big)$ and $O\big(K^2N_T + kN_TN_{RF}^2\big)$, respectively, where $k$ is the number of iterations. As an example, consider $N_T = 100$, $G = 3$, $K = 3$ and $N_{RF} = 3$, the complexity of the proposed HBF solution is only about 1 percent of the complexity of the HBF methods in [5, 7].

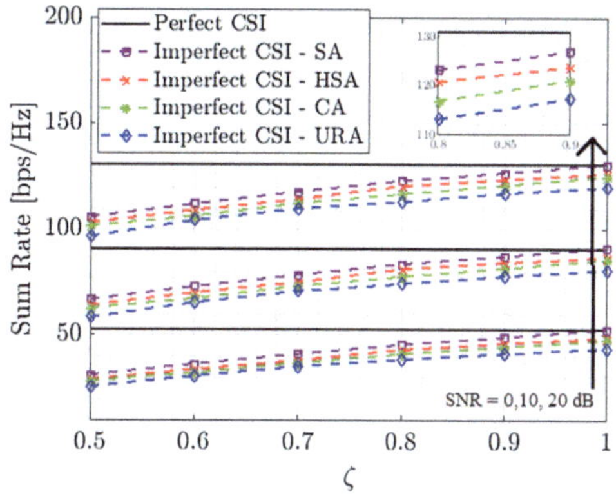


**Fig. 5.7** Sum-rate versus imperfect CSI $\zeta$ at SNR = 0, 10, 20 dB

### *5.4.4 Imperfect CSI*

In practical systems, CSI is contaminated by estimation error, therefore, we evaluate the impact of imperfect CSI on the performance of the proposed HBF solution. The estimated channel matrix $\hat{\mathcal{H}}$ is given as [8]:

$$\hat{\mathcal{H}} = \zeta \mathcal{H} + \sqrt{1 - \zeta^2}\mathbf{E}, \tag{5.16}$$

where $\mathcal{H}$ is the reduced-dimensional effective channel matrix at the baseband, $\zeta \in [0, 1]$ is the parameter that controls CSI accuracy and $\mathbf{E}$ is the error matrix with entities following the standard complex Gaussian distribution. In Fig. 5.7, we compare the effect of channel estimation error on the sum-rate for different array structures and observe that the proposed HBF scheme is not sensitive to the accuracy of CSI. In particular, SA can provide a higher sum-rate when compared to HSA, CA and URA with imperfect CSI (e.g., $\zeta = 0.7$).

## 5.5 Concluding Remarks

In this chapter, we have presented the mmWave MU-mMIMO HBF design and investigated the performance comparison of different 3D antenna array structures. In the low-complexity HBF design, the RF beamformer has been developed by using the ordered SVD of the channel matrix, whereas the baseband precoder is designed using the reduced-dimensional effective channel matrix via RZF. The simulation results show the high spectral and energy efficiencies of 3D antenna arrays when compared to URA for the proposed HBF solution as well as the robustness to the channel imperfections. Among different 3D configurations introduced in this letter, both HSA and SA can provide high uniform performance irrespective of the users' angular location around the BS.

## References


1. M. Mahmood, A. Koc, T. Le-Ngoc, 2D antenna array structures for hybrid massive MIMO precoding, in *Proceedings of the IEEE Global Communications Conference (GLOBECOM), Taipei, Taiwan, Dec. 7–11* (2020)
2. M. Mahmood, A. Koc, T. Le-Ngoc, Massive-MIMO hybrid precoder design using few-bit DACs for 2D antenna array structures, in *Proceedings of the IEEE International Conference on Communications (ICC), Montreal, Canada, Jun 14–23* (2021)
3. W. Ryoo, W. Sung, Beamforming using uniform spherical arrays: array construction, beam characteristics, and multi-rank transmission. IEEE Access **9**, 38731–38741 (2021)
4. E. Yaacoub, M. Al-Husseini, A. Chehab, K. Abualsaud, T. Khattab, M. Guizani, 3D beamforming with massive cylindrical arrays for physical layer secure data transmission. IEEE Commun. Lett. **23**(5), 830–833 (2019)
5. O. El Ayach, S. Rajagopal, S. Abu-Surra, Z. Pi, R.W. Heath, Spatially sparse precoding in millimeter wave MIMO systems. IEEE Trans. Wireless Commun. **13**(3), 1499–1513 (2014)
6. D. Kim, G. Lee, Y. Sung, Two-stage beamformer design for massive MIMO downlink by trace quotient formulation. IEEE Trans. Commun. **63**(6), 2200–2211 (2015)
7. X. Yu, J.C. Shen, J. Zhang, K.B. Letaief, Alternating minimization algorithms for hybrid precoding in millimeter wave MIMO systems. IEEE J. Sel. Topics Signal Process. **10**(3), 485–500 (2016)
8. A. Koc, A. Masmoudi, T. Le-Ngoc, 3D angular-based hybrid precoding and user grouping for uniform rectangular arrays in massive MU-MIMO systems. IEEE Access **8**, 84689–84712 (2020)
9. P. Zhao, Z. Wang, Joint user scheduling and hybrid precoding for multi-user mmWave systems with two-layer PS network, in *Proceedings of IEEE Global Communications Conference (GLOBECOM)* (IEEE, Piscataway, 2018), pp. 1–6
10. M. Mahmood, A. Koc, T. Le-Ngoc, Energy-efficient MU-massive-MIMO hybrid precoder design: low-resolution phase shifters and digital-to-analog converters for 2D antenna array structures. IEEE Open J. Commun. Soc. **42**, 1842–1861 (2021)
11. D. Zhang, P. Pan, R. You, H. Wang, SVD-based low-complexity hybrid precoding for millimeter-wave MIMO systems. IEEE Commun. Lett. **22**(10), 2176–2179 (2018)

# Chapter 6
# Metamaterial-Based Miniaturized Large-Scale Antenna Array Prototype for mMIMO

## 6.1 Introduction

The shift towards large-scale arrays represents a critical evolution in wireless technology, paving the way for next-generation networks capable of handling the immense and growing data traffic [1, 2]. These arrays facilitate the generation of highly directional beams with enhanced directivity and narrow beamwidth, improving power efficiency and reducing spatial correlation. Moreover, array beamforming dynamically focuses and shapes signal transmission and reception in specific directions, effectively boosting signal strength, expanding coverage, and improving interference management [3, 4].

To effectively design and comprehensively investigate the performance of large-scale arrays, this chapter begins with a quantitative examination of array beam radiation patterns as a function of array length. Various element radiation pattern (ERP) assumptions are incorporated to evaluate their influence on the overall array's radiation patterns. Theoretical expressions for array directivity and beamwidth as functions of array length are derived and analyzed, offering valuable insights into determining the effective array size based on beam resolution requirements. Subsequently, practical scenarios involving sub-arrays positioned at different locations within large-scale arrays are examined to evaluate their beamforming performance. The analysis focuses on the impacts of different surrounding environments on the achievable beamforming characteristics when using a maximum directivity beamformer. This study offers valuable insights into the beamforming performance and steerability of sub-arrays at different positions within a large array, providing a deeper understanding of their effectiveness in real-world deployments.

Based on the aforementioned studies, an $8\times8$ antenna array prototype utilizing the proposed dual-layer metamaterial-based patch antenna structure, as detailed in Chap. 3, is fabricated for mMIMO operation within the 3.35 GHz to 3.6 GHz band. Experimental measurements of element RL and inter-element MC are conducted and compared with the full-wave simulation results presented in the design process,



T. Le-Ngoc et al., *Half-Duplex/Full-Duplex MU-mMIMO*, Wireless Networks,
https://doi.org/10.1007/978-3-032-09195-6_6

verifying the proper functionality of the array prototype. Working with an RF switch network, a comprehensive experimental evaluation of all 64 elements in the CP 8×8 large-scale antenna array is conducted. This study provides a detailed analysis of ERP variations across different locations within the array. The information will facilitate the development and optimization of beamforming algorithms in subsequent chapters. Additionally, the beamforming performance of the array, integrated with a realistic beamforming network comprising quantized phase shifters and attenuators, is measured in an anechoic chamber. These measurements are conducted to assess the sub-array beam steering capabilities and evaluate the impact of surrounding environments on overall beamforming performance.

The main illustrative results and contributions of this chapter are summarized as follows:

- **Array length and beam radiation patterns:** A quantitative analysis of how the number of elements in an array shapes its beam pattern is conducted. Theoretical expressions for directivity and the 3 dB- and 10 dB-beamwidths as functions of the number of array elements are derived. To evaluate the element pattern impacts on the array beam patterns, we use three simulation scenarios: arrays with isotropic elements, arrays with homogeneous element patterns, and arrays with heterogeneous element patterns obtained from the $1 \times M$ HFSS full-wave simulations. The results demonstrate that the theoretical expressions accurately estimate the directivity and beamwidth values. Notably, as the number of antenna elements increases, the influence of individual element patterns on the linear array's main beam pattern diminishes. This provides insights into determining the effective length of the array in relation to directivity and beamwidth for large-scale array designs.
- **Surrounding environment impacts on sub-array beamforming performance:** The study explores practical scenarios where a sub-array within a large array is used for beamforming. It investigates how various surrounding environments affect the achievable beamforming performance when using an MDB beamformer, providing insights into the beamforming characteristics of sub-arrays located at different positions within the large array. The impacts on achieved beam steering angle, directivity, sidelobe level, and beamwidth variation due to different steering angles and asymmetrical sub-array locations are highlighted. The results illustrate the achievable beamforming performance and the potential beam steering angle deviations caused by sub-array surroundings.
- **8×8 CP metamaterial-based large-scale antenna array prototype and experimental measurements:** An 8×8 large-scale antenna array utilizing the proposed dual-layer patch antennas has been developed. Experimental measurements, including element RL, inter-element MC, and sample sub-array beamforming performance, were conducted to validate the full-wave simulation results presented in the previous chapter, demonstrating good agreement. A comprehensive experimental study was performed on all 64 elements of the CP 8×8 array, providing detailed information into variations in the ERPs across

different array positions. These measured ERPs are essential for assessing how ERP variations influence the performance of different beamforming techniques discussed in future chapters. Additionally, the study examines ERP categorization by analyzing symmetry and distortions within the array. The symmetry information derived from ERPs offers valuable guidance for enhancing beamforming optimization efficiency and improving overall array performance. The beamforming performance of the array, integrated with a realistic beamforming network comprising quantized phase shifters and attenuators, is measured in an anechoic chamber.

The rest of the chapter is structured as follows. Section 6.2 provides a quantitative analysis of how the number of elements in an array influences its beam pattern. Theoretical expressions for directivity, as well as 3 dB- and 10 dB-beamwidths, are derived as functions of the number of array elements. Section 6.3 discusses the effects of various surrounding environments on the achievable beamforming performance of sub-arrays using an MDB beamformer. This analysis offers insights into the beamforming characteristics of sub-arrays located at different positions within the large array. Then, Sect. 6.4 introduces the $8 \times 8$ CP antenna array prototype and describes the experimental measurement setup in the anechoic chamber. Empirical results of the measured antenna element radiation characteristics are presented in Sect. 6.5. This section includes an in-depth exploration of ERP variations, symmetry, and categorization, and evaluates their impact on beamforming performance. Various sub-array beamforming measurements are also presented and compared in Sect. 6.6. Section 6.7 concludes the chapter and provides possible future works.

## 6.2 Array Beamforming Performance Versus Number of Antenna Elements

The array length plays a crucial role in determining the beam resolution and beamforming capabilities of an antenna array. In this section, we analyze how the beam pattern of a ULA evolves as the number of elements increases. Understanding the beam characteristics, specifically the array directivity and beamwidth, provides designers with valuable insights to determine an effective array size for beamforming and to assess the impact of individual element patterns on the overall array performance. To provide a comprehensive analysis, we derive theoretical expressions for the array beam directivity and beamwidth. Three simulation scenarios are used to compare with the theoretically estimated values, which are

- **isotropic**: all elements assumed to be isotropic;
- **homogeneous**: all elements assumed to have the identical isolated Tx patch element radiation pattern shown in Fig. 3.5;

- **heterogeneous**: the HFSS full-wave simulation of $1 \times M$ array consisting of the proposed dual-layer patch antenna element structures is conducted, where each element has a different radiation pattern.

In all simulations, the elements are spaced approximately half a wavelength apart (4 cm) at an operating frequency of 3.5 GHz. Illustrative results show that the theoretical expressions can estimate the directivity and beamwidth values accurately. Notably, as the number of antenna elements grows, the linear array's main beam pattern becomes progressively less influenced by individual element patterns. In all the simulation scenarios, uniform steering weights across the antenna elements are employed.

### *6.2.1 Array Directivity vs Number of Antenna Elements*

In the context of a uniformly excited ULA comprising $M$-elements equally spaced at a half-wavelength distance the theoretical directivity (dir) can be articulated as [5]

$$\text{dir[dB]} = \text{dir}_{elem}\text{[dB]} + 10\log_{10}(M), \tag{6.1}$$

where $\text{dir}_{elem}$ is the single element directivity used in the array, and in our simulation, $\text{dir}_{elem}$=6.1 dB. The comparison between the theoretical values and the simulated results is shown in Fig. 6.1. Notably, all three plotted curves exhibit similar trends. The average and maximum differences between the homogeneous, heterogeneous, and theoretical values are less than 0.3 and 0.6 dB, respectively, indicating that the theoretical equation accurately estimates the realistic array directivity. Additionally, the average difference between the isotropic array and the

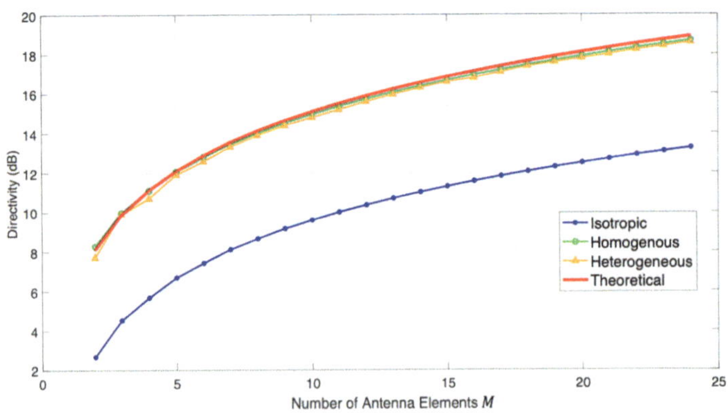


**Fig. 6.1** Linear array directivity vs number of antenna elements (simulation results)

heterogeneous results is 5.3 dB, primarily due to the 6.1 dB directivity difference between the proposed element and an isotropic element. The observation proves that the enhanced directivity of the antenna elements contributes to more concentrated array beams with higher overall array directivity.

### 6.2.2 Array Beamwidth vs Number of Antenna Elements

For a uniformly excited, equally half-wavelength spaced $M$-element linear isotropic array, the far-field radiation pattern in spherical coordinate $(\theta, \phi, r)$ can be represented as [5]

$$a(\theta,\phi,r)=\frac{e^{-j\beta r}}{r}\sum_{m=0}^{M-1} e^{-jm\pi\sin(\theta)}=\frac{e^{-j\beta r}}{r}e^{-j(M-1)\pi\sin(\theta)/2}\frac{\sin(M\pi\sin(\theta)/2)}{\sin(\pi\sin(\theta)/2)}. \tag{6.2}$$

As a result, the AdB-beamwidth is $2\theta_{AdB}$, where $\theta_{AdB}$ is the $\theta$ solution of the following equation

$$\frac{\sin(M\pi\sin(\theta)/2)}{\sin(\pi\sin(\theta)/2)}=10^{\frac{-A_{dB}}{10}}. \tag{6.3}$$

For $M \gg 2$, the 3 dB- and 10 dB-beamwidths in radians can be derived as

$$BW_{3dB}=\frac{1.77}{M},\ BW_{10dB}=\frac{3.1}{M}, \tag{6.4}$$

i.e., the 3 dB- and 10 dB-beamwidths are inversely proportional to the number of elements.

For the 3 dB beamwidth, Fig. 6.2 illustrates that all cases follow a similar trend. As the number of elements increases, the difference between all scenarios diminishes. Notably, when $M > 2$, differences among the simulation cases become less than 2°. This gap narrows to under 1° when $M > 10$. Similar observations can be found in the 10 dB-beamwidth simulation results, as shown in Fig. 6.3. When $M > 4$, the difference in 10 dB-beamwidth across the three simulation setups is less than 5°.

As the number of elements increases, the element pattern effects on the array main lobe pattern become negligible. The beamwidth decrease rate becomes slow when $M \geq 8$ and becomes nearly flat when $M > 14$, where the 3 dB- and 10 dB-beamwidths become smaller than 7° and 14°, respectively.

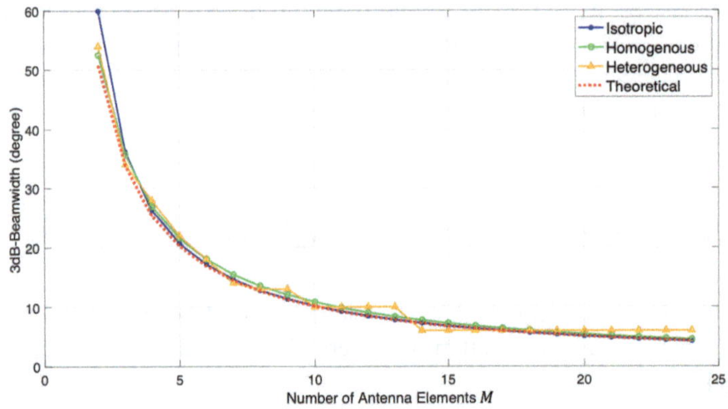


**Fig. 6.2** 3 dB-beamwidth vs number of antenna elements (simulation results)

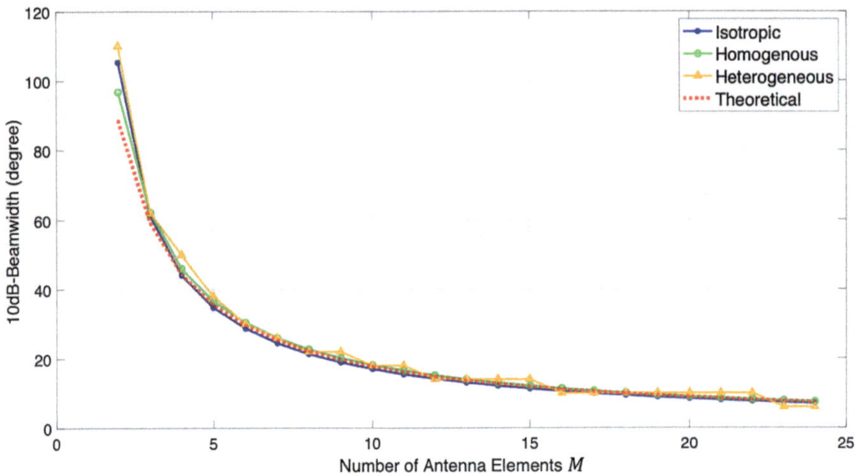


**Fig. 6.3** 10 dB-beamwidth vs number of antenna elements (simulation results)

## 6.3 Antenna Sub-array Beamforming Performance

The far-field radiation pattern of an $M$-element antenna array can be expressed as the summation of the response across all elements, modified by a beamforming vector, which can be expressed as

$$a(\theta, \phi, r) = \frac{e^{-j\beta r}}{r} \sum_{m=0}^{M-1} w_m(\theta, \phi) A_m(\theta, \phi) e^{-j\beta (r_m \mathbf{v}_m \mathbf{v}^T)}, \tag{6.5}$$

where $\beta = 2\pi/\lambda_0$ is the phase propagation constant and $r$ is the distance from the array origin to the far-field observation point. $A_m(\theta, \phi)$ represents the $m$th element gain in the direction $(\theta, \phi)$. $r_m \boldsymbol{v}_m$ denotes the position vector of the $m$th element, while $\boldsymbol{v}$ is the unit directional vector pointing from the array origin to the far-field observation point. $w_m$ is the complex beamforming weight applied to the $m^{\text{th}}$ element. In this section, an MDB vector is employed to align the phase across array elements and ensure constructive interference in the desired steering direction $(\theta, \phi)$. This equation highlights how the far-field radiation pattern of an antenna array is influenced by the individual element patterns, the beamforming weights, and the geometric layout of the elements. It illustrates the collective effect of array configuration on beam steering and pattern shaping, demonstrating the interplay between individual element characteristics and overall array performance.

### *6.3.1 Simulation Setup of Sub-array Beamforming*

The proposed dual-layer EBG-slotted antenna element is incorporated into a uniform rectangular array (URA) configuration. Due to its complex structure, accurate and realistic full-wave simulations demand substantial computational resources. Given the limitations of available computational resources, including a PC with 640 GB of RAM and 70 CPU cores, a 4×8 antenna array is simulated. This simulation is conducted to investigate how sub-array beamforming performance varies at different positions within a large array. For a comparative analysis, three sample 1×4 sub-arrays located in the left, right, and right-top edges are chosen, as illustrated in Fig. 6.4a. The beams are steering within a $\theta$ angle from $-50°$ to 50°, adjusted in increments of 5°. In the HFSS setup, lumped ports are used to control the feeding and steering weights for each antenna element, enabling precise beamforming configuration. The steering weights used in the simulations have a uniform magnitude, and phase shift values are changed based on the steering angle.

The simulated 2D cuts at $\phi = 0°$ and $\phi = 90°$ of three sub-arrays at steering angle $\theta = 0°$ are shown in Fig. 6.4b–d. Positioned in the array's top two rows, these sub-arrays produce beams with noticeable downward tilts, as evidenced in the $\phi = 0°$ cut plots. The Left 1×4 sub-array exhibits a beam tilted by 4°, while the TopRight 1×4 sub-array has a beam tilted by $-3°$ in the $\theta$ angle. Notably, all three beams maintain a normalized sidelobe level below $-11.9$ dB. The Right 1×4 sub-array demonstrates a 1.1 dB lower normalized sidelobe level compared to the TopRight 1×4 sub-array situated at the edge.

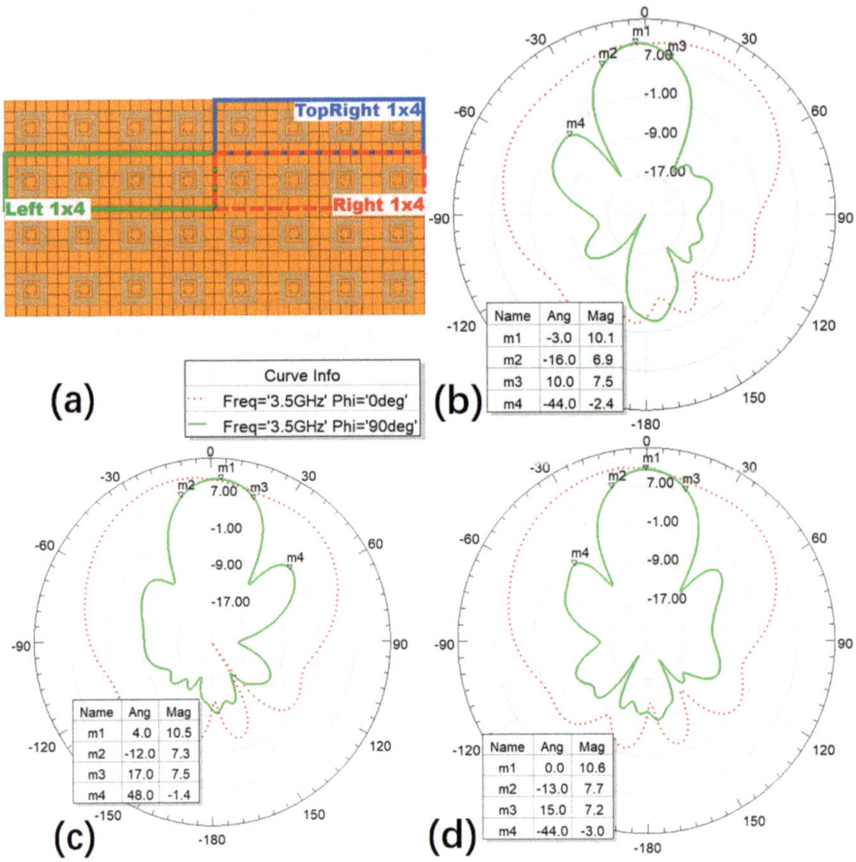


**Fig. 6.4** 1×4 sub-arrays on a 4×8 array: sub-array locations (**a**) and $\phi = 0^o$ and $\phi = 90^o$ 2D cuts beam pattern at steering angle $\theta = 0°$ of TopRight 1×4 (**b**), Left 1×4 (**c**), and Right 1×4 (**d**) (simulation results)

## 6.3.2 Sub-array Location vs Beam Peak Tilt

Within a $\theta$ steering range $-50°$ to $50°$, the average directivity for the Left, Right, and TopRight 1×4 sub-arrays are 11.2, 11.2, and 10.9 dB, respectively, demonstrating that location variation has a negligible effect on beamforming directivity.

Figure 6.5 depicts the beam peak directions achieved by the three sub-arrays relative to the intended steering angle. Notably, the Left 1×4 sub-array plot consistently exhibits a higher position compared to the results of the Right and TopRight sub-arrays. With identical steering weights, the Left 1×4 sub-arrays get a beam steering angle of 3.5° greater than the Right sub-array on average. The TopRight 1×4 sub-array has an average difference of $-3°$, in comparison to the Right sub-array.

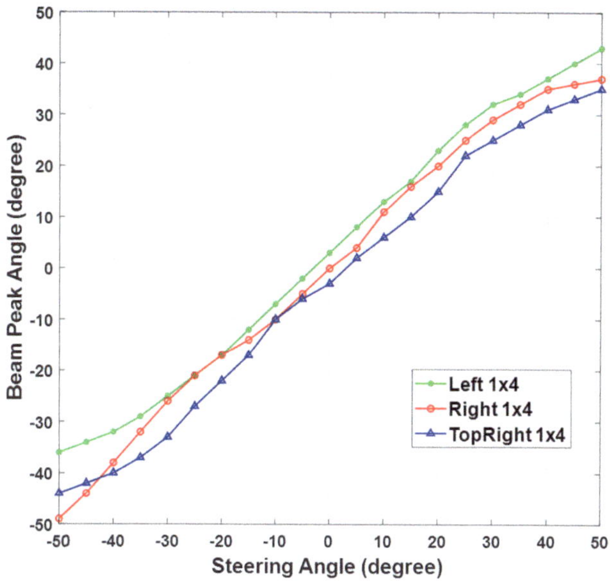


**Fig. 6.5** Sub-array beam peak angle vs target steering angle (simulation results)

Moreover, it can be observed that a larger beam peak tilt occurs when the beam is steering away from the array center. For the Left 1×4 sub-array, an average angle tilt of 5.7° is observed when steering within the range $\theta = -50°$ to $\theta = 0°$. In contrast, the Right 1×4 and TopRight 1×4 sub-arrays display average beam tilts of $-2.7°$ and $-6.5°$, respectively, when steering within the range $\theta = 0°$ to $\theta = 50°$. This trend indicates that the beam tilt angle increases as the steering angle moves further away from the array center. Consequently, the following observations can be made:

- Asymmetrically positioned sub-arrays tend to tilt their beams towards the array's center. The beam tilt gets larger as the steering angle increases.
- Edge sub arrays demonstrate radiation patterns more tilted towards the array center compared with the sub-arrays close to the array center.

### *6.3.3 Sub-array Location vs Sidelobe and Beamwidth*

The simulation results of the 3 dB-beamwidth and the normalized sidelobe level w.r.t. the main beam peak of the three sub-arrays across $\theta$ steering range $-50°$ to 50° are shown in Fig. 6.6.

When steering away from the array center, the sub-arrays with asymmetric placements get a relatively narrower 3dB-beamwidth. Additionally, Fig. 6.6b illustrates

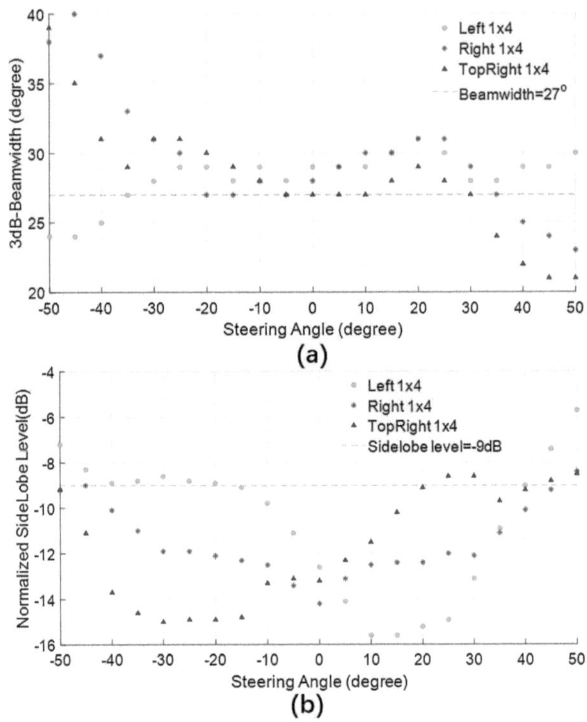


**Fig. 6.6** (**a**) 3dB-Beamwidth and (**b**) normalized sidelobe levels of 1×4 sub-arrays for steering angle from − 50° to 50° (simulation results)

that the sub-arrays with an asymmetric position typically exhibit quicker sidelobe level increases while steering away from the array center and the edge sub-array sidelobe level grows more quickly than the center sub-array sidelobe. The high sidelobe levels can potentially increase interference with other signals or systems operating in the sidelobe regions and also lead to a loss of signal power in the main lobe directions, leading to a degradation in the beamforming performance.

The presence of the ground plane and EBG patches acts as a reflective medium, leading to reflections and interference with the direct radiation from the patch elements. Therefore, when a sub-array is asymmetrically positioned on the array during the beamforming process, this non-uniform reflecting surface contributes to an extra amount of radiation interference. There is an amplified radiation on the side with a larger reflective surface. Therefore, the main lobe pattern becomes asymmetric and the radiation intensity is more pronounced in the direction of the larger ground plane, leading to the beam angle tilt. When the beam steers towards the array center (where the ground plane is larger), the extra reflection adds into the main lobe and contributes to a wider beamwidth. Meanwhile, when steering the beam away from the array center, the extra reflected waves become a main contributor to the sidelobe level increase.

To mitigate beam tilting and asymmetric steerability during the beamforming process, employing a sub-array with symmetric positioning relative to the steering plane can be beneficial. Additionally, if an asymmetrically located sub-array is used for beam focusing, a correction angle $\Delta\theta$ can be added to or subtracted from the target steering angle during the steering vector calculation. This adjustment, as discussed in Sect. 6.3.2, helps compensate for potential beam tilting and directivity degradation, ensuring improved beamforming accuracy and performance.

## 6.4 8×8 Dual-Layer Metamaterial-Based CP Antenna Array Prototype and Measurement Setup

### *6.4.1 8×8 Antenna Array Prototype*

An 8×8 dual-layer EBG array prototype featuring L-slotted CP patch antennas has been developed for mMIMO systems, operating within the frequency range of 3.35 to 3.6 GHz, as shown in Fig. 6.7. The array consists of a total of 64 antenna elements.

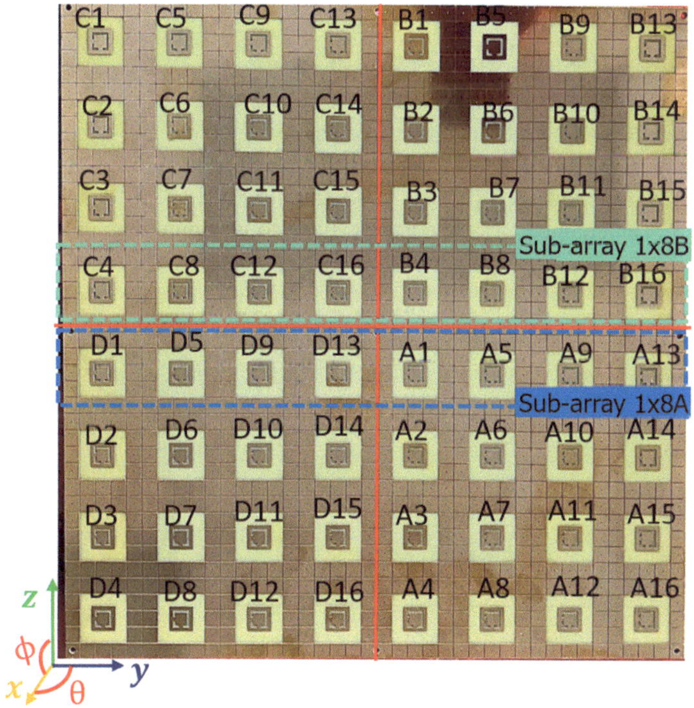


**Fig. 6.7** 8×8 CP antenna array prototype, and element and sub-array indices

Spatial aliasing can occur when the separation distance between elements exceeds half a wavelength, causing the coherent addition of plane waves at different incident angles, leading to grating lobes. These grating lobes degrade beam resolution and introduce significant interference in the undesired directions. To prevent this issue in our design, the separation distance between neighboring elements is set to 2.78 cm (0.32 $\lambda_0$ at 3.5 GHz) edge-to-edge or 4 cm (0.47 $\lambda_0$) center-to-center.

The dual-layer EBG structure incorporated into the array induces a slow-wave effect, resulting in a 65% reduction in the size of the radiating patches, as detailed in Chap. 3. This reduction leaves sufficient space for the upper layer's EBG slots, which serve as a band-pass filter, curtailing surface wave leakage and reducing inter-element MC within the array. All 64 antenna elements share the same FR-4 dielectric substrate with a dielectric constant $\epsilon_r = 4.6$ and the same ground plane. The array utilizes CP patch antenna elements, with the polarization direction controlled by the relative positions between the L-shaped slot and the feeding point on the central radiating patches. A single feeding probe is strategically placed 3 mm away from the center of each patch element. Mini-sized JSC coaxial RF connectors are soldered to each antenna element to ensure a secure, low-loss, and impedance-matched connection between the feeding coaxial cable and the feeding probes. The array dimensions are a width of $W_T = 32$ cm and a length of $L_T = 32$ cm.

### 6.4.2 Experimental Measurement Setup

Over-the-air (OTA) measurements of the antenna element and sub-array radiation characteristics were conducted in an anechoic chamber with internal dimensions of 20 ft (6.1 m) in length, 8 ft (2.4 m) in width, and 8 ft (2.4 m) in height, as shown in Fig. 6.8. The chamber was adequately sized to accommodate the antenna array

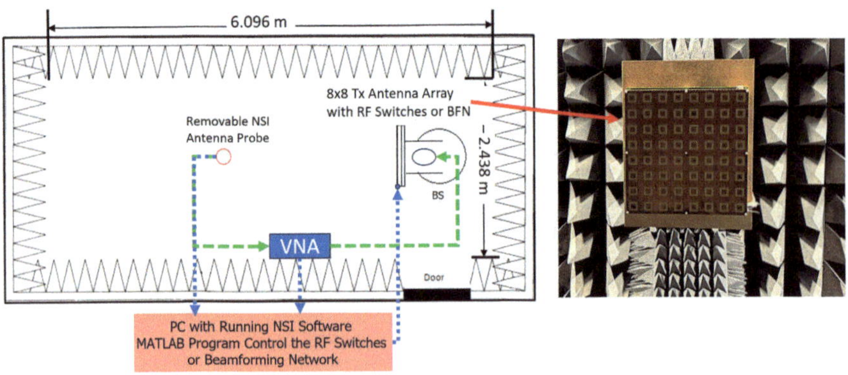


**Fig. 6.8** 8×8 circularly polarized antenna array prototype measurement setup in the anechoic chamber

prototype under test, which measured 32 × 32 cm (width × height). The array was positioned more than 4 ft away from the chamber walls, which were lined with C-RAM SFC-48 RF absorbers. These absorbers provided reflection levels below −45 dB at 3.5 GHz, effectively minimizing reflections and external interference to ensure accurate measurement of radiation characteristics.

#### 6.4.2.1 Antenna Element Measurements

To minimize the number of connections required during antenna element measurements and to protect the fragile JSC connectors, calibrated 1-to-16 RF switches controlled by a customized serial peripheral interface (SPI) module were employed. This setup enabled sequential switching and measurement of each of the 16 elements in a set. During element pattern and S-parameter measurements, only one antenna element was excited and measured at a time. The antenna element return losses were measured using a Keysight N5247A vector network analyzer (VNA). Radiation patterns were obtained using the NSI-2000 platform, equipped with a probe horn antenna, automatic rotation positioners, and the same VNA. The VNA provided a transmitting power of 10 dBm during these measurements.

For ERP scanning, an angular resolution of 1° was utilized for both the $\theta$ and $\phi$ angles. Specifically, the $\theta-$angle scan ranged from $-180°$ to 180°, while the $\phi-$angle spanned $-90°$ to 90°. Measurements were performed across the operational frequency range of 3.35 to 3.6 GHz, with frequency increments of 0.05 GHz. As a result, the 3D radiation pattern of the $m$th antenna element at each operating frequency was stored in a matrix $\boldsymbol{A}_m$ with dimension 361 × 181, capturing the full angular and frequency-dependent radiation characteristics.

#### 6.4.2.2 Antenna Sub-array Beamforming Measurements

The sample transmitting antenna sub-arrays, integrated with beamforming networks (BFNs), were used in the measurements to investigate the sub-array radiation patterns and the beam steerability. For sub-array consisting of $M$ antenna elements, where $M = 4, 8, 16$ in this section, a 1-to-$M$ power splitter was employed to evenly distribute the input power across $M$ paths. Then the $M$-path signals were modified by the BFN modules. The BFN setup comprises $M$ digitally controlled 6-bit phase shifters, capable of full 360° phase adjustment in 5.625° increments, and $M$ 7-bit digitally controlled attenuators, providing up to 32 dB of attenuation with fine adjustment steps of 0.25 dB. In this section, when the MDB vector is employed for beam steering, the attenuators are configured with a minimum attenuation of 0 dB, and the phase shifter values are set based on the calculated complex conjugate steering vectors of the sub-array.

To ensure an efficient and compact arrangement, these components were interconnected using short coaxial cables and compact JSC connectors, conveniently positioned behind the antenna arrays. The BFN was controlled through a customized

SPI module integrated with a Raspberry Pi 4. A PC running a MATLAB-based control program interfaced with the SPI module via a USB connection, enabling the generation of beamforming vectors and precise control of the BFN. On the receiving side, a horn probe antenna was used. The measurements employed the same NSI-2000 platform setup as the single-element tests, including automatic rotation positioners and a PNA. The radiation pattern scanning setup remained consistent, ensuring comparable data for evaluating the sub-array's beamforming performance.

## 6.5 Antenna Element Measurement Results

### *6.5.1 Antenna Element Return Loss*

Sixteen antenna elements in Sub-array B, located in the top-right quadrant of the array prototype, were utilized for impedance bandwidth measurements. The measured $S_{11}$ parameter, representing the return loss and impedance matching level, for these 16 elements over the frequency range of 3.1 to 3.8 GHz is illustrated in Fig. 6.9.

The results show that all 16 antenna elements exhibit similar $S_{11}$ trends. For each element, the $S_{11}$ parameter remains below −10 dB within the operating bandwidth of 3.35 to 3.80 GHz, indicating effective impedance matching. Additionally, $S_{11}$ values drop below −20 dB in the range of 3.39 to 3.43 GHz, pinpointing the center resonant frequencies of the antenna elements.

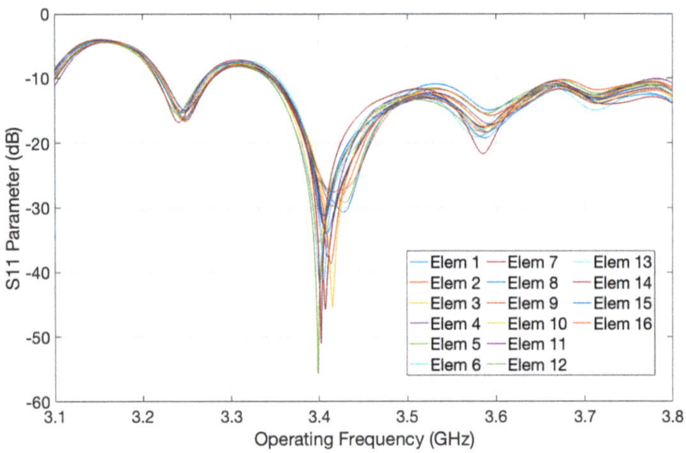


**Fig. 6.9** S11-parameters of antenna elements (measurement results)

The measured results align well with the simulated data, where the $S_{11}$ parameter is below $-10$ dB over the bandwidth of 3.34 to 3.78 GHz. This agreement demonstrates consistent RL performance within the target frequency band. A high RL signifies well-matched antenna impedance, ensuring that the majority of the input power is effectively transmitted rather than reflected, thereby improving signal transmission efficiency and overall system performance.

### 6.5.2 Antenna Inter-Element Mutual Coupling

To evaluate inter-element isolation performance, MC between the sixteen antenna elements in 4×4 Sub-array B was measured. In each test, a single antenna element was excited, and the receiving port was sequentially switched among the other 15 antenna elements. A total of $16 \times 15 = 240$ MC measurements were performed across the array.

The worst (highest) MC values measured within the frequency range of 3.35 to 3.6 GHz are presented in Fig. 6.10. Among the combinations, eight exhibited MC levels higher (i.e., worse) than $-15$ dB, while all combinations maintained MC levels below $-14.4$ dB. On average, neighboring antenna elements achieved an inter-element MC of $-18.7$ dB. The distance between antenna elements plays a critical role in determining the magnitude of MC. MC is generally higher between

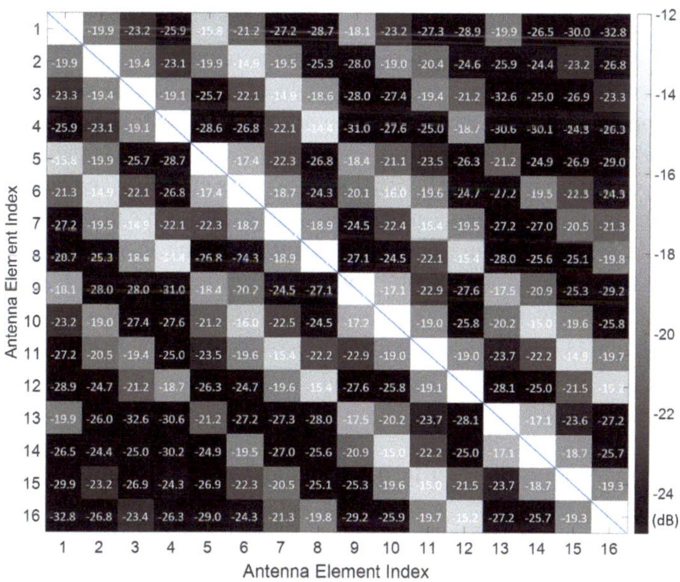


**Fig. 6.10** Worst MC magnitude (in dB) in the frequency range 3.35–3.6 GHz between antenna elements in sub-array B (measurement results)

neighboring elements compared to elements with greater separations along the same direction.

Among the eight possible directions, the MC between neighboring elements in the same row is stronger than that in other directions. This trend aligns with the HFSS simulation results discussed in Chap. 3. For all combinations of neighboring elements not in the same row, the MC magnitude remains below −17 dB. Additionally, the edge and corner elements exhibit isolation performance comparable to that of the elements located at the center of the array, demonstrating consistent inter-element isolation across the array configuration.

### 6.5.3 Element Radiation Patterns Across the 8 × 8 Array Prototype

The radiation patterns of the 64 antenna elements at 3.5 GHz, depicted in terms of directivity (in dB) with peak directivity values, are illustrated in Fig. 6.11. The ERP 2D cuts in both the $\theta$ and $\phi$ planes for 16 sample elements, 8 elements close to the array center and 8 elements at the array edge, are presented in Fig. 6.12.

#### 6.5.3.1 Achievable Element Directivity

Across the array's 64 elements, an average directivity of 7.0 dB is achieved. Meanwhile, antenna elements in the stronger inter-element MC direction (i.e., along the $y$−axis as shown in Sect. 6.5.2), exhibited a relatively lower directivity. For instance, the measured ERPs at the left and right edges exhibit an average directivity of 6.3 dB, approximately 1.1 dB lower than the average directivity of the four central elements in the array. Furthermore, the lowest directivity (Tx-C Element 2) of 5.5 dB is observed, which is about 1.9 dB less than the center elements' average directivity. Since the stronger inter-element interaction in this direction can potentially induce stronger coupling currents in adjacent elements, altering their radiation patterns. Directivity is diminished as the MC causes energy to be radiated in unintended directions or coupling by neighboring elements. Conversely, the elements along the $z$-axis, which is the weaker coupling direction, showed a negligible impact on directivity.

#### 6.5.3.2 Element Radiation Pattern Distortion and Tilt

Notably, elements situated near the array's center demonstrate more uniform radiation patterns, characterized by their smoother surface in the plot. This uniformity is attributed to a more consistent ground plane influence and a balanced distribution of surrounding elements. This pattern uniformity is particularly evident in the central

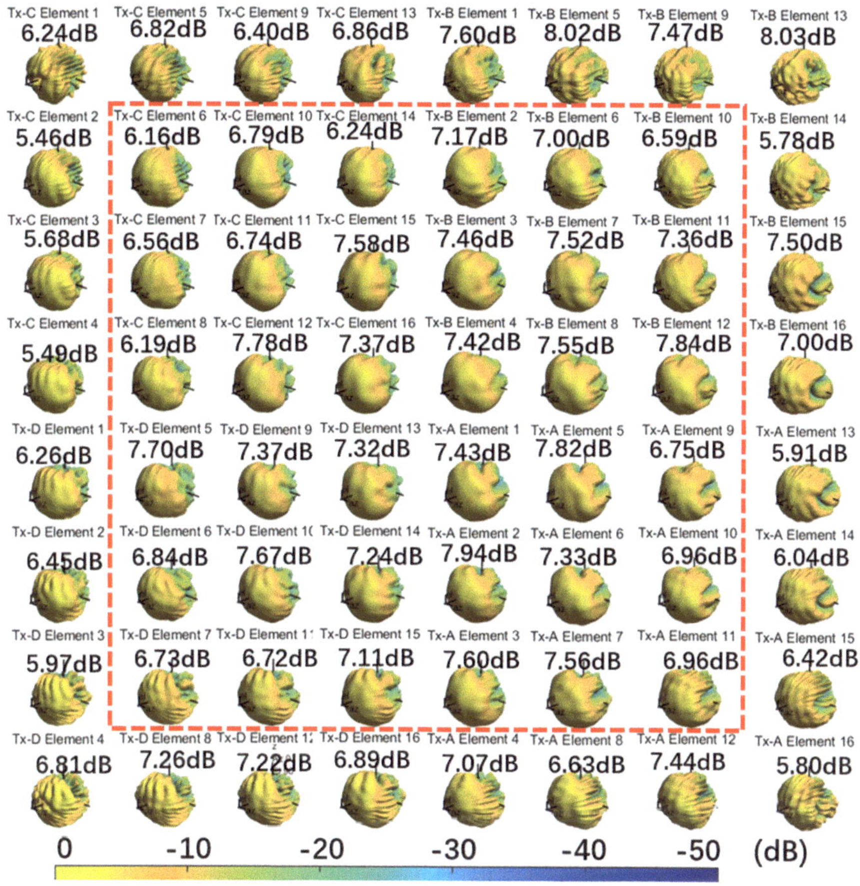


**Fig. 6.11** 3D radiation patterns (directivity in dB) of the 64 elements at 3.5 GHz (measurement results)

6×6 section of the 8×8 array, as highlighted by the red square in Fig. 6.11. Moving towards the edges of the array, however, leads to a noticeable increase in ERP distortion, with the surfaces of these patterns becoming coarser. The most significant pattern distortions are observed in the corner elements, C1, D4, B13, and A16.

Additionally, radiation patterns of elements on the array's edges are observed to bias towards the array center. For instance, patterns of elements on the left edge (C1-C4 and D1-D4) display a rightward inclination, whereas those on the right edge (A13-A16 and B13-B16) show a leftward shift. Top-edge elements exhibit patterns that tilt downwards, while bottom-edge elements have patterns that tilt upwards. The ERP tilted behaviors are attributed to the uneven ground plane and surrounding EBG/radiating patches, which act as a reflective medium. Therefore, when an element is asymmetrically positioned on the array, this uneven reflecting surface contributes to a non-uniform amount of radiation interference. A stronger

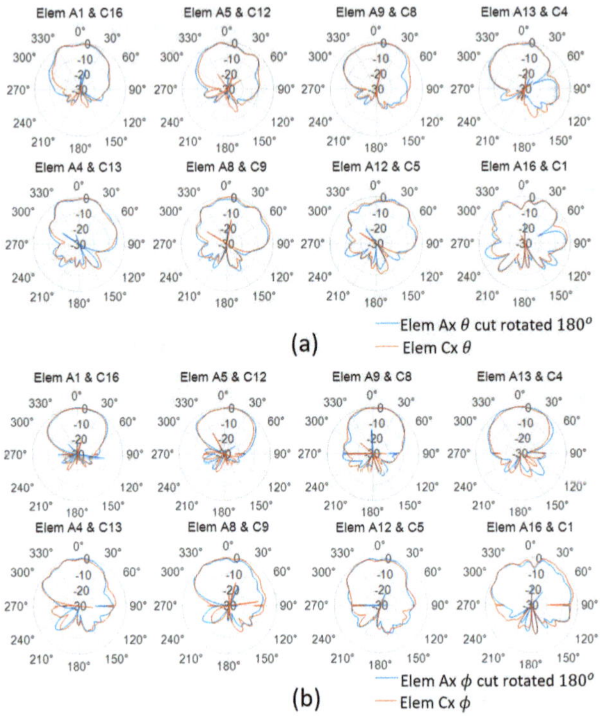


**Fig. 6.12** Symmetric ERP comparison (**a**) $\theta$ cuts. (**b**) $\phi$ cuts (measurement results)

radiation on the side with a larger reflective surface can be observed. Consequently, the ERP becomes asymmetric and the radiation intensity is greater in the direction of the larger ground plane, leading to the radiation pattern tilt.

### *6.5.4 Element Radiation Patterns Categorization and Symmetry*

The 2D $\theta$ and $\phi$ radiation patterns for 16 sample antenna elements: elements A1, A5 A9 A13 with their diagonally symmetric elements C16, C12, C8, C4, which are close to the array center; elements A4, A8, A12, A16 (located at right-bottom edge) with their symmetric located elements C13, C9, C5, C1 (located at left-top edge), are plotted in Fig. 6.12. To facilitate a direct comparison of ERPs, we rotated the EPRs in sub-array-A by 180° within the array $yz$-plane and then plotted the 2D cuts.

The radiation patterns of two elements situated at diagonally symmetric positions w.r.t. the array center align in both $\theta$ and $\phi$ dimensions, upon rotating one of the

ERPs by 180° in the array plane. The observation suggests that within an antenna array when elements are situated in symmetric environments along both the E-plane and H-plane, their radiation patterns demonstrate centrosymmetry. Consequently, the radiation pattern of a specific element can be precisely determined by rotating the pattern of its diagonally opposite element by 180°. For each element pair, the radiation pattern $\boldsymbol{A}_m$ and its symmetric ERP $\boldsymbol{A}_m^{sym}$ have the relationship

$$A_m(\theta, \phi) = A_m^{sym}(-\theta, -\phi). \tag{6.6}$$

Leveraging the symmetry of ERPs in an array, the heterogeneous ERPs can be grouped, effectively reducing the number of unique ERPs to half the total number of elements. In the case of our 8×8 antenna array prototype, this principle allows the 64 elements to be efficiently paired into 32 pairs. The ERP symmetry information can also help create comparable sub-arrays, with elements having identical radiation patterns towards opposite $\theta$ and $\phi$ directions. For symmetric sub-arrays, composed of elements symmetrically positioned within the array, the sub-array response vector exhibits symmetry, resulting in nearly identical steering vectors toward opposite directions. Classifying ERPs can significantly improve the efficiency of beamforming analysis and the development of optimization techniques by enabling the reuse of optimal beamforming vectors. Optimized beamforming vectors derived from one sub-array can be directly applied to its symmetric counterpart to generate a mirrored beam pattern. This mirrored pattern features main lobe steering and sidelobe patterns oriented in opposite directions. Further details and discussions on the beamforming performance of symmetric sub-arrays will be presented in Chap. 7.

## 6.6 Sub-array Beamforming Measurement Results

### 6.6.1 4×4 Antenna Sub-array Beamforming Measurement

The 32 elements on the top half of the 8×8 mMIMO testbed are divided into two 4×4 sub-arrays, which are 4×4-C and 4×4-B. Each sub-array is driven by a beamforming vector with uniform magnitude and phase shift values. The 3D beam patterns of the sub-arrays are shown in Fig. 6.13, while the normalized 2D cut plots along the azimuth ($\theta$ angle) and elevation ($\phi$ angle) directions are presented in Fig. 6.14. Both 4×4 sub-arrays exhibit similar narrow main beams with four low sidelobes in both the azimuth and elevation directions.

For the Tx-4×4-C sub-array, the main lobe's 3 dB-beamwidth is 28.3°, and the 10 dB-beamwidth is 51.1° in the azimuth direction. In the elevation direction, the 3 dB-beamwidth is 26.7°, and the 10 dB-beamwidth is 47.0°. The measured gain is 13.3 dB, and the directivity is 15.9 dB. For the Tx-4×4-B sub-array, the 3 dB-beamwidth is 27.1°, and the 10 dB-beamwidth is 47.4° in the azimuth direction. In the elevation direction, the 3 dB-beamwidth is 27.7°, and the 10 dB-beamwidth

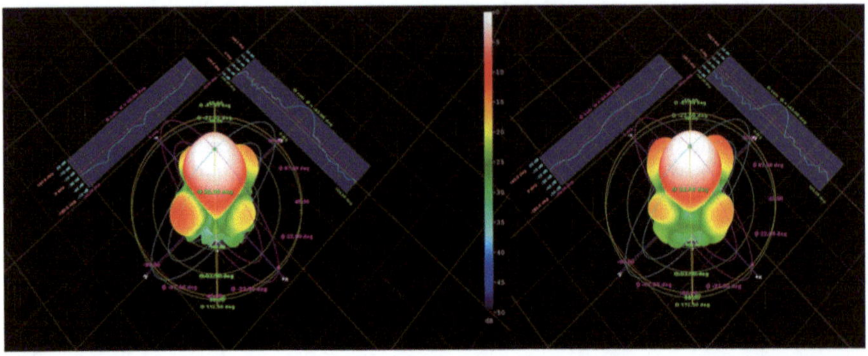

**Fig. 6.13** Tx-4×4-C (left)/-B (right) antenna sub-array 3D beams (measurement results)

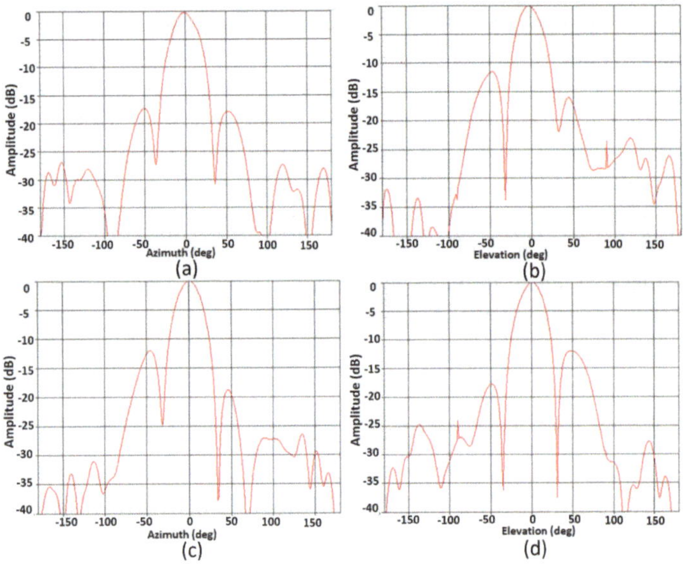


**Fig. 6.14** Beam 2D cuts for Tx-4×4-C sub-array. (**a**) Azimuth plane. (**b**) Elevation plane; for Tx-4×4-B sub-array. (**c**) Azimuth plane. (**d**) Elevation plane (measurement results)

is 49.0°. The gain is 13.6 dB, with a directivity of approximately 15.9 dB. The measured sub-array beam gain and directivity closely match the HFSS simulation results of an isolated 4×4 array, with differences of less than 0.3 dB.

The two 4×4 sub-arrays show consistent beamforming characteristics. The difference in the 3 dB-beamwidth between the sub-arrays is less than 0.6°, while the gain difference is 0.3 dB. The measured efficiencies are −2.6 dB (55.0%) for the Tx-4×4-C sub-array and −2.3 dB (58.9%) for the Tx-4×4-B sub-array. For both sub-arrays, the normalized sidelobe level is lower (better) than −11 dB in both azimuth and elevation directions. The narrow beamwidth and high gain enable the sub-arrays to achieve effective beamforming performance when steering beams in both horizontal and vertical directions.

### 6.6.2 1×4 and 1×8 Sub-array Beamforming Measurement

The array's capability for beam steering and nulling is determined by the number of antenna elements available in that direction. For instance, our 8×8 array has a maximum of 8 elements in either the $\theta$ or $\phi$ direction, meaning that the narrowest beamwidth, effective steerability, and nulling capability are determined by the 8-element sub-array length. Two sample ULAs, a 1×4 sub-array located on the right side of the fourth row and 1×8 sub-array spanning the entire fourth row of the array prototype, are used to investigate the radiation patterns of ULAs with different array lengths and their beam steerability. The measured 3D radiation patterns at 3.5 GHz, obtained with uniform feeding power and zero phase shift, are shown in Fig. 6.15. The corresponding 2D cuts along the azimuth and elevation directions are presented in Fig. 6.16.

Both the 1×4 and 1×8 sub-arrays produce a plate-shaped beam pattern characterized by sidelobes on either side of the main beam. For the Tx 1×4 sub-array, a directivity of 11.3 dB is achieved, with a 3 dB-beamwidth of 27.3°. Additionally, the normalized sidelobe level of − 14.1 dB and a first null beamwidth of 66° are observed. In contrast, the Tx 1×8 sub-array demonstrates a higher directivity of 14.0 dB, with a narrower 3 dB-beamwidth of 14.2° in the azimuth direction. The normalized sidelobe level for the Tx 1×8 sub-array is − 13.6 dB.

The comparison between the two sub-arrays provides valuable insights into how the radiation pattern evolves with increasing array length. Doubling the array size

(a) (b) (c) (d)

**Fig. 6.15** 3D beam patterns of the 1×4 and 1×8 sub-arrays. (**a**) 1×4 side view. (**b**) 1×4 top view. (**c**) 1×8 side view. (**d**) 1×8 top view (measurement results)

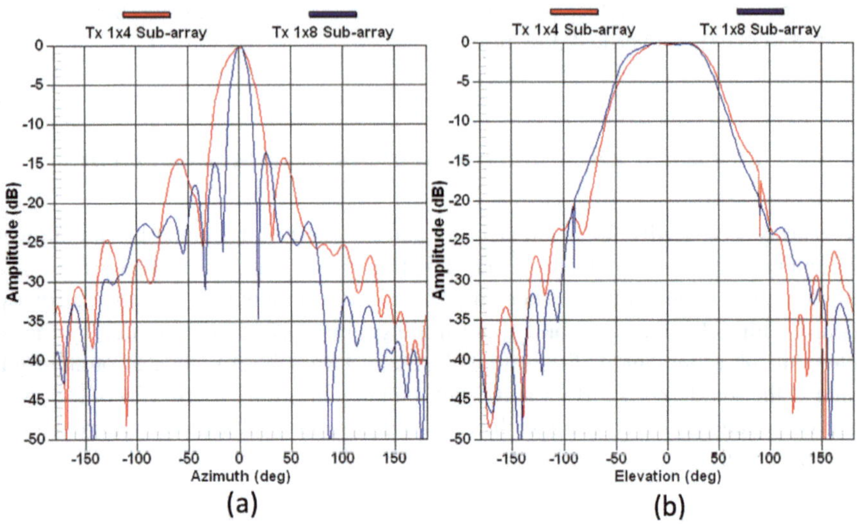


**Fig. 6.16** 2D beam pattern cuts of the 1×4 and 1×8 sub-arrays: (**a**) in azimuth. (**b**) elevation directions (measurement results)

from 1×4 to 1×8 enhances directivity, increasing from 11.3 to 14.0 dB. This gain is 0.3 dB less than the theoretical 3 dB directivity improvement expected when doubling the array length, demonstrating the array's enhanced ability to focus energy in the desired direction. Similarly, doubling the array size reduces the 3 dB-beamwidth by nearly half, consistent with the observed directivity increase. A narrower beamwidth improves the array's spatial resolution, allowing it to better distinguish between closely spaced targets or users while reducing interference from undesired directions.

The measured results of the $1 \times M$ sub-arrays for $M = 4$ and $M = 8$ closely align with the simulation results presented in Sect. 6.2, with an average directivity difference of less than 0.5 dB. The experimentally measured beamwidths also show good agreement, with an average discrepancy of 1°. This consistency validates the performance of the fabricated sub-arrays and supports the theoretical predictions.

### *6.6.3 Sample 1×4 Sub-array Beam Steering Performance*

The 2D cuts of the measured sample 1×4 sub-array beamforming results for steering angles $\theta$ ranging from $-50°$ to $+50°$ are shown in Fig. 6.17 and the beam radiation characteristics are summarized in Table 6.1. Uniform magnitude steering weights were employed, with phase shift values based on the target steering angles.

The measured results show an average directivity of 11.0 dB, with a directivity variation of 0.9 dB across the entire steering range. The average 3 dB-beamwidth

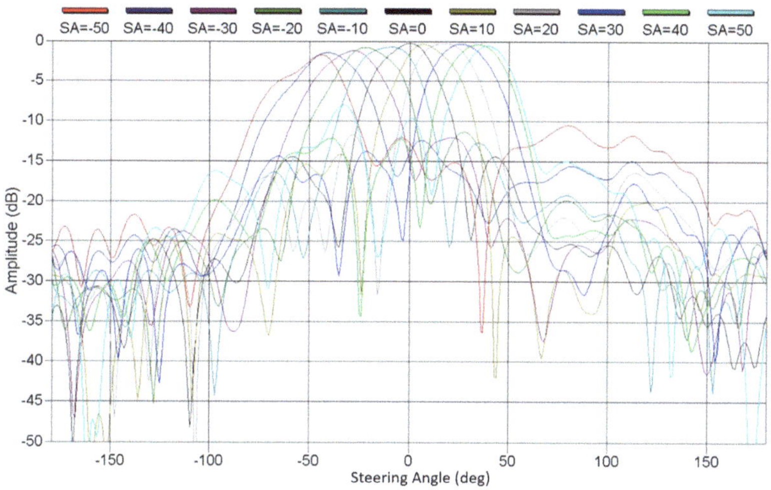


**Fig. 6.17** Sample 1×4 sub-array beam steering 2D cuts with steering angle (SA) $-50°$ to $50°$ (measurement results)

and 10 dB-beamwidth are 28.8° and 51.2°, respectively. The normalized sidelobe level remains below $-9.0$ dB across the steering range and drops below $-10.3$ dB for steering angles between $-50°$ and $+40°$. With uniform magnitude steering vectors, the difference between the achieved and target beam directions is less than 5° within this range.

Compared to the simulation results, presented in Sect. 6.3, of a similar Right 1×4 sub-array with the same relative location, the Tx 1×4 sub-array exhibits comparable trends in beam peak angle tilt, directivity, beamwidth, and sidelobe level variations. When the beam is steered away from the array center, the asymmetric location of the sub-array results in narrower 3 dB- and 10 dB-beamwidths and an average beam tilt of $-3.8°$, which is $-1.1°$ more tilted than the simulation results. A faster sidelobe level increase can be observed when steering away from the array center, i.e., in the positive steering directions. The measured average directivity and sidelobe differ from the HFSS simulated values by 0.2 and 1.0 dB, respectively.

The ground plane and conductive surfaces act as reflective media, leading to reflections and interference with the direct radiation from the patch elements. When the sub-array is asymmetrically positioned on the array, this non-uniform reflecting surface introduces additional radiation interference in the beamforming process. There is amplified radiation on the side with a larger reflective surface. Consequently, the main lobe pattern becomes asymmetric, and the radiation intensity is more pronounced in the direction of the larger ground plane, leading to beam angle tilt. When the beam is steered toward the array center (where the reflective surface is larger), the extra reflections constructively add to the main lobe, leading to a wider beamwidth. Conversely, steering the beam away from the array center reduces main lobe reinforcement but increases sidelobe levels due to the influence of reflected

**Table 6.1** Tx 1×4 sub-array beam steering measurement results

| Steering angle (deg) | Achieved angle (deg) | Directivity (dB) | 3 dB BW(deg) | 10 dB BW (deg) | SLL (dB) |
|---|---|---|---|---|---|
| −50 | −51.2 | 10.8 | 29.1 | 57.5 | −10.3 |
| −45 | −49.0 | 10.8 | 26.8 | 59.8 | −11.2 |
| −40 | −40.6 | 10.7 | 30.0 | 61.0 | −11.0 |
| −35 | −35.3 | 10.9 | 33.0 | 60.4 | −10.7 |
| −30 | −31.0 | 11.0 | 30.8 | 57.1 | −10.8 |
| −25 | −27.2 | 10.7 | 32.9 | 58.9 | −10.9 |
| −20 | −21.0 | 11.0 | 30.1 | 54.3 | −10.6 |
| −15 | −16.8 | 11.1 | 31.1 | 50.8 | −10.5 |
| −10 | −12.4 | 11.1 | 30.8 | 48.7 | −11.9 |
| −5 | −7.5 | 11.2 | 29.0 | 48.0 | −12.4 |
| 0 | −1.2 | 11.3 | 27.3 | 47.9 | −14.1 |
| 5 | 3.5 | 11.5 | 24.5 | 48.0 | −16.4 |
| 10 | 7.0 | 11.5 | 26.6 | 48.3 | −13.8 |
| 15 | 10.1 | 11.3 | 28.7 | 47.4 | −11.8 |
| 20 | 14.9 | 10.9 | 29.3 | 47.2 | −11.4 |
| 25 | 22.7 | 10.9 | 29.0 | 48.8 | −11.6 |
| 30 | 26.6 | 10.8 | 29.0 | 49.1 | −13.4 |
| 35 | 31.7 | 10.8 | 28.8 | 48.1 | −12.9 |
| 40 | 38.0 | 11.0 | 28.0 | 46.5 | −11.9 |
| 45 | 39.0 | 11.2 | 25.8 | 44.5 | −10.0 |
| 50 | 41.0 | 11.3 | 23.8 | 43.0 | −9.0 |

waves. These findings, consistent with the simulation results, underscore the significant impact of the ground plane and sub-array positioning on beamforming performance. They emphasize the importance of careful design considerations to mitigate beam asymmetry and optimize performance in asymmetric sub-array configurations.

### *6.6.4 Sample 1×8 Sub-array Beam Steering Performance*

The Tx 1×8 sub-array, symmetrically located along the $\theta$ axis, was utilized for experimental beam steering investigations. The measured beam characteristics within the steering range of $-60^\circ$ to $+60^\circ$ at 3.5 GHz are shown in Fig. 6.18 and summarized in Table 6.2. An effective beam peak steering range from $-58.1^\circ$ to $59.3^\circ$ was achieved.

Within the range of $-50^\circ$ to $+50^\circ$, the 1×8 sub-array demonstrated an average directivity of 13.4 dB, an average 3 dB-beamwidth of $16.3^\circ$, and an average 10 dB-beamwidth of $27.8^\circ$. The 3 dB- and 10 dB-beamwidths remained narrower than 19.2

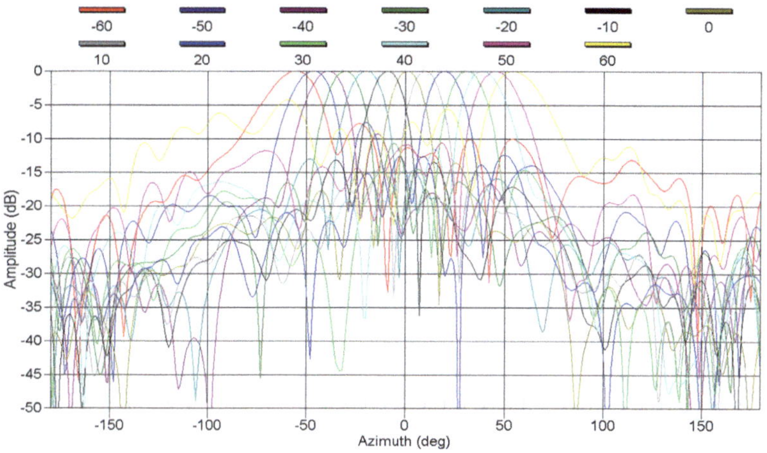


**Fig. 6.18** Sample 1×8 sub-array beam steering 2D cut for steering angle from $-50^o$ to $+50^o$ (measurement results)

**Table 6.2** Sample 1×8 sub-array beam measurement results (BW: Beamwidth; SLL: Sidelobe Level)

| Steering angle (deg) | Achieved angle(deg) | Directivity (dB) | 3 dB BW(deg) | 10 dB BW (deg) | SLL (dB) |
|---|---|---|---|---|---|
| −60 | −58.1 | 11.6 | 23.4 | 46.1 | −7.8 |
| −50 | −48.8 | 12.9 | 18.6 | 31.9 | −7.6 |
| −40 | −41 | 13 | 17.2 | 29.6 | −9.3 |
| −30 | −30.2 | 13 | 15.9 | 26.9 | −10.1 |
| −20 | −21.3 | 13.3 | 16 | 26.7 | −13 |
| −10 | −9.2 | 13.9 | 14.7 | 24.9 | −13.2 |
| 0 | 0.9 | 14 | 14.2 | 24 | −13.6 |
| 10 | 9.7 | 14 | 13.8 | 23.5 | −14.1 |
| 20 | 20.3 | 13.4 | 15.4 | 26.2 | −12.4 |
| 30 | 30.6 | 13.2 | 16.4 | 27.6 | −14.5 |
| 40 | 42.3 | 13.1 | 17.8 | 30.9 | −11.6 |
| 50 | 49 | 13.2 | 19.2 | 34 | −10.6 |
| 60 | 59.3 | 11.5 | 22.8 | 41.7 | −5.6 |

and 34°, respectively, within this range. The HFSS simulation and experimental results are in good agreement, showing average differences in directivity and 3 dB-beamwidth of 0.1 dB and 0.8°, respectively. The normalized sidelobe level is lower than the main lobe by at least − 9.3 dB in the steering range of − 40° to + 50°. The sidelobe increases rapidly when the steering angle exceeds 50°. As the steering angle increases, the apparent spacing between elements also increases due to the projection of element spacing along the beam direction. This can result in the effective element spacing exceeding half a wavelength, leading to the formation of grating lobes, which contributes to the rapid increase in sidelobe levels.

The symmetric ground plane and conductive surfaces on the left and right sides of the $1 \times 8$ array prototype result in more balanced reflections and interference with the direct radiation from the patch elements. Consequently, the $1 \times 8$ sub-array exhibits symmetric beamforming characteristics when steering the beam in both positive and negative $\theta$ directions. The achieved steering angles, directivities, and observed increases in beamwidth and sidelobe levels follow similar trends in both directions, highlighting the benefits of symmetric sub-array positioning for achieving a more uniform beamforming performance across the steering range.

## 6.7 Concluding Remarks

This chapter has presented a comprehensive study of large-scale antenna array design, from theoretical derivations to experimental validations. First, a quantitative analysis was conducted to characterize the influence of array length on beam radiation patterns, yielding theoretical expressions for directivity and the 3dB- and 10dB-beamwidths. Simulations under various element pattern assumptions confirmed that the derived formulas effectively predict beam behavior, while also demonstrating that the impact of individual element patterns diminishes as the array size increases. Then, attention turned to practical scenarios in which sub-arrays are positioned at different positions within large-scale arrays. The resulting investigations highlight how surrounding environments, asymmetries, and steering angles affect beamforming performance when using an MDB. These findings underscore the potential deviations in beam steering angle, directivity, and sidelobe characteristics depending on sub-array location and design considerations.

Building on these insights, an $8 \times 8$ metamaterial-based patch antenna array was designed and fabricated for mMIMO operation in the 3.35 GHz to 3.6 GHz band. Measurements of element RL, inter-element MC, and sub-array beamforming performance agreed well with simulations, validating the array's design. A thorough examination of ERPs across all 64 elements further revealed how ERP variations, particularly symmetry and distortion, can guide more efficient beamforming optimizations. Moreover, beamforming experiments with a realistic network of quantized phase shifters and attenuators confirmed the array's steering capabilities and highlighted the effects of environmental factors on performance.

Overall, this chapter provides both theoretical and practical insights into the design and evaluation of large-scale arrays. The results and the array prototype serve as a foundation for subsequent chapters, where the measured ERPs and design considerations will help in the development of beamforming and interference management techniques, and the prototype will be used for experimental measurements of various MU-mMIMO beamforming algorithms.

## References

1. Z. Zhang, Y. Xiao, Z. Ma, M. Xiao, Z. Ding, X. Lei, G. K. Karagiannidis, P. Fan, 6G wireless networks: vision, requirements, architecture, and key technologies. IEEE Veh. Technol. Mag. **14**(3), 28–41 (2019)
2. E.G. Larsson, O. Edfors, F. Tufvesson, T.L. Marzetta, Massive MIMO for next generation wireless systems. IEEE Commun. Mag. **52**(2), 186–195 (2014)
3. P. Xingdong, H. Wei, Y. Tianyang, L. Linsheng, Design and implementation of an active multibeam antenna system with 64 RF channels and 256 antenna elements for massive MIMO application in 5G wireless communications. China Commun. **11**(11), 16–23 (2014)
4. S. Malkowsky, J. Vieira, L. Liu, P. Harris, K. Nieman, N. Kundargi, I.C. Wong, F. Tufvesson, V. Öwall, O. Edfors, The world's first real-time testbed for massive MIMO: design, implementation, and validation. IEEE Access **5**, 9073–9088 (2017)
5. W.L. Stutzman, G.A. Thiele, *Antenna Theory and Design* (Wiley, Hoboken, 2012)

# Chapter 7
# Far-field Nulling Control Beamforming with Measured Element Patterns for MU-mMIMO

## 7.1 Introduction

The deployment of large-scale antenna arrays in mMIMO enables the generation of steerable and directional beams characterized by high gain and narrow beamwidth. This capability of mMIMO beamforming significantly enhances the MU capacity and spectrum efficiency of wireless networks by facilitating the simultaneous directional transmission or reception of multiple data streams [1]. By dividing the array into sub-arrays, the BS can communicate with multiple users in different directions simultaneously, substantially improving overall throughput and spectral efficiency [2]. However, with the advancement of MU operations in mMIMO systems, MUI emerges as a critical concern. Multiple users sharing the same frequency resources often lead to significant interference, which poses a major challenge by degrading the SIR. Effective interference management is critical, as a lower SIR results in higher bit error rates and reduced system data throughput.

In such scenarios, NCB plays a pivotal role in enhancing signal reception quality. By strategically adjusting the magnitudes and phase shifts of feedings within the antenna array, NCB acts as a spatial filter, maximizing the beam gain in the target user directions and minimizing the radiation power in the potential interference directions. Consequently, the array beam main lobe can be focused toward the intended signal's direction of arrival and radiation nulls can be created to eliminate unwanted interference. This dual functionality not only maintains signal strength from the designated user but also efficiently suppresses interference from other users, significantly enhancing the efficacy of the MU-mMIMO system [3, 4]. Popular beamforming algorithms can be divided into two main categories, which are the non-iterative scheme and the iterative optimization scheme, as presented in Sect. 2.3.1.

For the non-iterative beamforming approach, the feeding weights can be derived from instant calculations based on the estimation of the inverse covariance matrix



T. Le-Ngoc et al., *Half-Duplex/Full-Duplex MU-mMIMO*, Wireless Networks,
https://doi.org/10.1007/978-3-032-09195-6_7

of the desired and interference channels. The performance is critically dependent on the precision of the channel's statistical data, particularly the estimated SV. However, real-world conditions introduce challenges like wavefront distortions, variations in ERPs, directional inaccuracies, and incoherent local scattering, all of which can cause notable discrepancies in the SV estimation. The simulation and evaluations of the proposed algorithms [5–19] predominantly assume isotropic or uniform ERPs, where the variations of the ERPs across a large-scale antenna array are ignored. The theoretical isotropic SV encapsulates solely the phase differences induced by an incoming signal at the inputs of isotropic sources, attributable to their distinct spatial positions. Consequently, the non-iterative methods usually generate suboptimal beamforming vectors, characterized by insufficient null depths and imprecise null positioning. While numerous beamforming techniques [20–22] have been devised to account for inter-element MC effects on beamforming performance by incorporating the MC matrix in the algorithms, it is important to recognize that element interactions are not the sole influential factor of beam patterns. Factors such as element positioning, surrounding environmental conditions, and asymmetries in the ground plane also play significant roles in creating varied ERPs throughout the array. As demonstrated in [23], the presence of heterogeneous ERPs within an antenna array can markedly compromise the efficacy of beamforming techniques. To address existing limitations, [4] introduces enhanced null steering beamforming and MVDR methods, tailored for non-isotropic radiation patterns generated from microstrip planar antenna array simulations. The schemes are designed for adaptability across various antenna array configurations and optimized for 3D spatial beamforming, showcasing accurate null and main lobe placements, resulting in high SIR values and robustness against increasing numbers of users. Nonetheless, a comprehensive analysis of realistic ERP variations in large-scale arrays remains unexplored, raising questions about how ERP insights could further refine NCB optimization effectiveness and efficiency. Additionally, obtaining accurate ERP information presents significant challenges in the development of non-iterative algorithms, as even minor inaccuracies can impact algorithmic performance and degrade the overall system efficiency. Furthermore, there is a pressing need for experimental validations to assess the NCB scheme's efficacy in mitigating interference with practical BFNs, which consist of quantized phase shifters and attenuators.

The iterative beamforming scheme is a search and optimization strategy that continuously adjusts feeding weights while evaluating the resulting objective metrics such as nulling depth, SIR, or achievable data rate. By introducing strategic perturbations to the beamforming weights, the scheme progressively enhances overall beamforming performance. This approach often leverages techniques such as successive FFT [24, 25] or sophisticated optimization algorithms like PSO [23, 26–28] and GA[27, 29, 30]. Nevertheless, similar to the existing research for the non-iterative approaches, we notice that most of the existing research for the iterative schemes [24–30] focuses on using isotropic or uniform ERPs in the algorithm design. Performance evaluation predominantly utilizes simulation-based methods, underpinned by the assumption of ideal characteristics for phase shifters and attenuators. The study of beamforming optimization, particularly through the integration of measured ERPs and the validation of these methods via experimental measurements, remains an important area to be studied. Additionally, the utilization of optimization algorithms such as GA and PSO in beamforming underscores the crucial importance of selecting initial search points. This decision markedly impacts the convergence rate and the quality of the optimal solution. An inadequately chosen starting point can result in slower convergence, potentially trapping the algorithm in sub-optimal solutions. Most existing algorithms initiate the search with the conventional MDB approach. Meanwhile, initiating vectors derived from non-iterative NCB schemes could expedite the optimization process, offering a more efficient pathway to achieving optimal beamforming configurations.

In this chapter, a perturbation-based nulling control beamforming (PNCB) scheme is developed and evaluated with the experimentally measured ERPs from our $8 \times 8$ large-scale antenna array, as presented in Chap. 6. The PNCB approach leverages the LCMV algorithm to generate an initial beamforming vector, with PSO applied to further refine nulling performance. The proposed algorithm effectively manages multiple null controls and facilitates null width adjustment. Experimental validation using realistic modulated signals and a quantized BFN substantiates the initial simulation findings, incorporating practical considerations of phase shifter and attenuator quantization. Both simulation and experimental results confirm the proposed PNCB approach's capability to create deep radiation nulls with precise control over null locations and adjustable nulling width. The main contributions of this work are summarized as follows:

- **Perturbation-based nulling control beamforming optimization for MU-mMIMO:** A PNCB strategy integrating PSO with the LCMV beamformer for initial search vector generation is developed. The scheme is tailored to create deep nulls with customizable widths, taking into account the impacts of realistic ERPs. Simulation with measured ERPs and sample $1 \times 8$ sub-array illustrates a normalized nulling level[1] better than $-52.4$ dB. To achieve wide nulls with adjustable widths, our approach strategically generates multiple adjacent nulls in close proximity. Illustrative results show that the technique can extend the null width to exceed 6° while maintaining an average nulling level better than $-45.3$ dB. The illustrative results underscore the technique's effectiveness in precisely shaping radiation nulls while minimally affecting the main lobe patterns, resulting in an average interference mitigation improvement of 30.9 dB compared to scenarios using the MDB scheme. By leveraging the pattern symmetry information achieved from Chap. 6, we effectively halve the total number of required optimization processes. This efficiency improvement is achieved by applying the optimal beamforming vector to its symmetric sub-array, thereby ensuring similar beamforming performance. Additionally, the impact of

[1] All null levels and nulling values are reported as normalized radiation power values in the target nulling directions, relative to the power level of the main lobe in the desired user directions. The power towards the desired user direction is taken as the reference point, corresponding to 0 dB.

phase shifter and attenuator quantization levels on beamforming performance is studied to provide valuable insights for selecting hardware in a realistic BFN setup, highlighting the potential trade-off between hardware cost and nulling effectiveness.

- **Experimental measurements of the nulling performance with realistic modulated signals and beamforming network:** Experimental measurements conducted in an anechoic chamber for two-user scenarios validate the effectiveness of the proposed PNCB technique, utilizing practical quantized phase shifters and attenuators. The results highlight the PNCB scheme's capability to achieve a significant gain difference, averaging over 35.6 dB, between the main and the interference users. This translates to a notable nulling depth enhancement of 22 dB on average over the MDB approach. When employing two transmitting User Equipments (UEs) with distinct 20 MHz (3.49–3.51 GHz) 64-QAM or 16-QAM modulated signals, the proposed PNCB scheme results in a considerable SIR enhancement of 22 dB and an EVM improvement of 21.3 dB on average, compared to the MDB approach. The proposed scheme demonstrates its capability to effectively suppress the interference signal, bringing it close to the noise level. The efficacy of the PNCB scheme is evidenced in the constellation diagrams, where a high-quality received signal is manifested by distinct, well-clustered symbols closely aligned with their ideal locations, showcasing minimal scatter and a lack of discernible interference patterns.

The rest of the chapter is structured as follows. Section 7.2 introduces the system model for the MU-mMIMO setup employed in developing the PNCB algorithm. It also outlines the LCMV scheme, which serves as the foundation for generating the initial search vector in our beamforming optimization process. In Sect. 7.3, we elaborate on the proposed PNCB schemes, providing comprehensive details on their formulation and implementation. Then, Sect. 7.4 utilizes the measured ERPs, presented in Sect. 6.5.3, to construct a customized antenna array simulation in MATLAB, showcasing the efficacy of our proposed PNCB algorithm in achieving precise nulling control with a consideration of realistic ERPs and the impacts of potential hardware impairments in the BFN on the nulling performance. Section 7.5 presents experimental measurements involving two users with the use of realistic modulated signals, which confirm the effectiveness of the proposed PNCB scheme in reducing MUI. Section 7.6 concludes the chapter and provides possible future work.

## 7.2 System Model

In the system setup, we consider a partially connected HBF scheme, which enhances hardware efficiency by reducing the number of connections between RF chains and antenna elements [2]. In this architecture, each RF chain is connected to a subset of the total antenna elements, grouped into antenna sub-arrays. This configuration

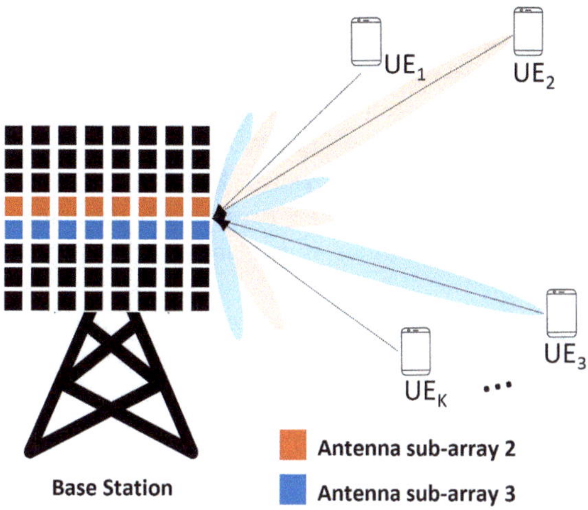


**Fig. 7.1** Nulling control beamforming with two sample sub-array beam patterns for MU-mMIMO

decreases hardware complexity while maintaining the effective beamforming capabilities of each sub-array. Each antenna sub-array features its own beam pattern, controlled by a BFN. As illustrated in Fig. 7.1, our study explores an MU-mMIMO operation scenario, where the BS serves $K$ users simultaneously. In the problem formulation, we focus on the receiving (uplink) array scenarios. The BS is equipped with antenna sub-arrays, each comprising $M$ antenna elements. The signal received by the $i$th sub-array can be represented as

$$y_i(\theta_1, \phi_1, \theta_2, \phi_2, \ldots, \theta_K, \phi_K,) = \underbrace{\boldsymbol{w}_i^H \mathbf{h}_i(\theta_i, \phi_i) s_i}_{\text{Desired Signal}} + \underbrace{\sum_{k \neq i}^{k=1,\ldots,K} \boldsymbol{w}_i^H \mathbf{h}_i(\theta_k, \phi_k) s_k}_{\text{Multi-user Interference}} + \underbrace{\boldsymbol{w}_i^H \boldsymbol{n}_i}_{\text{Modified Noise}}, \tag{7.1}$$

where $\boldsymbol{n}_i \sim \mathcal{CN}(0, \sigma_n^2 \mathbf{I}_M)$ is the Gaussian thermal noise at the receiver of sub-array $i$. The first term in the expression denotes the intended received signal of the $i$th sub-array from the desired user $i$, located in the direction of $(\theta_i, \phi_i)$. The symbol $s_i$ is the transmitted signal symbol from $i$th user, adhering to the power constrain $\mathbb{E}\{s_i s_i^H\} = 1$. The $i$th sub-array beamforming vector, $\boldsymbol{w}_i \in \mathbb{C}^{M\times 1}$, plays a pivotal role in manipulating the phase and amplitude of the incoming signal, which is capable of controlling the constructive and destructive summation of the received signal. The channel vector $\mathbf{h}_i(\theta_i, \phi_i) \in \mathbb{C}^{M\times 1}$ delineates the desired signal channel from user $i$ to the receiver of sub-array $i$. The channel can be randomly varying according to the realistic propagation under consideration, including multipath propagation. For robust beamforming, we assume that the BS has the knowledge of the channel sample and this channel sample remains almost unchanged during the beamforming computation. The SV is used to estimate the array response. Moreover, the SV can be modified to account for realistic ERPs or multi-path propagation effects by incorporating ERP gains or an equivalent complex multi-path gain to formulate a weighted SV. Since one of the main focuses of this work is to investigate the impact of realistic ERPs on beamformer optimization, the normalized pattern-weighted SV which encapsulates the antenna array's configuration, ERPs, and the user's locations, is used and can be expressed as

$$\mathbf{h}_i(\theta,\phi) = \begin{bmatrix} A_{i,1}(\theta,\phi)\exp\left(j\beta r_{i,1}\boldsymbol{v}(\theta,\phi)\boldsymbol{v}(\theta_{i,1},\phi_{i,1})^T\right) \\ A_{i,2}(\theta,\phi)\exp\left(j\beta r_{i,2}\boldsymbol{v}(\theta,\phi)\boldsymbol{v}(\theta_{i,2},\phi_{i,2})^T\right) \\ \vdots \\ A_{i,M}(\theta,\phi)\exp\left(j\beta r_{i,M}\boldsymbol{v}(\theta,\phi)\boldsymbol{v}(\theta_{i,M},\phi_{i,M})^T\right) \end{bmatrix}, \tag{7.2}$$

with $\beta = 2\pi/\lambda_0$ being the propagation phase constant and $\lambda_0$ is the free-space wavelength. The matrix $\boldsymbol{A}_{i,m}$ encapsulates the normalized radiation patterns for the $m$th antenna element of the $i$th antenna sub-array, characterizing its directional magnitude response across various angles. Each entry, denoted by $A_{i,m}(\theta,\phi)$, represents the normalized magnitude response of the antenna element $m$ when the signal arrives in the direction $(\theta,\phi)$. Under the assumption of isotropic elements, the ERPs exhibit uniform magnitude responses, i.e., $A_{i,1}(\theta,\phi) = A_{i,2}(\theta,\phi) = \ldots = A_{i,M}(\theta,\phi) = 1$. The terms $r_{i,m}$ and $\boldsymbol{v}(\theta_{i,m},\phi_{i,m})$ represent the distance and the unit position vector of the $m$th antenna element within the $i$th antenna sub-array relative to the sub-array center or phase reference point. $\boldsymbol{v}(\theta,\phi)$ denotes the unit position vector pointing in the direction $(\theta,\phi)$.

The second term in (7.1) highlights the challenge posed by MUI in scenarios where $K$ users are simultaneously served. The signals from $K-1$ non-desired users are perceived by the $i$th sub-array as MUI, significantly impacting the sensitivity of the receiver and the overall performance of the MU-mMIMO system. The transmitted symbol vector for the $k$th interference user is represented by $s_k$, which satisfies the power constraint $\mathbb{E}\{s_k s_k^H\} = 1$, where $k = 1,\ldots,K$ and $k \neq i$. The interference channel vector $\mathbf{h}_i(\theta_k,\phi_k) \in \mathbb{C}^{M\times 1}$ maps the interference paths from the $k$th user who has an angular location $(\theta_k,\phi_k)$ relative to the $i$th receiver. Accordingly, the SIR at the $i$th receiver in the system can be calculated as

$$\mathrm{SIR}_i = \frac{\mathbb{E}\{|\boldsymbol{w}_i^H\mathbf{h}_i(\theta_i,\phi_i)s_i|^2\}}{\mathbb{E}\{|\sum_{k\neq i}^{k=1,\ldots,K}\boldsymbol{w}_i^H\mathbf{h}_i(\theta_k,\phi_k)s_k|^2\}}. \tag{7.3}$$

### 7.2.1 Maximum-Directivity and Nulling Control Beamforming

The beamforming process plays a pivotal role in managing the gain of signals transmitted or received in various directions, improving the SIR. In the context of MDB, the feeding weight vector, denoted as $\boldsymbol{w}_0(\theta_i, \phi_i)$, is designed to optimize the directivity of the beam towards the target user direction $(\theta_i, \phi_i)$. Therefore, the power of the desired signal term in (7.1) is maximized. To steer the main lobe effectively, the beamforming vector is designed to have phases aligned with the phase shifts across antenna elements, as shown in the (7.2), leading to a constructive summation of the signals across all elements.

For the sake of simplicity,[2] in the following analysis, we consider a ULA configuration, with elements aligned along the $y-$axis, which primarily facilitates beam steering in the $\theta$ angle directions with a constant elevation angle. To simplify the notation, we eliminate the $\phi$ angle in the later discussions. The corresponding MDB vector for target steering angle $\theta_i$ can be expressed as

$$\boldsymbol{w}_0(\theta_i) = [1, e^{j\beta d \sin(\theta_i)}, \ldots, e^{j(M-1)\beta d \sin(\theta_i)}]^T, \tag{7.4}$$

where $d$ is the distance between the neighbouring antenna elements in the $1 \times M$ ULA.

The covariance matrix of the received signal at the $i$th sub-array, comprising multi-user signals and thermal noise prior to beamforming, is given by

$$\mathbf{R}_i = \sum_{k=1}^{K} \mathbf{h}_i(\theta_k, \phi_k)\, \mathbf{h}_i^H(\theta_k, \phi_k) + \sigma_n^2 \mathbf{I}_M, \tag{7.5}$$

where the diagonal entries represent the total received power at each antenna element, while the off-diagonal entries capture the spatial correlations induced by the multi-user signals.

To enhance the SIR and improve the system's performance, NCB serves as an effective strategy for increasing spatial discrimination between users. This method adjusts the intensity of the beam's radiation across various angles, thereby establishing a differential gain for users located in disparate directions. Operating within linear optimization constraints, LCMV is a widely used non-iterative beamforming technique that adjusts array feeding weights to minimize the variance of the received signal. The approach is effective in spatial filtering, maintaining the integrity of desired signals while simultaneously diminishing interference from other sources. Additionally, the LCMV approach's proficiency in managing multiple constraints

---

[2] The array's capabilities for beam resolution, steering, and nulling are primarily determined by the number of antenna elements available along the desired direction. While this study focuses on a ULA configuration, the algorithms developed are versatile and can be applied to arrays of any other geometric configuration without limitations.

allows for the flexible control of the beam pattern, ensuring the main lobe beam pattern remains optimally shaped without significant distortion.

With the LCMV algorithm, the optimum weights are selected to minimize the undesired signal power received in direction[3] $\boldsymbol{\theta}_N$, while maintaining the power gain relative to the MDB, in the intended user direction $\theta_i$. This ensures that the $i$th sub-array's response adheres to specified constraints expressed as

$$\boldsymbol{C}_i^H \boldsymbol{w}_i = \boldsymbol{d}, \tag{7.6}$$

where the response matrix $\boldsymbol{C}_i \in \mathbb{C}^{M \times K}$ for the $i$th sub-array includes array response vectors for both the desired and interfering directions, represented as

$$\boldsymbol{C}_i(\theta_1, \ldots, \theta_K) = [\mathbf{h}_i(\theta_1), \ldots, \mathbf{h}_i(\theta_K)], \tag{7.7}$$

and the target response constraint vector $\boldsymbol{d} \in \mathbb{R}^{K \times 1}$ specifies the sub-array response at each user location with expression

$$\boldsymbol{d} = [d_1, \ldots, d_K]^T. \tag{7.8}$$

In this context, $d_i = 1$ imposes a unit gain constraint at the target user's direction $\theta_i$ for the $i$th sub-array, ensuring $\boldsymbol{w}_i^H \mathbf{h}_i(\theta_i) = 1$. To minimize interference at the locations of other users, all other constraint elements $d_k = 0$, enforcing $\boldsymbol{w}_i^H \mathbf{h}_i(\theta_k) = 0$ for $k = 1, \ldots, K$ and $k \neq i$.

The closed-form solution can be derived using the Lagrange multiplier method with multiplier vector $\boldsymbol{\lambda} \in \mathbb{C}^{K \times 1}$. The Lagrangian function, incorporating the constraints, can be expressed as

$$\mathcal{L}(\boldsymbol{w}_i, \boldsymbol{\lambda}) = \boldsymbol{w}_i^H \mathbf{R}_i \boldsymbol{w}_i - 2\,\mathrm{Re}\{\boldsymbol{\lambda}^H (\boldsymbol{C}_i^H \boldsymbol{w}_i - \boldsymbol{d})\}, \tag{7.9}$$

where $\mathbf{R}_i \in \mathbb{C}^{M \times M}$ denotes the covariance matrix of the received signal for the $i$th sub-array before being combined by the beamforming vector $\boldsymbol{w}_i$. The covariance matrix encapsulates the effects of the desired signal, interference, and noise prior to beamforming.

The solution to the optimization problem can be found by taking the gradient of the Lagrangian with respect to $\boldsymbol{w}_i$ and setting it to zero, while also considering the constraints

$$\nabla_{\boldsymbol{w}_i^H} \mathcal{L} = \mathbf{R}_i \boldsymbol{w}_i - \boldsymbol{C}_i \boldsymbol{\lambda} = 0, \tag{7.10}$$

$$\boldsymbol{\lambda} = (\boldsymbol{C}_i^H \mathbf{R}_i^{-1} \boldsymbol{C}_i)^{-1} \boldsymbol{d}. \tag{7.11}$$

[3] To simplify the representation, the $1 \times (K - 1)$ angle vector $\boldsymbol{\theta}_N$ is defined as $[\theta_1, \ldots, \theta_{i-1}, \theta_{i+1}, \ldots, \theta_K]$, representing the set of interference user directions.

Substituting $\boldsymbol{\lambda}$ back into Eq. (7.10), the closed-form LCMV beamformer weights can be obtained as

$$\boldsymbol{w}_{\mathrm{lcmv}}(\theta_i, \boldsymbol{\theta}_N) = \mathbf{R}_i^{-1}\boldsymbol{C}_i(\boldsymbol{C}_i^H\mathbf{R}_i^{-1}\boldsymbol{C}_i)^{-1}\boldsymbol{d}. \tag{7.12}$$

where given the constraints on degrees of freedom in a $1 \times M$ array, the algorithm is possible to control up to a maximum of $M - 1$ arbitrary nulls.

While the LCMV utilizes spatial filtering to preserve the intensity of the intended signal and diminish interference, its effectiveness critically depends on accurately aligning the array's response with the spatial characteristics of both the desired and interfering signals. As outlined in Eq. (7.2), the desired signal channel vector and the interference channel vector are shaped by the radiation patterns of individual elements. Under the assumption of isotropic elements, the array's response is solely a function of the collective arrangement. In scenarios with isotropic and uniform ERPs, optimal beamforming vectors can be achieved with negligible deviations, resulting in deep nulls in the potential interference directions, as illustrated by the results from [23].

## 7.3 Perturbation-Based Nulling Control Beamforming

In practical scenarios, various factors contribute to heterogeneous ERPs, including inter-element MC, variations in the surrounding environments of antenna elements, and asymmetries in the dielectric substrate and ground plane, as noted in [4, 31]. As presented in Chap. 6, these heterogeneous ERPs exhibit tilted and distorted main lobes across elements within the large-scale antenna array. This variation introduces non-uniformities in the magnitude responses $\mathbf{h}_i(\theta, \phi)$ across different directions and elements, impacting the accuracy of the estimated array response and potentially causing deviations between the expected and actual spatial characteristics of the signal. The deviations in the numerical values can be amplified during the inversion of the covariance matrix $\mathbf{R}_i$ and array response matrix $\boldsymbol{C}_i$. As a result, the LCMV algorithm may produce suboptimal steering coefficients, which impair interference suppression and misalign the intended and achieved nulling angles. As demonstrated in [23], larger variations in ERPs exacerbate degradation in both nulling depth and the accuracy of nulling angle positioning.

To enhance the effectiveness of NCB while accounting for realistic ERPs, finely-tuned perturbations can be introduced to the $\boldsymbol{w}_{\mathrm{lcmv}}$, which helps to search for new optimal NCB vectors that improve the nulling performance. However, as the number of antenna elements $M$ increases, the complexity of deriving an effective beamformer rises significantly. For example, determining the optimal result in a set of $\chi$-level uniformly quantized magnitude perturbations and $\gamma$-level uniformly quantized phase perturbations for $M$-element array using exhaustive search needs $\chi^M\gamma^M$ iterations. To circumvent the prohibitive computational burden associated with exhaustive search, PSO is adopted. PSO utilizes a swarm of intelligent particles, each representing a beamforming vector candidate, which collectively explores the search domain under predefined boundaries to optimize a given objective function. The movements of the particles are guided primarily by their best-known positions in the search space (personal best) as well as the entire swarm's best-known positions (global best). The stochastic nature of PSO is especially suited for tackling non-convex optimization challenges, making it a robust tool in complex problem-solving scenarios.

The computational complexity of PSO mainly depends on three key factors in the optimization problem formulation and settings: (1) the number of particles ($N_p$): representing the total number of candidate solutions being evaluated and updated in each iteration, (2) the number of iterations ($N_n$): representing the number of times the algorithm iterates over the updating process of particles, (3) the number of elements ($M$) in the antenna array. Therefore, the complexity of the PSO for the proposed PNCB can be represented as $O(N_p N_n M)$. The scalability of PSO allows for easy adjustments in the number of particles or iterations, enabling a trade-off between optimization performance and cost-effectiveness.

Additionally, a threshold for the maximum number of steady iterations can be set in the optimization process: if the global best solution does not improve after a specified number of consecutive iterations, the optimization will stop, outputting the current global best candidate. This flexibility in setting the iteration threshold is helpful in practical scenarios where the beamforming optimization performance can be limited by realistic hardware impairments, such as phase shifter or attenuator quantization levels. This approach prevents over-exploration of the search space and ensures efficient optimization within practical constraints. In our proposed PNCB scheme, we integrate PSO to refine the beamforming vector $\boldsymbol{w}_{\mathrm{lcmv}}(\theta_i, \boldsymbol{\theta}_N)$, significantly boosting the efficiency of identifying new optimal beamforming solutions. By initializing the search with $\boldsymbol{w}_{\mathrm{lcmv}}(\theta_i, \boldsymbol{\theta}_N)$, we not only expedite the convergence process but also mitigate the risk of getting trapped in suboptimal solutions in the optimization process. Details of the proposed algorithms are presented in the following sections.

### 7.3.1 Multiple-User Nulling Control Beamforming

Let $\boldsymbol{\zeta}_p^n \in \mathbb{R}^{2M}$ denote the search vector for the $p$th particle at the $n$th iteration, defined as

$$\boldsymbol{\zeta}_p^n = [\mathrm{mag}(\boldsymbol{w}_p^n), \mathrm{ang}(\boldsymbol{w}_p^n)]^T, \tag{7.13}$$

where $\text{mag}(\boldsymbol{w}_p^n) \in \mathbb{R}^{1\times M}$ denotes the amplitudes of the beamforming weights $\boldsymbol{w}_p^n$, and the subsequent $\text{ang}(\boldsymbol{w}_p^n) \in \mathbb{R}^{1\times M}$ corresponds to the phases of $\boldsymbol{w}_p^n$. The initial search vector, derived from the LCMV beamformer, is expressed as

$$\boldsymbol{\zeta}_{\text{lcmv}}(\theta_i, \boldsymbol{\theta}_N) = [\text{mag}(\boldsymbol{w}_{\text{lcmv}}(\theta_i, \boldsymbol{\theta}_N)), \text{ang}(\boldsymbol{w}_{\text{lcmv}}(\theta_i, \boldsymbol{\theta}_N))]^T. \tag{7.14}$$

The proposed PNCB optimization algorithm aims to minimize beam radiation towards the $K-1$ interference user directions $\boldsymbol{\theta}_N$, and maintain the directivity towards the intended user's direction $\theta_i$. Therefore, the objective cost function to be minimized can be defined as

$$\Omega(\boldsymbol{\zeta}_p^n) = \boldsymbol{\alpha}_1 \text{dir}(\boldsymbol{\zeta}_p^n, \boldsymbol{\theta}_N) - \alpha_2 \text{dir}(\boldsymbol{\zeta}_p^n, \theta_i) + Q, \tag{7.15}$$

where the $\boldsymbol{\alpha}_1 \in \mathbb{R}^{1\times(K-1)}$ and the $\alpha_2 \in \mathbb{R}$ are arbitrary constant vector and scalar, and $\boldsymbol{\alpha}_1 > \mathbf{0}_{(K-1)}$ and $\alpha_2 > 0$. The function $\text{dir}(\boldsymbol{\zeta}_p^n, \boldsymbol{\theta})$ denotes the normalized evaluated directivity values at the angle $\boldsymbol{\theta}$ for a given beamforming search vector $\boldsymbol{\zeta}_p^n$ at the $p$th particle during the $n$th iteration. The outputs of the function dir(.) have the same dimension as the input variable $\boldsymbol{\theta}$. For example, $\text{dir}(\boldsymbol{\zeta}_{\text{lcmv}}, \theta_i) \in \mathbb{R}$ and the vector $\text{dir}(\boldsymbol{\zeta}_{\text{lcmv}}, \boldsymbol{\theta}_N) \in \mathbb{R}^{K-1}$ give the baseline of the directivity in the target angle and the arbitrary nulling levels when the LCMV beamformer is used. Hence, in the cost function, an increase in target directivity or a decrease in radiation power in the interference directions both contribute to reducing the overall cost. In the search process, $N_p$ particles are used in each iteration and total $N_n$ iterations are considered.

A penalty factor $Q$ is implemented to assess the efficacy of the current search vector $\boldsymbol{\zeta}_p^n$ relative to the initial $\boldsymbol{\zeta}_{\text{lcmv}}$ and to guarantee that the directivity towards the intended user remains within a predefined directivity reduction threshold, denoted by $\epsilon$. This assessment can be formulated as

$$\text{Condition1:}\ \text{dir}(\boldsymbol{\zeta}_p^n, \boldsymbol{\theta}_N) < \text{dir}(\boldsymbol{\zeta}_{\text{lcmv}}(\theta_i, \boldsymbol{\theta}_N), \boldsymbol{\theta}_N), \tag{7.16}$$

$$\text{Condition2:}\ \text{dir}(\boldsymbol{\zeta}_p^n, \theta_i) \geq \text{dir}(\boldsymbol{\zeta}_{\text{lcmv}}(\theta_i, \boldsymbol{\theta}_N), \theta_i) - \epsilon, \tag{7.17}$$

and based on condition (7.16) and (7.17), the $Q$ has the values

$$Q = \begin{cases} 0, & \text{if both conditions are met,} \\ Q_1, & \text{if one of the two conditions is met,} \\ Q_2, & \text{if neither of the conditions are met,} \end{cases} \tag{7.18}$$

where $0 < Q_1 < Q_2$ to differentiate between the fulfillment of the conditions.

Then, the optimization problem for the proposed PNCB can be formulated as

$$\begin{aligned}
&\min_{\hat{\boldsymbol{\zeta}}} \quad \Omega(\hat{\boldsymbol{\zeta}}),\\
\text{s.t.} \quad &\text{Constraint 1: } \forall q \in \text{mag}(\hat{\boldsymbol{w}}), 0 \leq q \leq 1,\\
&\text{Constraint 2: } \text{ang}(\hat{\boldsymbol{w}}) \geq \text{ang}(\boldsymbol{w}_{\text{lcmv}}(\theta_i, \boldsymbol{\theta}_N)) - \Delta p,\\
&\quad \text{and } \text{ang}(\hat{\boldsymbol{w}}) \leq \text{ang}(\boldsymbol{w}_{\text{lcmv}}(\theta_i, \boldsymbol{\theta}_N)) + \Delta p,
\end{aligned} \tag{7.19}$$

where Constraint 1 establishes the search boundary for the magnitude of each element in the beamforming vector, limiting it to the range [0, 1]. Constraint 2 delineates the phase perturbation limits, with $\Delta p$ specifying the maximum allowable variation from the initial LCMV beamformer phase $\text{ang}(\boldsymbol{w}_{\text{lcmv}}(\theta_i, \boldsymbol{\theta}_N))$. $\Delta p$ is determined based on the maximum phase alteration that results in the largest acceptable mainlobe steering angle change.

In the $n$th iteration, the personal best beamforming vector $\boldsymbol{\zeta}^n_{\text{best},p}$ for the $p$th particle and the current global best beamforming vector $\boldsymbol{\zeta}^n_{\text{best}}$ among all the iterated particles can be expressed as

$$\boldsymbol{\zeta}^n_{\text{best},p} = \underset{\boldsymbol{\zeta}_p^{(n*)}, \forall n*=0,1,\cdots,N_n}{\arg\min} \quad \Omega\left(\boldsymbol{\zeta}_p^{(n*)}\right), \tag{7.20}$$

$$\boldsymbol{\zeta}^n_{\text{best}} = \underset{\boldsymbol{\zeta}^n_{\text{best},p}, \forall p=1,\cdots,N_p}{\arg\min} \quad \Omega\left(\boldsymbol{\zeta}^n_{\text{best},p}\right). \tag{7.21}$$

The shifting factor at $n$th iteration and $p$th particle $\boldsymbol{\delta}^n_p \in \mathbb{C}^{2M}$ is dynamically updated based on the value of the $\boldsymbol{\zeta}^n_{\text{best},p}$ and $\boldsymbol{\zeta}^n_{\text{best}}$. This update is crucial for modulating the beamforming vector's rate of change, enhancing convergence probability. The shifting factor initializes as a zero-vector. The update can be expressed as

$$\boldsymbol{\delta}_p^{(n+1)} = \mathbf{J}\boldsymbol{\delta}^n_p + \mathbf{K}(\boldsymbol{\zeta}^n_{\text{best},p} - \boldsymbol{\zeta}^n_p) + \mathbf{L}(\boldsymbol{\zeta}^n_{\text{best}} - \boldsymbol{\zeta}^n_p), \tag{7.22}$$

where $\mathbf{K}, \mathbf{L}$ are positive random diagonal matrices for self-adjustment and social adjustment, respectively. These matrices govern particle interactions and the tendency of each particle to move towards its personal best and global best positions, respectively. The $\mathbf{J}$ denotes the diagonal inertia weight matrix, essential for maintaining a balance between exploration and exploitation while searching for the optimal solution. Subsequently, the beamforming search vector will be updated by

$$\boldsymbol{\zeta}_p^{(n+1)} = \text{clip}(\boldsymbol{\delta}_p^{(n+1)} + \boldsymbol{\zeta}^n_p, \boldsymbol{\zeta}_L, \boldsymbol{\zeta}_U), \tag{7.23}$$

**Algorithm 7.1:** PNCB Scheme

**Input**: $\theta_i, \boldsymbol{\theta}_N, \epsilon$.
**Output**: $\hat{\boldsymbol{\zeta}} = \boldsymbol{\zeta}^n_{\text{best}}$
Define the array based on $\boldsymbol{A}_{i,1}, \boldsymbol{A}_{i,2}, \ldots, \boldsymbol{A}_{i,M}$, and the array configuration.
**for** $n$ =*0:*$N_n$ **do**
  **for** $p$=*1:*$N_p$ **do**
    **if** $n$=*0* **then**
      Initialize the shifting factor $\boldsymbol{\delta}^{(0)}_p \leftarrow 0$.
      **if** $p$=*1* **then**
        Initialize $\boldsymbol{\zeta}^n_p \leftarrow \boldsymbol{\zeta}_{\text{lcmv}}(\theta_i, \boldsymbol{\theta}_N)$ based on (7.12) and (7.14).
      **else**
        Initialize $\boldsymbol{\zeta}^n_p$ uniformly distributed in the search region $[\boldsymbol{\zeta}_L, \boldsymbol{\zeta}_U]$.
      **end**
    **else**
      Update shifting factor $\boldsymbol{\delta}^n_p$ using (7.22). Update vector $\boldsymbol{\zeta}^n_p$ using (7.23).
    **end**
    Update penalty factor $Q$ via (7.16), (7.17), (7.18)
    and evaluate cost function $\Omega(\boldsymbol{\zeta}^n_p)$ via (7.15).
    Find the personal best $\boldsymbol{\zeta}^n_{\text{best},p}$ using (7.20).
  **end**
  Find the global best $\boldsymbol{\zeta}^n_{\text{best}}$ using (7.21).
**end**

with $\boldsymbol{\zeta}_L, \boldsymbol{\zeta}_U \in \mathbb{R}^{2M}$ defining the lower-bound and upper-bound of perturbations, respectively. $\boldsymbol{\zeta}_L$ and $\boldsymbol{\zeta}_U$ are derived from Constraints 1 and 2 in expression (7.19). The clip function rigorously enforces the specified bounds by nudging the perturbed search vector's magnitudes and phases toward their closest boundary values whenever they stray outside the predefined search area. If the shifting factor propels the updated search vector beyond these limits, it is reined into zero, ensuring that the search remains confined to the feasible region. The detailed steps of the algorithm are outlined in Algorithm 7.1.

### 7.3.2 *Nulling-Width Control Beamforming*

Interference in communication systems often exhibits dynamic spatial characteristics due to factors like changing propagation paths and the relative motion between receivers and interference sources. This non-stationary nature can cause the interference direction of arrival change, potentially missing the arbitrary beamforming nulls. Such misalignment can significantly impair the nulling ability to suppress interference, highlighting the need for null widening in scenarios where pinpointing the precise direction of interference is challenging. By widening the nulls, a safeguard is established against the unpredictable nature of interference directions, thereby bolstering the system's resilience. This strategy also proves advantageous in signal-dense environments where interference might originate from several proximate directions. Broadened nulls ensure extensive coverage, effectively mitigating interference from a wide angular spectrum, and thus preserving the integrity of the communication channel.

In the proposed PNCB framework, we introduce a strategic method to enlarge the nulling width by creating a sequence of adjacent nulls around the target nulling angle $\theta_k$, where $k \in \{1, 2, \ldots K\}$ and $k \neq i$. This is achieved by setting a predetermined nulling width $\theta_W$ and creating a series of adjacent nulls in the range. Centered around $\theta_k$, the expansion of nulling angles is formulated as

$$\begin{aligned}\boldsymbol{\theta}'_k = [\theta_k - \frac{\theta_W}{2}, \theta_k - \frac{\theta_W}{2} + \Delta\theta, \theta_k - \frac{\theta_W}{2} + 2\Delta\theta, \ldots, \theta_k + \frac{\theta_W}{2}], \\ k = 1, \ldots, K, \text{and } k \neq i,\end{aligned} \tag{7.24}$$

where $\Delta\theta$ represents the angular increment between consecutive nulls. Therefore, this expansion leads to the enhancement of the original nulling direction vector, $\boldsymbol{\theta}_N$, which encapsulates the central angular positions of the targeted nulls. By incorporating multiple proximate nulls, the expanded nulling direction vector is represented as

$$\boldsymbol{\theta}'_N = [\boldsymbol{\theta}'_1, \ldots, \boldsymbol{\theta}'_{(i-1)}, \boldsymbol{\theta}'_{(i+1)}, \ldots, \boldsymbol{\theta}'_K], \tag{7.25}$$

thereby broadening the nulling coverage to more effectively counteract interference from various directions. For an array with $M$ degree of freedom, the size of the expanded nulling angle vector $\boldsymbol{\theta}'_N$ must be equal to or less than $M-1$. Similar to the previous section, the LCMV generated $\boldsymbol{\zeta}_{\text{lcmv}}(\theta_i, \boldsymbol{\theta}'_N)$, is used as the beamforming vector at the initial search point.

To thoroughly assess nulling performance across the entire target angular span, we utilize the geometric mean of directivity ($\text{dir}_{GM}(.)$) within the target nulling width. The nulling depth geometric mean function can be expressed as

$$\text{dir}_{GM}(\boldsymbol{\zeta}^n_p, \theta_k, \theta_W) = \frac{1}{2N_S+1}\sum_{q=-N_S}^{q=+N_S} \text{dir}(\boldsymbol{\zeta}^n_p, \theta_k + q\frac{\theta_W}{2N_S}), k=1, \ldots, K, \text{and } k \neq i \tag{7.26}$$

where $(2N_S + 1)$ represents the predefined total number of angular sample power levels evaluated within the nulling range $[\theta_k - \frac{\theta_W}{2}, \theta_k + \frac{\theta_W}{2}]$ for the nulling depth evaluation. Subsequently, the collective geometric mean nulling depth vector, which is output of $\text{dir}_{GM}(\boldsymbol{\zeta}^n_p, \boldsymbol{\theta}_N, \theta_W) \in \mathbb{R}^{K-1}$ across all the target nulling angles within the vector $\boldsymbol{\theta}_N$, can be represented as

$$\begin{aligned}\text{dir}_{GM}(\boldsymbol{\zeta}^n_p, \boldsymbol{\theta}_N, \theta_W) = [&\text{dir}_{GM}(\boldsymbol{\zeta}^n_p, \theta_1, \theta_W), \ldots, \text{dir}_{GM}(\boldsymbol{\zeta}^n_p, \theta_{(i-1)}, \theta_W), \\ &\text{dir}_{GM}(\boldsymbol{\zeta}^n_p, \theta_{(i+1)}, \theta_W), \ldots, \text{dir}_{GM}(\boldsymbol{\zeta}^n_p, \theta_K, \theta_W)].\end{aligned} \tag{7.27}$$

**Algorithm 7.2:** PNCB Scheme for Null-width Control

**Input**: $\theta_i$,$\boldsymbol{\theta}_N$, $\theta_W$, $\epsilon$.
**Output**: $\hat{\boldsymbol{\zeta}} = \boldsymbol{\zeta}^n_{\text{best}}$
Define the array based on $\boldsymbol{A}_{i,1}, \ldots, \boldsymbol{A}_{i,M}$, and the array configuration.
Calculate the expanded nulling vector $\boldsymbol{\theta}'_N$ using Eqs. (7.24) and (7.25).
**for** $n$ =0:$N_n$ **do**
  **for** $p$=1:$N_p$ **do**
    **if** $n$=0 **then**
      Initialize the shifting factor $\boldsymbol{\delta}^{(0)}_p \leftarrow 0$.
      **if** $p$=1 **then**
        Initialize $\boldsymbol{\zeta}^n_p \leftarrow \boldsymbol{\zeta}_{\text{lcmv}}(\theta_i, \boldsymbol{\theta}'_N)$ based on (7.12) and (7.14).
      **else**
        Initialize $\boldsymbol{\zeta}^n_p$ uniformly distributed in the search region $[\boldsymbol{\zeta}_L, \boldsymbol{\zeta}_U]$.
      **end**
    **else**
      Update shifting factor $\boldsymbol{\delta}^n_p$ using (7.22). Update vector $\boldsymbol{\zeta}^n_p$ using (7.23).
    **end**
    Update penalty factor $Q$ via (7.28), (7.29), (7.18)
    and evaluate cost function $\Omega(\boldsymbol{\zeta}^n_p)$ via (7.30).
    Find the personal best $\boldsymbol{\zeta}^n_{\text{best},p}$ using (7.20)
  **end**
  Find the global best $\boldsymbol{\zeta}^n_{\text{best}}$ using (7.21)
**end**

Therefore, the nulling depth comparison condition in (7.16) can be modified as

$$\text{Condition1: } \text{dir}_{GM}(\boldsymbol{\zeta}^n_p, \boldsymbol{\theta}_N, \theta_W) < \text{dir}_{GM}(\boldsymbol{\zeta}_{\text{lcmv}}(\theta_i, \boldsymbol{\theta}'_N), \boldsymbol{\theta}_N, \theta_W), \tag{7.28}$$

$$\text{Condition2: } \text{dir}(\boldsymbol{\zeta}^n_p, \theta_i) \geq \text{dir}(\boldsymbol{\zeta}_{\text{lcmv}}(\theta_i, \boldsymbol{\theta}'_N), \theta_i) - \epsilon. \tag{7.29}$$

The cost function for the $p$th particle and $n$th instance in the PSO will be evaluated based on the $\text{dir}_{GM}(\boldsymbol{\zeta}^n_p, \boldsymbol{\theta}_N, \theta_W)$ and can be expressed as

$$\Omega(\boldsymbol{\zeta}^n_p) = \boldsymbol{\alpha}_{w,1}\text{dir}_{GM}(\boldsymbol{\zeta}^n_p, \boldsymbol{\theta}_N, \theta_W) - \alpha_{w,2}\text{dir}(\boldsymbol{\zeta}^n_p, \theta_i) + Q_w, \tag{7.30}$$

where the predefined vector $\boldsymbol{\alpha}_{w,1} \in \mathbb{R}^{1\times(K-1)}$ and scalar $\alpha_{w,2}$ are arbitrary scalar constants, satisfying $\boldsymbol{\alpha}_{w,1} > \mathbf{0}_{(K-1)}$ and $\alpha_{w,2} > 0$. The penalty factor $Q_w$ will be determined based on (7.18) and condition (7.28) and (7.29). The detailed steps of the algorithm are outlined in Algorithm 7.2.

## 7.4 Simulation Results of Beam Nulling with Measured Element Radiation Pattern

In this section, measured ERPs $\boldsymbol{A}_m$ at 3.5 GHz in Sect. 6.5.3, are used in beamforming simulations to define the customized antenna array with element spacing 4 cm, which has a consistent layout as our antenna array prototype. In the simulation, 1×8B sub-array is used, as shown in Fig. 6.7. MATLAB with the Phased Array Toolbox and Adaptive Beamformer Toolbox is used in the simulation. The Algorithms 7.1 and 7.2 are also implemented in MATLAB. The threshold of the directivity degradation in the intended user direction $\epsilon$ is set to 3 dB. The comparison and analysis encompass three distinct beamforming approaches:

1. **MDB**: This scenario employs a maximum-directivity beamformer, defined by the beamforming formulation in Eq. (7.4). This approach serves as a baseline, focusing on maximizing directivity without implementing nulling techniques.
2. **LCMV**: This setup employs the LCMV scheme to compute the beamformer with the assumption of isotropic ERPs. The process, as detailed in Eq. (7.12), aims to introduce radiation nulls in specific directions while preserving the desired signal strength.
3. **PNCB**: This scenario incorporates our proposed perturbation-based NCB approach, generating vectors through the iterative process detailed in Algorithms 7.1 or 7.2. The integration of LCMV with PSO aims to improve the beamforming vectors, optimizing the nulling depth with an accurate nulling direction for improved interference mitigation.

### *7.4.1 Nulling Control Beamforming for Multiple Users*

To assess the effectiveness of the proposed PNCB technique in MU environments, a four-user scenario with one desired user and three potential interference users was simulated, where the four users are located at $-35°, -10°, 10°$, and $35°$, respectively. The users are separated with an angular difference greater than 20°, which is more than half the amount of the first-null beamwidth[4] of 1×8 array, to avoid the nulling process impacting on the main lobe beam pattern. Figure 7.2 presents the simulation results of the normalized beam radiation patterns, w.r.t. mainlobe peak values, showcasing the four combinations of beam steering and nulling angles.

[4] The measurement results from the sample 1×8 arrays show a mainlobe (first-null) beamwidth of $BW_{FN} = 30°$.

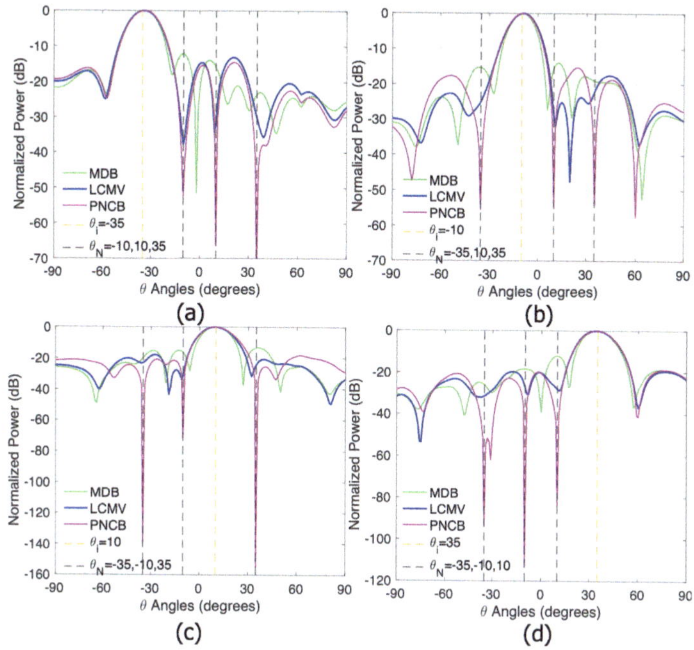


**Fig. 7.2** 1 × 8 sub-array nulling beamforming $\theta$-cuts for the four-user case, with the desired user located at (**a**) − 35°, (**b**) − 10°, (**c**)10°, (**d**)35° (simulation results)

#### 7.4.1.1 No Nulling or LCMV Scheme

With an MDB strategy, an average normalized radiation power level of −15.7 dB across the interference directions is observed, with a power level ranging from −12.1 to −26.1 dB. Across the tested scenarios, an average directivity of 14.3 dB and an average 3dB-beamwidth of 15.4° are achieved, respectively. When the LCMV approach is applied, only marginal improvement in terms of interference mitigation can be observed. The level of achieved arbitrary nulls ranged from −22.0 to −37.6 dB. An average nulling value of −27.6 dB can be achieved, indicating an average of 11.9 dB deeper (better) nulls compared with the no-nulling case. Under the LCMV approach, the achieved nulling angles showed an average deviation of 2.5° from the intended nulling directions, with a maximum angle deviation of 8°.

#### 7.4.1.2 Proposed PNCB Scheme

The proposed PNCB scheme lowers the power radiating in the potential interference direction notably. The achieved null levels are consistently lower than −52.4 dB, with deviations from the target nulling angles of less than 0.5° in all four evaluated scenarios. The results indicate a 40.3 and 30.4 dB interference mitigation enhancement in the worst cases compared with the no-nulling case and the LCMV scheme, respectively. The achieved null levels impressively range from −52.4 to −154.9 dB. Moreover, compared to the MDB case and the LCMV scheme, the PNCB approach realized nulls that are, on average, 53.3 and 41.4 dB lower (better) at the angles of potential interference, respectively. These deep nulls provide a substantial margin for practical implementation considerations, such as losses due to phase shifter and attenuator quantization.

Additionally, with the PNCB scheme, an average directivity and 3dB-beamwidth of 13.9 dB and 16.9° can be achieved, respectively. Simulations show that the PNCB approach results in a slight increase in the 3dB-beamwidth, ranging from 0.6° to 2.2°, compared to beam patterns using the MDB scheme. Meanwhile, the impact on the main lobe directivity is minimal, with less than 0.6 dB degradation observed in all test cases. This demonstrates the PNCB scheme's efficacy in maintaining robust main lobe characteristics while significantly improving nulling performance.

### *7.4.2 Nulling Width Optimization*

To assess the effectiveness of the proposed PNCB scheme in widening the radiation nulls, simulations were carried out utilizing a 1×8B sub-array with Algorithm 7.2. The simulation framework was designed with a steering angle set at $\theta_i = 0°$ and iterative nulling center $\theta_N \in [-50°, -40°, -30°, -20°, 20°, 30°, 40°, 50°]$. The target nulling width, $\theta_W = 6°$, approximately equates to half the 3dB-beamwidth of the 1×8B antenna array configuration. The angular increment between consecutive nulls $\Delta\theta = 3°$ is used in the simulation. The simulation results are shown in Fig. 7.3 with details of the average[5] and the highest (worst) normalized nulling level within the nulling angular span from $\theta_N - \frac{\theta_W}{2}$ to $\theta_N + \frac{\theta_W}{2}$ summarized in Table 7.1.

Without any nulling considerations, an average radiation power level of −22.1 dB can be observed, with the normalized nulling levels fluctuating between −15.6 and −26.6 dB in the target nulling range. When the LCMV scheme is employed, an average nulling level of −28.7 dB can be achieved in the target 6-degree nulling width. Specifically, the nulling level ranges from −24.9 to −34.0 dB. The LCMV methodology, influenced by discrepancies in the heterogeneous ERPs, yields only a marginal enhancement in potential interference mitigation. Compared to the results based on the MDB, the LCMV scheme improves the average nulling depth by 6.6 dB. This modest improvement underscores the LCMV approach's challenges in fully counteracting the inconsistencies inherent in ERPs, highlighting its limitations in achieving optimal nulling efficacy.

Meanwhile, the proposed PNCB strategy markedly enhances the nulling depth with an accurate nulling width and direction control. Across test scenarios, the

[5] This study utilizes arithmetic means to characterize the average nulling depth within an angular span.

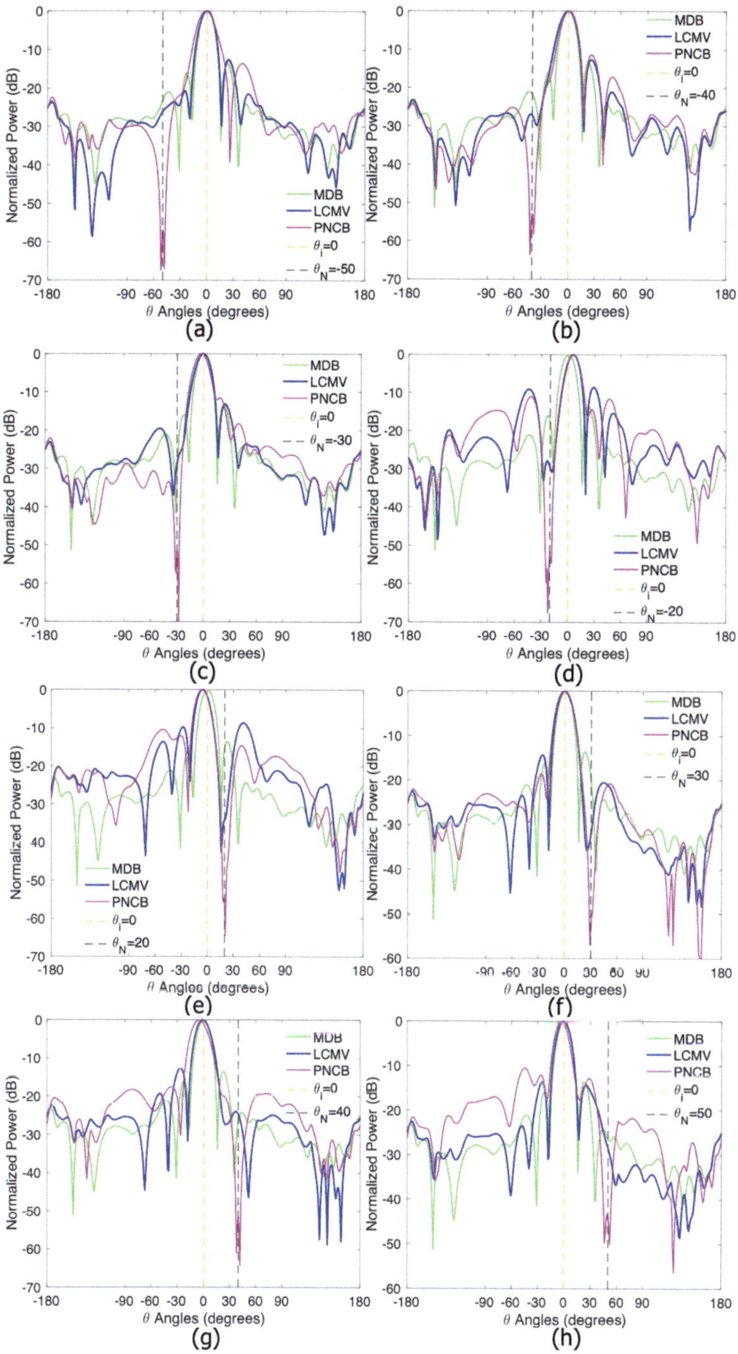


**Fig. 7.3** 1×8B sub-array nulling control beamforming $\theta$ cuts with steering angle $\theta_i = 0°$, target nulling width $\theta_W = 6°$ and nulling center $\theta_N$ at (**a**) $-50°$, (**b**) $-40°$, (**c**) $-30°$, (**d**) $-20°$, (**e**) $20°$, (**f**) $30°$ , (**g**) $40°$ , (**h**) $50°$ (simulation results)

**Table 7.1** Simulation results with nulling width optimization

| | Average nulling level (dB) | | | Worst nulling level (dB) | | |
|---|---|---|---|---|---|---|
| $\theta_N$ | MDB | LCMV | PNCB | MDB | LCMV | PNCB |
| − 50° | −23.2 | −26.2 | −57.7 | −21.8 | −25.4 | −53.4 |
| − 40° | −21.8 | −27.5 | −53.7 | −21.0 | −26.9 | −49.5 |
| − 30° | −25.6 | −28.6 | −50.5 | −20.3 | −26.1 | −46.4 |
| − 20° | −18.1 | −29.2 | −48.2 | −16.0 | −27.8 | −40.6 |
| 20° | −15.6 | −34.0 | −49.0 | −13.5 | −31.0 | −43.0 |
| 30° | −19.3 | −30.8 | −46.8 | −15.8 | −27.4 | −41.0 |
| 40° | −26.6 | −24.9 | −51.4 | −24.3 | −24.0 | −46.9 |
| 50° | −26.1 | −28.4 | −45.3 | −25.0 | −26.6 | −43.4 |

average nulling levels ranging from −45.3 to −57.7 dB with a 6-degree nulling width are achieved. Compared to MDB and LCMV schemes, the PNCB approach yields average improvements of 28.3 and 21.6 dB, respectively. When comparing the limit (highest nulling values) in the target nulling span, the proposed PNCB scheme achieves suppression values better than 40.6 dB, which indicates a nulling depth improvement of 27.1 and 16.7 dB than the MDB and LCMV schemes, respectively. The results highlight the proficiency of the PNCB scheme in attenuating interference from wide directions, demonstrating its robustness against fluctuations in interference directionality.

With the PNCB scheme, an average directivity of 13.1 dB and a 3dB-beamwidth of 15.9° can be observed. Additionally, the normalized sidelobe levels consistently remain below −10.4 dB. Moreover, the consistency observed in the main lobe patterns between the MDB and PNCB schemes suggests that the proposed nulling beamforming approach has a negligible impact on the radiation patterns in the target direction. Specifically, when comparing the beam patterns using the PNCB scheme with the beam patterns using the MDB scheme, we observe an average directivity decrease of 1.7 dB and an average 3dB-beamwidth increase of 2.1°. This finding attests to the nulling strategies' ability to maintain the structural integrity of the main lobes while effectively counteracting interference, thereby ensuring optimal signal transmission and reception.

### *7.4.3 Element Symmetry on the Beam Nulling Optimization*

As presented in Sect. 6.5.4, antenna elements located at diagonally symmetric locations on a large-scale array perform symmetric EPRs. Leveraging the symmetry of ERPs in an array, the heterogeneous ERPs can be grouped, effectively reducing the number of unique ERPs to half the total number of elements. In the case of our 8×8 antenna array prototype, this principle allows the 64 elements to be efficiently paired into 32 pairs. The ERP symmetry information can also help create comparable sub-arrays, with elements having identical radiation patterns towards opposite $\theta$ and $\phi$ directions. For symmetric sub-arrays which are composed of elements positioned symmetrically within the array, the $i$th array response vector $\mathbf{h}_i$ and its symmetric sub-arrays channel vector $\mathbf{h}_i^{sym}$ have the relationship

$$\mathbf{h}_i(\theta, \phi) = \mathbf{h}_i^{sym}(-\theta, -\phi), \tag{7.31}$$

which demonstrates the symmetric sub-arrays possess nearly identical SVs towards opposite directions.

The ERP classification can significantly enhance the efficiency of beamforming analysis and the development of optimization techniques, facilitating the reuse of the optimal beamforming vector. For a given sub-array characterized by a steering angle at $\theta_i$ and nulling angles at $\boldsymbol{\theta}_N$, the optimized beamforming vector $\hat{\boldsymbol{w}}(\theta_i, \boldsymbol{\theta}_N)$ can be directly applied to its symmetric sub-array to generate a mirrored beam pattern. This mirrored pattern achieves the main lobe steering and nulling in opposite orientations. Consequently, incorporating the optimized beamforming vector $\hat{\boldsymbol{w}}(\theta_i, \boldsymbol{\theta}_N)$ into the symmetric sub-array during the NCB optimization process enhances computational efficiency while preserving beamforming and nulling performance, making it a valuable approach for large-scale array beamformer designs. The relationship between the optimized beamforming vector for symmetric sub-array pairs can be represented as

$$\hat{\boldsymbol{w}}^{sym}(-\theta_i, -\boldsymbol{\theta}_N) = \hat{\boldsymbol{w}}(\theta_i, \boldsymbol{\theta}_N). \tag{7.32}$$

To explore the potential benefits of exploiting ERP symmetry within the PNCB optimization process, we engaged two symmetric sub-arrays, 1×8B and 1×8A which are the fourth- and fifth- rows in the 8 × 8 array as shown in Fig. 6.7, for our simulations. We applied Algorithm 7.1 to determine the optimal PNCB vector for the 1×8B sub-array, which was then applied to its symmetric sub-array, the 1×8A sub-array. The simulations were conducted with steering angles $\theta_i \in [-30°, -20°, -10°, 30°]$ and a fixed nulling angle $\theta_N = 5°$. A distance greater than 15° is kept between the steering angle and the designated nulling angle to avoid the nulling process impacting the main lobe radiation patterns.

The simulation results for the symmetric sub-array pair 1×8B and 1×8A radiation pattern $\theta$ cuts are depicted in Fig. 7.4. By applying the optimal PNCB beamforming vector, the symmetric sub-array 1×8A can achieve comparable but mirrored beam patterns compared with the 1×8B. When comparing the radiation power in the mainlobe and sidelobe, a comparable magnitude and beamwidth at the opposite angular locations can be achieved. The results are consistent with the analysis illustrated in Sect. 6.5.4. In terms of nulling efficacy, the symmetric sub-array 1 × 8A exhibits pronounced nulls at precise locations, with nulling levels ranging between −41.3 and −51.0 dB. The angular difference between the realized and intended nulling direction is smaller than 1°, indicating high accuracy in null placement.

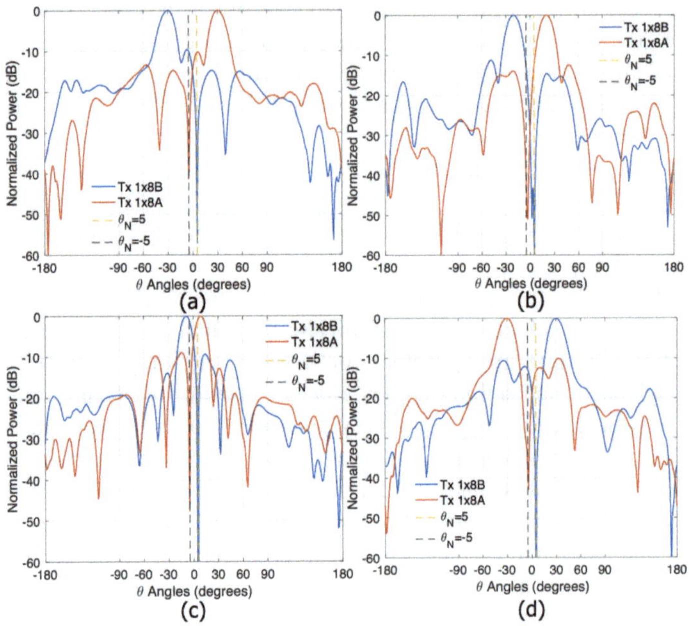


**Fig. 7.4** Symmetric sub-array 1×8B and 1×8A with nulling angle $\theta_N = 5°$ and steering angle (**a**) $\theta_i = -30°$, (**b**) $\theta_i = -20°$, (**c**) $\theta_i = -10°$, (**d**) $\theta_i = 30°$ (simulation results)

At the same time, a reduction in the nulling depth for symmetric sub-arrays has been noted. This degradation in nulling efficacy is attributed to the small difference existing in the radiation patterns of symmetric elements. These discrepancies are a result of the inherent limitations in the precision of ERP measurements and the manufacturing tolerances of the antenna array. Such factors introduce slight imperfections in the optimal PNCB vector, subtly impacting its performance. In scenarios that require additional refinement of the beamforming vector performance, utilizing the optimal beamforming vectors from symmetric sub-arrays as initial vectors for optimization can potentially lead to considerable time savings in the process.

### 7.4.4 *Phase-Shifter and Attenuator Quantization Influence on the Beam Nulling Depth Measurements*

The quantization levels in phase shifters and attenuators constrain beamforming precision, directly impacting both the accuracy of null placement and the achievable null depth. In the previous sections, simulations assumed ideal, infinitely precise phase shifters and attenuators within the beamforming network. This section delves

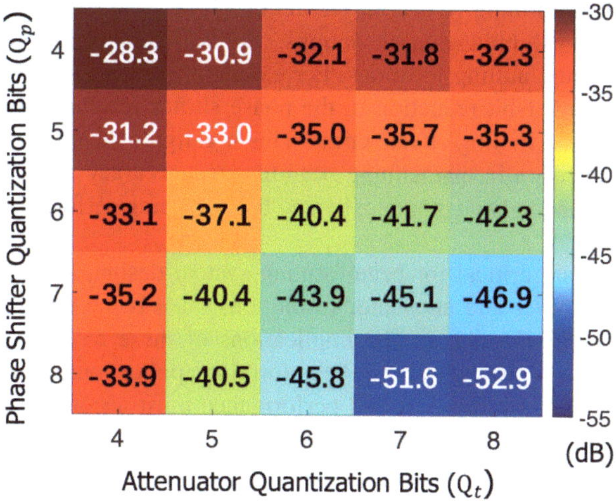


**Fig. 7.5** Average achievable nulling level (dB) versus the number of phase shifter and attenuator quantization bits (simulation results)

into simulations conducted with the 1×8B sub-array, incorporating practical $Q_p$-bit phase shifters, and $Q_t$-bit attenuators. For an affordable iterative simulation time and insightful comparison, we focus on a single arbitrary nulling scenario. Twenty randomly chosen $(\theta_i, \theta_N)$ pairs are considered, where $|\theta_i - \theta_N| > 15°$ to prevent interference from the nulling process on the main lobe pattern. The quantization levels $Q_p$ and $Q_t$ are in the range from 4 to 8. The average achievable nulling level across the 20 tested cases for each $(Q_p, Q_t)$ pair is presented in Fig. 7.5, offering insights into the impact of quantization levels on the effectiveness of nulling in beamforming applications.

The results highlight a significant correlation between the reduction in quantization bits and the deterioration of the achievable null depth in the beamforming process. As the resolution of phase shifters $Q_p$ and attenuators $Q_t$ decreases, we observe a marked degradation in nulling performance. Specifically, reducing the quantization levels from $Q_p$=$Q_t$=8 to $Q_p$=$Q_t$=4 leads to a substantial decrease in nulling depth, averaging a 24.6 dB loss. The rate of decline in nulling depth diminishes as $Q_p$ and $Q_t$ decrease. With a 7-bit attenuator and a 6-bit phase shifter,[6] the normalized null level experienced a drop from −52.9 to −41.7 dB, reflecting a decrease of nearly 11.2 dB.

Additionally, the impact of reducing $Q_p$ is more pronounced than that of $Q_t$. Phase shifters play a pivotal role in orchestrating phase alignment across antenna elements, enabling the constructive and destructive interference required

[6] The available attenuator and phase shifter setup for our BFN in the Lab and experimental results will be presented in the following section.

for effective beamforming. When the resolution of the phase shifters decreases, the beam pattern can exhibit angular deviations, resulting in misalignment between the intended and actual nulling locations. For example, in the range where $Q_p \geq 5$ and $Q_t \geq 7$, each 1-bit reduction in the phase shifter resolution leads to a 3.4–7 dB degradation in nulling effectiveness. This highlights the critical role of phase quantization precision in maintaining beamforming accuracy and underscores the trade-offs between hardware complexity and beamforming performance in practical implementations.

The findings are critical for beamforming system design, as they suggest that both the phase shifter and attenuator resolutions must be carefully considered to maintain nulling performance. The implications of these results are particularly relevant when selecting hardware for beamforming applications, indicating that the trade-off between hardware cost and performance must account for the potential degradation in nulling effectiveness due to quantization. This balance is essential for ensuring that beamforming systems can reliably achieve the desired signal integrity and interference suppression in real-world operations.

## 7.5 Experimental Measurements

To assess the effectiveness of the proposed PNCB scheme in mitigating MUI, experimental measurements were carried out in the anechoic chamber equipped with C-RAM SFC-48 absorbers. The placement of the 8×8 antenna array prototype, along with the two monopole antenna probes in the chamber, is depicted in Fig. 7.6. Each monopole antenna probe, connected to either a vector signal generator (VSG) or a spectrum analyzer (SA), represents a UE. The two fixed monopole antenna probes are separated with an angular separation of 22°, i.e., $|\theta_i - \theta_N| = 22°$, and have the same distance from the antenna array center. The 8×8 array is affixed to a motorized positioner capable of $\theta$-angle rotation, allowing for precise adjustments of the array's orientation relative to the probes, thereby enabling a thorough evaluation of the PNCB scheme's capability of interference mitigation across various $(\theta_i, \theta_N)$ combinations.

The sample 1×8B antenna sub-array, integrated with BFN, was employed in the measurements. The BFN setup comprises 8 digitally controlled 6-bit phase shifters, which allow for full 360° phase adjustment in 5.625° increments, and 8 7-bit digitally controlled attenuators, providing up to 32 dB of attenuation in fine adjustment steps of 0.25 dB. To ensure efficiency and a compact arrangement, these components are interconnected using short coaxial cables and compact JSC connectors and are conveniently located behind the antenna arrays. The BFN is managed through a customized SPI module, seamlessly integrated with a Raspberry Pi 4. Additionally, a PC equipped with a MATLAB-based control program interfaces with the SPI module through a USB cable connection, facilitating the generation of the beamforming vectors based on the $(\theta_i, \theta_N)$ and controlling the BFN.

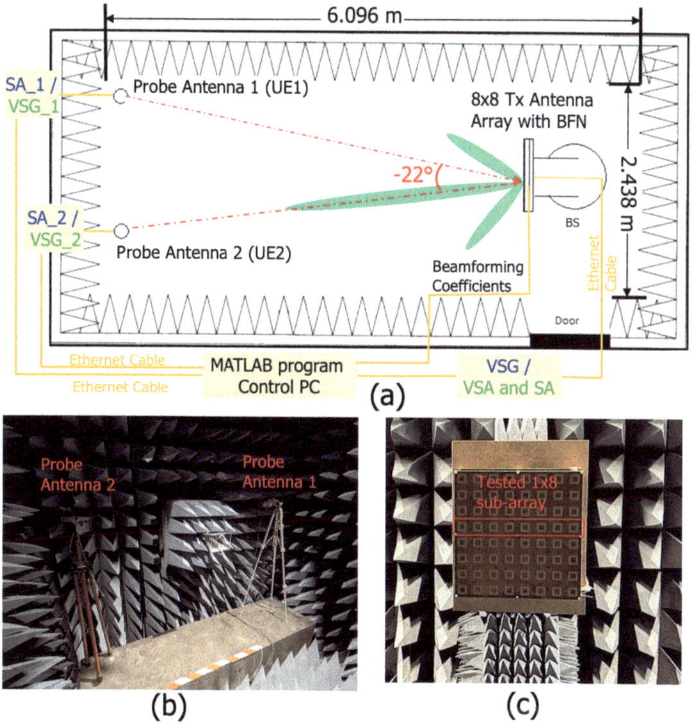


**Fig. 7.6** Experimental setup of the proposed nulling control beamforming scheme (**a**) 8×8 antenna array prototype and antenna probes placement in the anechoic chamber; (**b**) the two probe antennas; (**c**) the 8×8 antenna array prototype

The experimental evaluations of the proposed PNCB schemes are divided into two parts. In Sect. 7.5.1, the measurements aim to evaluate the achievable nulling depth, quantified as the gain difference between the received signals of two users. Additionally, the generation of the optimal PNCB vector is explored to optimize beamforming performance and enhance nulling capabilities. Given the importance of accurately monitoring power levels at two angles simultaneously, i.e., the desired signal direction $\theta_i$ and the target nulling direction $\theta_N$, the sample 1×8B sub-array is used for transmitting. Additionally, each of the two probe antennas is connected to an SA, enabling the simultaneous tracking of power levels. The difference in the readings between the two SAs serves as an indicator of the system's efficacy in mitigating potential interference with respect to the primary user's signal, i.e., the normalized null depth. The optimal PNCB vector which provides the deepest null is recorded.

In Sect. 7.5.2, we delve into a setup where two transmitting users are distinguished by their unique modulation schemes. Each user transmitter is represented by one antenna probe connected to one VSG. One user is designated as the intended user, while the other serves as a source of potential interference. The 1×8B sub-array is used for receiving, where the received signal will be analyzed by the vector signal analyzer (VSA) and the SA. The measurement results of the SIR and the EVM are presented.

In the evaluation process, the SIR quantifies the level of the desired user signal relative to the interfering signals, which is essential in MU communication systems. Nulling beamforming enhances the SIR by adjusting beam gain across angles, thereby increasing beam spatial gain discrimination and reducing the MUI. Additionally, EVM measures the difference between the expected symbol and the actual received sample in the constellation diagram and is directly related to the received SINR [32]. A lower EVM indicates higher received SINR, and hence, better system performance, i.e., lower transmission error rate, leading to higher data integrity and system reliability. EVM is particularly important in MU scenarios, where interference significantly impacts and limits the system's performance and a low EVM represents effective interference mitigation.

## *7.5.1 Nulling Control Beamforming Measurements*

### 7.5.1.1 Measurement Setup

At the beginning of the optimization process, the initial beamforming vector, $\boldsymbol{w}_{\text{lcmv}}(\theta_i, \theta_N)$, is derived using Eq. 7.12. The $\boldsymbol{\zeta}_{\text{lcmv}}(\theta_i, \theta_N)$ vector, along with its measured power level at the two probes, i.e., $\text{dir}(\boldsymbol{\zeta}_{\text{lcmv}}(\theta_i, \theta_N), \theta_N)$ and $\text{dir}(\boldsymbol{\zeta}_{\text{lcmv}}(\theta_i, \theta_N), \theta_i)$, is used as the initial searching vector and the baseline in the optimization. Algorithm 7.1 implemented in MATLAB is executed on the controller PC, leveraging readings from the SAs connected to the two antenna probes. The radiation power measured in the directions of User Equipment 1 (UE-1) and User Equipment 2 (UE-2) corresponds to the intended signal power $\text{dir}(\boldsymbol{\zeta}_p^n, \theta_i)$ and the interference signal power $\text{dir}(\boldsymbol{\zeta}_p^n, \theta_N)$, respectively. These measurements are transmitted to the local PC through USB cables and are utilized in the execution of Algorithm 7.1. Then $\boldsymbol{\zeta}_p^n$ will be updated accordingly. The threshold of the directivity degradation in the intended user direction $\epsilon$ is set to 3 dB. In the experimental setup, it is crucial to note that each SA can only evaluate one test case of $\boldsymbol{\zeta}_p^n$ at a time, preventing the parallel evaluation of $N_p$ particles within each iteration. This sequential measurement of $N_p$ particles, coupled with the communication overhead between the SA and the PC executing the optimization algorithm, significantly extends the overall time required for experimentation. To manage this and ensure the experiments remain within a reasonable timeframe, we have reduced the number of iterations and the size of the particle swarms, setting $N_n$ and $N_p$ to 4 and 100, respectively.

The VSG produces a modulated NR-FR1-TM3.1 test signal [33], encompassing a narrow-band scenario where a 20 MHz bandwidth from 3.49 to 3.51 GHz is considered. This signal is then distributed into eight equal paths using a 1-to-8 power splitter. Subsequently, the 8 paths are connected to 8 digitally controlled 6-bit phase shifters and 8 7-bit digitally controlled attenuators, setting the complex-valued coefficients (of the beamforming vector). After multiplying by the corresponding beamforming coefficients, the 8 signal paths are fed to 8 transmit power amplifiers prior to the 8 transmit antenna elements of the designated $1 \times 8$ sub-array. Measurements for 16 $(\theta_i, \theta_N)$ pairs are conducted with the SAs at UE-1 and UE-2.

#### 7.5.1.2 MUI Suppression Measurement Results

The experimental findings, presented in Table 7.2, illustrate the effectiveness of beamforming strategies in suppressing interference power relative to the main user's signal power. When the MDB scheme is employed, average normalized radiation power of −13.6 dB in the direction of interference is observed, with values fluctuating between −7.7 and −19.3 dB. The LCMV approach offers a marginal enhancement in diminishing the power in the interference direction, achieving null levels ranging from −14.3 to −25.2 dB, and an average normalized nulling level of −21.5 dB. This represents an average improvement of 7.9 dB over scenarios without nulling.

Meanwhile, the proposed PNCB scheme demonstrates a substantial reduction in radiation towards potential interference directions across 16 measured $(\theta_i, \theta_N)$ scenarios, achieving an average radiation nulling depth of 35.6 dB. Specifically, the PNCB scheme results in a power differential between the two users that spans

**Table 7.2** Comparison of the interference suppression measurements

| | | Normalized null level (dB) | | | Nulling improvement (dB) |
|---|---|---|---|---|---|
| $\theta_i$ | $\theta_N$ | MDB | LCMV | PNCB | $Null_{PNCB}-Null_{MDB}$ |
| −40 | −18 | −7.7 | −19.1 | −35.4 | −27.7 |
| −34 | −12 | −9 | −14.4 | −37.8 | −28.8 |
| −20 | 2 | −10.7 | −19.5 | −39 | −28.3 |
| −18 | −40 | −19.3 | −23.8 | −35.2 | −15.9 |
| −12 | 10 | −10 | −23.5 | −33.9 | −23.9 |
| −12 | −34 | −17.6 | −21.9 | −35.1 | −17.6 |
| −6 | 16 | −10.4 | −25.2 | −39.3 | −28.8 |
| −4 | 18 | −12 | −23.2 | −35.5 | −23.4 |
| 2 | −20 | −16.8 | −19.4 | −35.6 | −18.8 |
| 8 | −14 | −17.1 | −25.1 | −36.1 | −19 |
| 10 | −12 | −17.9 | −24.9 | −38.7 | −20.8 |
| 14 | 36 | −15.9 | −22.1 | −32.1 | −16.1 |
| 16 | −6 | −16.1 | −25.2 | −38 | −21.9 |
| 18 | −4 | −15.7 | −23.4 | −34.9 | −19.3 |
| 30 | 8 | −11.9 | −18.2 | −32.4 | −20.5 |
| 36 | 14 | −9.7 | −14.3 | −30.9 | −21.2 |
| Average | | −13.6 | −21.5 | −35.6 | −22 |

from −30.9 to −39.3 dB. Compared to the MDB and LCMV methods, the proposed PNCB scheme shows a considerable enhancement, with an average increase in the power differential between the target user and potential interference sources of 22 and 14.2 dB, respectively. The most significant improvements observed in nulling depth, compared to scenarios without nulling and the results utilizing the LCMV beamformer, are 28.8 and 23.4 dB, respectively.

## *7.5.2 SIR and EVM Measurements*

### 7.5.2.1 Measurement Setup

In this section, we delve into a realistic setup where two transmitting UEs are distinguished by their unique modulation schemes. The UEs, each connected to a VSG, transmitted 20 MHz NR-FR1-TM3.1 test signals concurrently with a transmitting power of 7 dBm per UE through a single monopole antenna. One user was designated as the intended user, while the other served as a source of potential interference. The two users were differentiated by their modulation schemes: 16-QAM for UE-1 and 64-QAM for UE-2. UE-2 was positioned at $\theta = 0°$ and UE-1 at $\theta = 22°$. Therefore, two $(\theta_i, \theta_N)$ measurement scenarios are created $(0°, 22°)$ and $(22°, 0°)$. For both cases, three beamforming schemes were measured and compared: MDB (no nulling control), LCMV nulling approach, and proposed PNCB scheme (LCMV+PSO).

On the receiver side, the 1×8B antenna sub-array was directly connected to LNAs, followed by the BFN comprising eight power attenuators and eight phase shifters. The LNAs were utilized to boost measurement sensitivity. Subsequently, the signals from these eight paths were merged using an 8-to-1 power combiner and directed to an SA for measuring the received signal power. Moreover, the received signals undergo analysis with a Keysight VSA for signal demodulation and EVM evaluation.

### 7.5.2.2 Measurement Results

The measured received signal constellation diagrams are illustrated in Figs. 7.7 and 7.9. Detailed characteristics are summarized in Table 7.3. It can be observed that for both cases a considerable SIR and the EVM improvement can be achieved with the proposed PNCB scheme.

**UE-1 16-QAM as Main User** When UE-1 (16-QAM) serves as the primary user, the absence of the nulling strategy results in a very noisy received constellation diagram characterized by scattered symbols, as depicted in Fig. 7.7a. As shown in Fig. 7.8a, the interference signal power of −55.5 dBm is observed in the target 20 MHz bandwidth. The received power from the intended user equals −34.4 dBm.

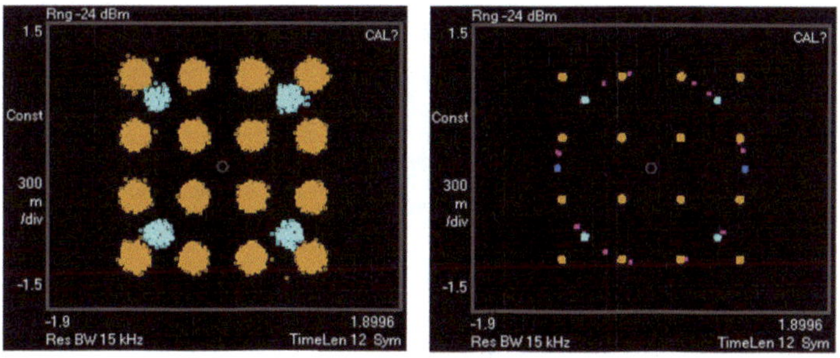


(a) (b)

**Fig. 7.7** Received signal constellations: measurement with 16-QAM UE-1 as the main user, i.e. $(\theta_i, \theta_N) = (22°, 0°)$ with (**a**) MDB, (**b**) proposed PNCB scheme

**Table 7.3** Measurement results of beamforming nulling techniques in the two-user scenario

| | Main received power (dBm) | Interference power (dBm) | SIR (dB) | EVM (dB) |
|---|---|---|---|---|
| *UE-1 16-QAM as main user* | | | | |
| MDB | −34.4 | −55.5 | 21.1 | −18.9 |
| LCMV | −36.6 | −56.9 | 20.3 | −20.2 |
| PNCB | −36.5 | −79.2 | 42.7 | −36.5 |
| *UE-2 64-QAM as main user* | | | | |
| MDB | −34.5 | −45 | 10.5 | −6 |
| LCMV | −35.8 | −54.4 | 18.6 | −8.8 |
| PNCB | −37.1 | −69.9 | 32.8 | −30.9 |

This leads to a SIR of 21.1 dB and an EVM of −18.9 dB. A high EVM can lead to substantial signal distortion and challenge the demodulation processes, potentially increasing communication error rates.

The LCMV nulling scheme, hindered by misalignment between the intended and actual nulling angles due to variations in ERPs, offers very limited improvement in received signal quality. Specifically, it achieves a modest EVM improvement of 1.3 dB. Despite reducing interference by 1.4 dB, the LCMV scheme inadvertently decreases the desired signal power by 2.2 dB, resulting in a slight 0.8 dB degradation in the SIR.

In contrast, the employment of the proposed PNCB scheme yields significant enhancements. It suppresses interference to −79.2 dBm, as shown in Fig. 7.8b. The approach boosts the SIR to 42.7 dB and markedly improves (lowers) the EVM to −36.5 dB. Compared with the MDB approach, these improvements translate to a 23.7 dB decrease in interference power, a 21.6 dB rise in the SIR, and a 17.6 dB better EVM, signifying a considerable improvement in the received signal quality.

The effectiveness of the PNCB in suppressing interference signals can also be demonstrated by comparing the achieved EVM with the operating scenario without interference. When only UE-1 is active and UE-2 is turned off, an EVM value

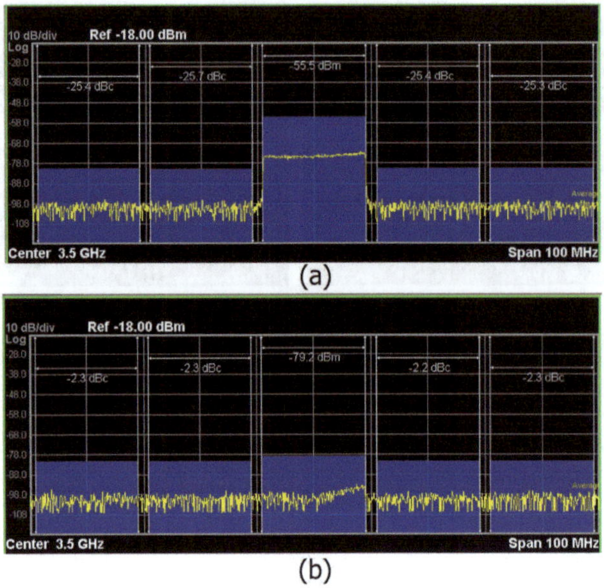


**Fig. 7.8** Received interference power with 16-QAM UE-1 as the main user, i.e. $(\theta_i, \theta_N) = (22°, 0°)$ with (**a**) MDB, (**b**) proposed PNCB scheme

of −37.8 dB is achieved. When both UEs are active, the proposed PNCB scheme achieves an EVM of −36.5 dB, which is only 1.3 dB higher than the EVM in the absence of the interfering user. This result indicates that the proposed PNCB scheme effectively suppresses the interference signal, bringing it close to the noise level. This minimal difference also suggests that instead of the MUI, after applying the PNCB, the receiver's effective noise floor becomes the limiting factor.

Additionally, as shown in Fig. 7.7b, the constellation diagram reveals a high-quality received signal after implementing the PNCB scheme. Symbols are distinctly and tightly grouped around their ideal constellation points, with minimal scatter and no discernible patterns, reflecting a low EVM and high signal integrity. The clear separation of symbols underscores the minimal interference and distortion present, which is essential for precise demodulation and decoding in wireless communication systems.

#### 7.5.2.3 UE-2 64-QAM as Main User

As illustrated in Fig. 7.9a, when only the MDB approach is applied, a very noisy and interfered received signal constellation diagram is observed. Due to the high interference from the UE-1, distinguishing the desired signal constellations from UE-2 becomes almost impossible. The high MUI hinders the successful demodulation of the received signal, potentially leading to a high bit error rate.

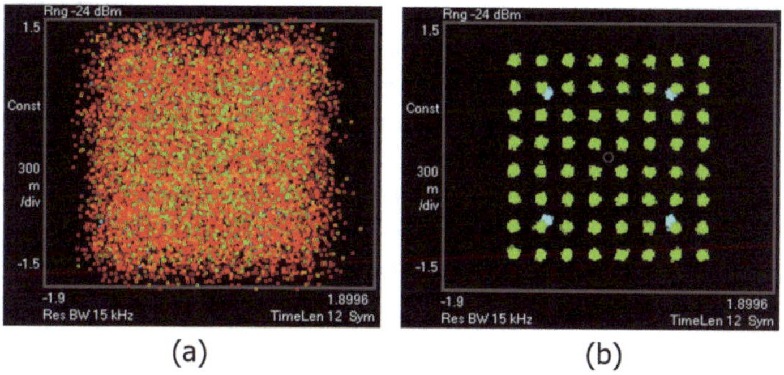


**Fig. 7.9** Received signal constellations: measurement with 64-QAM UE-2 as the main user, i.e. $(\theta_i, \theta_N) = (0°, 22°)$ with (**a**) MDB, (**b**) proposed PNCB scheme

Without specific nulling consideration, the received power from the main user UE-2 is −34.5 dBm, with interference power at −45 dBm, leading to an SIR of 10.5 dB and an EVM of −6 dB. The LCMV nulling approach leads to a slight improvement in received signal quality, where the interference power is reduced to −54.4 dBm, indicating a 9.4 dB lower interference power. The SIR is improved to 18.6 dB, and the EVM decreases to −8.8 dB, indicating a SIR and EVM improvement of 8.1 and 2.8 dB, respectively.

The implementation of the proposed PNCB scheme brings substantial gains. Interference is further diminished to −69.9 dBm, elevating the SIR to 32.8 dB and significantly improving the EVM to −30.9 dB. Compared with the LCMV scheme, the results represent a 15.5 dB reduction in interference power, a 14.2 dB increase in SIR, and a 22.1 dB enhancement in EVM. When comparing with the MDB scheme, a 24.9 dB reduction in interference power, a 22.3 dB increase in SIR, and a 24.9 dB enhancement in EVM, marking a significant leap in signal quality.

With the proposed PNCB scheme, a high-quality received signal is evident in a constellation diagram through distinct, tightly grouped symbols around their ideal points, with minimal scatter and no discernible patterns in the scatter, indicating a very low EVM and high signal integrity, as shown in Fig. 7.9b. Clear separation between symbols and symmetrical scatter around each point suggest minimal noise and distortion, crucial for accurate demodulation and decoding in digital communication systems.

## 7.6 Concluding Remarks

A PNCB scheme is proposed to effectively suppress MUI by creating deep and wide radiation nulls precisely aligned with potential interference directions. The symmetry information of the ERPs is leveraged to streamline the optimization process via reusing optimized beamforming vectors for symmetric sub-arrays. The proposed PNCB approach employs weights generated by the LCMV method as an initial point for optimization. Subsequent weight perturbations are applied to identify the actual optimal beamforming vectors, compensating for the impacts caused by heterogeneous ERPs. PSO is implemented to improve search efficiency. Illustrative results demonstrate the proposed PNCB capability to achieve an effective interference mitigation level of 52.4 dB in scenarios involving multiple users with a negligible influence on the main lobe pattern. The proposed PNCB can consistently maintain an average suppression ranging from −45.3 to −57.7 dB over a 6-degree angle span. Experimental measurements conducted in an anechoic chamber, featuring two users distinguished by 16-QAM and 64-QAM modulation signals, reveal that the proposed PNCB techniques can attain an EVM improvement of 21.3 dB and a SIR improvement of 22.0 dB. Furthermore, the study investigates the influence of attenuator and phase-shifter quantization bits on the nulling depth, providing valuable insights into the achievable performance with various attenuator and phase-shifter setups.

The PNCB algorithm-generated beamforming vectors offer a robust and practical dataset for training various neural network or learning-based models in future studies, providing an effective and real-time MUI suppression mechanism. Additionally, as the size of antenna array apertures continues to grow, the current angular gain-differential NCB algorithm can be extended to a location gain-differential NCB scheme, accommodating the effects of near-field spherical wavefronts. This extended scheme has the potential to enhance the gain disparity between users in both collinear and non-collinear scenarios, laying the groundwork for further discussions in the next chapter.

## References

1. F. Rusek, D. Persson, B.K. Lau, E.G. Larsson, T.L. Marzetta, O. Edfors, F. Tufvesson, Scaling up MIMO: opportunities and challenges with very large arrays. IEEE Signal Process. Mag. **30**(1), 40–60 (2012)
2. E.G. Larsson, O. Edfors, F. Tufvesson, T.L. Marzetta, Massive MIMO for next generation wireless systems. IEEE Commun. Mag. **52**(2), 186–195 (2014)
3. I.P. Gravas, Z.D. Zaharis, T.V. Yioultsis, P.I. Lazaridis, T.D. Xenos, Adaptive beamforming with sidelobe suppression by placing extra radiation pattern nulls. IEEE Trans. Antennas Propag. **67**(6), 3853–3862 (2019)
4. Z.D. Zaharis, I.P. Gravas, P.I. Lazaridis, T.V. Yioultsis, T.D. Xenos, Improved beamforming in 3D space applied to realistic planar antenna arrays by using the embedded element patterns. IEEE Trans. Veh. Technol. **71**(6), 6145–6157 (2022)
5. M. Khosravi, M. Fakharzadeh, M.H. Bastani, Large array null steering using compressed sensing. IEEE Signal Process. Lett. **23**(8), 1032–1036 (2016)
6. R. Qian, M. Sellathurai, D. Wilcox, A study on MVDR beamforming applied to an ESPAR antenna. IEEE Signal Process. Lett. **22**(1), 67–70 (2014)
7. Z. Liu, S. Zhao, G. Zhang, B. Jiao, Robust adaptive beamforming for sidelobe canceller with null widening. IEEE Sensors J. **19**(23), 11213–11220 (2019)
8. Z. Liu, S. Zhao, C. Zhang, G. Zhang, Flexible robust adaptive beamforming method with null widening. IEEE Sensors J. **21**(9), 10579–10586 (2021)

9. J.A. Srar, K.-S. Chung, A. Mansour, Adaptive array beamforming using a combined LMS-LMS algorithm. IEEE Trans. Antennas Propag. **58**(11), 3545–3557 (2010)
10. L. Zhang, W. Liu, R.J. Langley, Adaptive beamforming with real-valued coefficients based on uniform linear arrays. IEEE Trans. Antennas Propag. **59**(3), 1047–1053 (2010)
11. J. Capon, High-resolution frequency-wavenumber spectrum analysis. Proc. IEEE **57**(8), 1408–1418 (1969)
12. L. Yang, M.R. McKay, R. Couillet, High-dimensional MVDR beamforming: optimized solutions based on spiked random matrix models. IEEE Trans. Signal Process. **66**(7), 1933–1947 (2018)
13. X. Wang, M. Amin, X. Cao, Analysis and design of optimum sparse array configurations for adaptive beamforming. IEEE Trans. Signal Process. **66**(2), 340–351 (2017)
14. O.L. Frost, An algorithm for linearly constrained adaptive array processing. Proc. IEEE **60**(8), 926–935 (1972)
15. B.R. Breed, J. Strauss, A short proof of the equivalence of LCMV and GSC beamforming. IEEE Signal Process. Lett. **9**(6), 168–169 (2002)
16. J. Xiong, H. Hong, H. Zhang, N. Wang, H. Chu, X. Zhu, Multitarget respiration detection with adaptive digital beamforming technique based on SIMO radar. IEEE Trans. Microw. Theory Technol. **68**(11), 4814–4824 (2020)
17. Y. Chen, T. Le-Ngoc, B. Champagne, C. Xu, Recursive least squares constant modulus algorithm for blind adaptive array. IEEE Trans. Signal Process. **52**(5), 1452–1456 (2004)
18. R.C. de Lamare, M. Haardt, R. Sampaio-Neto, Blind adaptive constrained reduced-rank parameter estimation based on constant modulus design for CDMA interference suppression. IEEE Trans. Signal Process. **56**(6), 2470–2482 (2008)
19. Y. Cai, R.C. de Lamare, Low-complexity variable step-size mechanism for code-constrained constant modulus stochastic gradient algorithms applied to CDMA interference suppression. IEEE Trans. Signal Process. **57**(1), 313–323 (2008)
20. Z. Ye, C. Liu, Non-sensitive adaptive beamforming against mutual coupling. IET Signal Process. **3**(1), 1–6 (2009)
21. B. Liao, S.-C. Chan, Adaptive beamforming for uniform linear arrays with unknown mutual coupling. IEEE Antennas Wirel. Propag. Lett. **11**, 464–467 (2012)
22. Z. Zheng, K. Liu, W.-Q. Wang, Y. Yang, J. Yang, Robust adaptive beamforming against mutual coupling based on mutual coupling coefficients estimation. IEEE Trans. Veh. Technol. **66**(10), 9124–9133 (2017)
23. Y. Gong, A. Yaseen, R. Morawski, T. Le-Ngoc, Perturbation-based adaptive beamforming for MU-mMIMO, in *2023 IEEE 98th Vehicular Technology Conference (VTC2023-Fall)* (IEEE, 2023), pp. 1–5
24. G. Mahanti, V.B. Dikkani, S. Mandal, Placement of wide nulls in the radiation pattern of a linear array antenna using iterative fast Fourier transform, in *2012 IEEE International Conference on Signal Processing, Communication and Computing (ICSPCC 2012)* (IEEE, 2012), pp. 552–555
25. W.P. Keizer, Fast low sidelobe synthesis for large planar array antennas utilizing successive fast Fourier transforms of the array factor. IEEE Trans. Antennas Propag. **55**(3), 715–722 (2007)
26. M.M. Khodier, C.G. Christodoulou, Linear array geometry synthesis with minimum sidelobe level and null control using particle swarm optimization. IEEE Trans. Antennas Propag. **53**(8), 2674–2679 (2005)
27. S.K. Goudos, V. Moysiadou, T. Samaras, K. Siakavara, J.N. Sahalos, Application of a comprehensive learning particle swarm optimizer to unequally spaced linear array synthesis with sidelobe level suppression and null control. IEEE Antennas Wirel. Propag. Lett. **9**, 125–129 (2010)
28. Y. Liao, J. Wang, Q.H. Liu, Transmit beampattern synthesis for frequency diverse array with particle swarm frequency offset optimization. IEEE Trans. Antennas Propag. **69**(2), 892–901 (2020)
29. A. Hamza, H. Attia, Fast beam steering and null placement in an adaptive circular antenna array. IEEE Antennas Wireless Propag. Lett. **19**(9), 1561–1565 (2020)

30. M.R.R. Khan, V. Tuzlukov, Null steering beamforming for wireless communication system using genetic algorithm, in *2011 IEEE International Conference on Microwave Technology & Computational Electromagnetics* (IEEE, 2011), pp. 289–292
31. C.M. Schmid, S. Schuster, R. Feger, A. Stelzer, On the effects of calibration errors and mutual coupling on the beam pattern of an antenna array. IEEE Trans. Antennas Propag. **61**(8), 4063–4072 (2013)
32. W.B. Hasan, P. Harris, H. Bromell, A. Doufexi, M. Beach, Adaptive user grouping based on EVM prediction for efficient & robust massive MIMO in TDD. IEEE Access **7**, 162683–162696 (2019)
33. 3GPP, NR; Base Station (BS) conformance testing; Part 1: conducted conformance testing. 3rd Generation Partnership Project (3GPP), TS 38.141-1, v19.2.0 (Release 19) (2025). https://portal.3gpp.org/desktopmodules/Specifications/SpecificationDetails.aspx?specificationId=3367

# Chapter 8
# Near-Field Nulling Control Beamfocusing for Multi-User Interference Suppression

## 8.1 Introduction

To enhance spatial resolution and address the increasing demands for greater spectral efficiency and capacity, antenna arrays are being developed with progressively larger apertures [1–3]. With the growth in aperture size, the Rayleigh distance ($r_{FF}$)—which serves as the boundary between near-field and far-field operating regions—scales quadratically, resulting in an expanded radiative near-field region characterized by a spherical wavefront [4, 5]. While spherical wavefronts introduce challenges in channel modeling and very large-scale antenna arrays increase hardware complexity for beamforming, near-field propagation offers exciting opportunities [1, 6, 7]. Unlike conventional far-field beamforming, which operates solely in the angular domain, near-field beamfocusing introduces an additional degree of freedom by shaping beam patterns in both angular and distance domains [8, 9]. As demonstrated in [8], near-field LDMA leverages beamfocusing properties and spherical-domain sampling methods to achieve up to a 10-fold increase in spectral efficiency. As presented in Chap. 2, this innovative approach significantly enhances flexibility in interference mitigation, providing a promising pathway for future 6G extremely large antenna array systems.

Notably, existing investigations of near-field radiation patterns and evaluations of proposed algorithms predominantly rely on isotropic or simplified modeling assumptions for the element patterns, such as cosine elements [1–4, 4, 6, 7, 9–13]. The theoretical isotropic SV used in these evaluations accounts only for the magnitude and phase differences induced at the element sources by an incoming signal, due to their distinct spatial positions relative to observation points. Nevertheless, realistic antenna element-wise interactions and heterogeneous element radiation patterns, affected by factors like the surrounding environment, material EM properties, and structural geometric complexities, are often ignored. These EM interactions significantly impact the performance and robustness of proposed near-field algorithms in practice. Therefore, antenna array full-wave simulations,



T. Le-Ngoc et al., *Half-Duplex/Full-Duplex MU-mMIMO*, Wireless Networks,
https://doi.org/10.1007/978-3-032-09195-6_8

employing the finite element method (FEM) to solve Maxwell's equations, can be crucial for understanding variations in realistic array radiation patterns at different observation distances. These full-wave simulations also provide insightful and practical knowledge on the achievable array radiation patterns, facilitating the evaluation of various beamforming algorithms.

Moreover, the distance-dependent beam gain in the near-field offers significant opportunities for beamfocusing by maximizing the desired signal gain and suppressing potential interference through nulling radiation at interference locations. This approach leads to substantial improvements in robustness and performance in MU-mMIMO systems. As presented in Chap. 7, NCB schemes can generally be divided into two types: (1) Non-iterative Schemes: In these methods, the feeding weights are derived from instantaneous calculations based on the estimation of the inverse covariance matrix of the desired, where popular NCB algorithms including MVDR [14–18], LCMV [19–21], CMT [22, 23], LMS [24–27]. (2) Iterative Approaches: These methods continuously adjust the feeding weights while evaluating objective metrics such as nulling depth, SIR, or achievable data rate. In the iterative search process, various stochastic optimization algorithms like PSO [28–31] and GA [29, 32, 33] have been explored. However, most existing research on NCB focuses on far-field angular-domain scenarios, where users are differentiated only by their angular locations. The development of near-field NCB algorithms, which consider both angular and distance domains, remains a crucial area of study. Furthermore, the lack of near-field NCB algorithms validated through full-wave simulation studies represents a significant research gap. Addressing this gap is essential to realize and understand the potential of near-field beamfocusing techniques for mitigating MUI in practical applications.

Building on the aforementioned observations, this study has two primary objectives. Firstly, comprehensive full-wave simulations incorporating realistic element radiation patterns and EM wave characteristics are performed to investigate how the near-field array radiation pattern varies with different field observation distances. The results offer valuable insights into the variations of spherical wavefronts and the achievable beamfocusing gain within the near-field region. Secondly, two near-field NCB optimization schemes are developed to effectively suppress MUI in near-field MU-mMIMO systems. The effectiveness of the proposed algorithms is validated also through full-wave simulation results. The main contributions of this work are summarized as follows:

- **Full-wave simulation studies of the array near-field radiation patterns:** Full-wave simulations were conducted using Ansys HFSS software on a sample $1 \times 24$ array composed of our previously designed dual-layer metamaterial-based patch antenna elements. These simulations aimed to investigate the variation of the array radiation pattern at different EM field observation distances. The results reveal the transition from the reactive near-field region, where evanescent waves dominate, to the radiative near-field region, characterized by distance-dependent radiation patterns, and ultimately to far-field radiation patterns. Additionally, the study presents the achievable radiation beamwidth at varying distances and the intensity of leakage waves for both uniform-phase beamformers and maximum-directivity beamformers.

- **Constant and non-constant modulus near-field nulling control beamfocusing optimization for MU-mMIMO:** This study proposes two near-field NCB optimization schemes. The first is a non-iterative, non-constant-modulus approach based on the LCMV algorithm. This method nullifies radiation at interference user locations through a closed-form calculation that incorporates the array response constraint vector and the inverse of the signal covariance matrix. The second scheme is an iterative, constant-modulus technique developed using PSO algorithms, which exclusively controls the phase-shift values in the beamfocusing weights. Full-wave simulations conducted on the sample $1 \times 24$ patch antenna array demonstrate that both optimization schemes achieve interference suppression exceeding 26.1 dB in two-user scenarios. Furthermore, comprehensive simulations across various user locations evaluate the differences in achievable beamfocusing gain between desired and interference users in both the colinear and non-colinear scenarios. These results underscore the effectiveness of the proposed NCB optimization schemes in enhancing interference suppression and beamforming performance in near-field MU-mMIMO systems.

The rest of the chapter is structured as follows: Sect. 8.2 reviews and compares the differences in channel models and SV characteristics between the near-field and far-field operating regions of large-scale antenna arrays. Comprehensive full-wave simulation results illustrate the transition of array radiation patterns from the near-field to the far-field propagation as distance increases. Then, Sect. 8.3 presents the system model for the near-field MU-mMIMO setup, employing a partially connected HBF scheme in the NCB algorithm designs. Section 8.4 outlines the proposed LCMV beamforming algorithm using the array's near-field SV, which effectively controls radiation nulling and creates a significant beamfocusing gain difference between interference and desired users. Full-wave simulation results demonstrate the performance of the proposed approach and its nulling effectiveness for various colinear user locations. Furthermore, in Sect. 8.5, a constant-modulus NCB scheme called perturbation-based NCB (PNCB), employing a PSO algorithm, is developed and investigated. Section 8.6 concludes the chapter and suggests possible directions for future work.

## *8.1.1 Full-Wave Simulation and Evaluation Setup*

Figure 8.1 illustrates the steps from NCB vector generation to full-wave simulation and evaluation employed in this study. The proposed near-field NCB algorithms are implemented in MATLAB, generating the optimal beamfocusing vectors, which are saved in a .csv file format. To evaluate the effectiveness of the proposed nulling algorithm with a practical antenna array setup, Ansys HFSS simulation software is employed. HFSS accurately models practical hardware antenna arrays

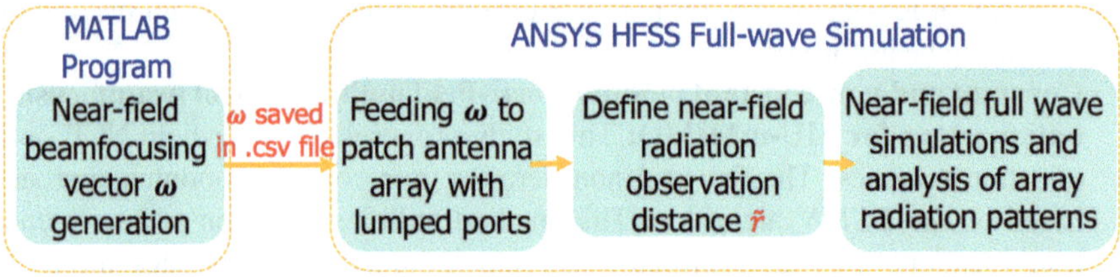


**Fig. 8.1** Near-field beamfocusing vector evaluation flow diagram

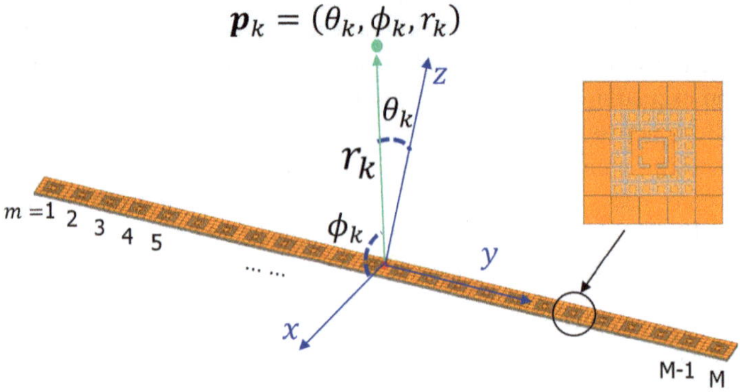


**Fig. 8.2** $1 \times M$ Patch antenna array simulation setup in HFSS

by adhering to the physical laws of EM wave propagation. By using the FEM approach to solve Maxwell's equations, HFSS ensures that simulations account for real-world EM interactions, including mutual coupling, conductor and dielectric material properties, and EM geometric complexities of the antenna array. The result radiation patterns resemble those observed in practical scenarios, providing realistic and reliable evaluations for the performance.

As demonstrated in Fig. 8.2, a $1 \times 24$ antenna array[1] is built in Ansys HFSS, consisting of our previously designed metamaterial-based patch antenna elements operating around 3.5 GHz as presented in Chap. 3. Each antenna element has a dimension $4 \times 4$ cm. The dual-layer metamaterial structure is employed to miniaturize the center radiating patch by introducing slow-wave effects. This design

[1] Due to the high complexity of the metamaterial-based patch antenna structure in full-wave simulations and limitations in computational resources, particularly RAM (the available system has 640 GB of RAM and 70 CPU cores), a $1 \times 24$ uniform linear array (ULA) is employed for simulations. This configuration utilizes approximately 90% of the computational resources. For the sample $1 \times 24$ array, the aperture size is approximately 0.92 m, with a wavelength of 0.086 m (calculated at 3.5 GHz). Therefore, the far-field distance $r_{FF}$ is approximately 19.7 m. All the illustrative results of the array radiation pattern, E-field strength, and beam focusing values are generated from HFSS full-wave simulations.

leaves sufficient space for the upper layer's EM bandgap slots, which act as a band-pass filter to curtail surface wave leakage and reduce interactions between elements within the array. An isolation level better than 14.3 dB between neighboring antenna elements is achieved. To achieve circular polarization, an L-slot center patch with a probe-feeding mechanism is developed. In HFSS, the array is configured with lumped port setups. This setup allows precise control over the magnitude and phase of the feeding coefficient at each antenna element. The beamfocusing vectors are loaded and applied to the pre-built antenna array, and a predefined near-field observation distance $\tilde{r}$ is used to generate the electric field (E-field) radiation pattern at that distance in spherical coordinates. In the presentation, the antenna array is oriented along the $y-$axis, with its primary radiation direction perpendicular to the array plane in the $z-$axis which is defined as the angles $\theta = 0°$ and $\phi = 0°$. The angle $\theta$ is measured within the $zy-$plane, and the angle $\phi$ is measured within the $zx-$plane, as illustrated in Fig. 8.2.

## 8.2 Near-Field Channel Model and Full-Wave Simulation

### *8.2.1 Near-Field and Far-Field Radiation Patterns*

The radiation characteristics of an antenna can be classified into three regions according to the distance between the antenna and the observation point [34]. The first region is known as the reactive near-field region, which occurs when the observation point is within a few wavelengths of the antenna. In this region, reactive fields dominate over radiating fields, meaning that the EM fields primarily store and release energy rather than propagate as radiating waves. Evanescent waves that do not escape but oscillate between the antenna and the immediate surroundings are predominant here.

As the distance from the antenna increases greater than a few wavelengths, the wave characteristics transition into the radiating near-field region, also known as the Fresnel region. In this intermediate region, radiative waves become dominant, and due to phase differences across the elements in a large array, the angular field distribution varies with distance from the antenna. In the Fresnel region, the fields have not yet fully developed into planar waves; instead, they exhibit spherical wavefronts.

Further increasing the observation distance beyond the Rayleigh distance brings the EM waves into the far-field region, also known as the Fraunhofer region. In this region, the angular field distribution remains constant regardless of distance. It is characterized by fully developed planar waves, where the wavefronts are essentially flat, and the EM waves propagate away from the antenna with minimal divergence.

To visualize the radiation pattern characteristics at different distances from the array, Fig. 8.3 shows full-wave simulation results of the E-field radiation pattern (in dB) for the sample $1 \times 24$ patch array operating at 3.5 GHz, with a field observation

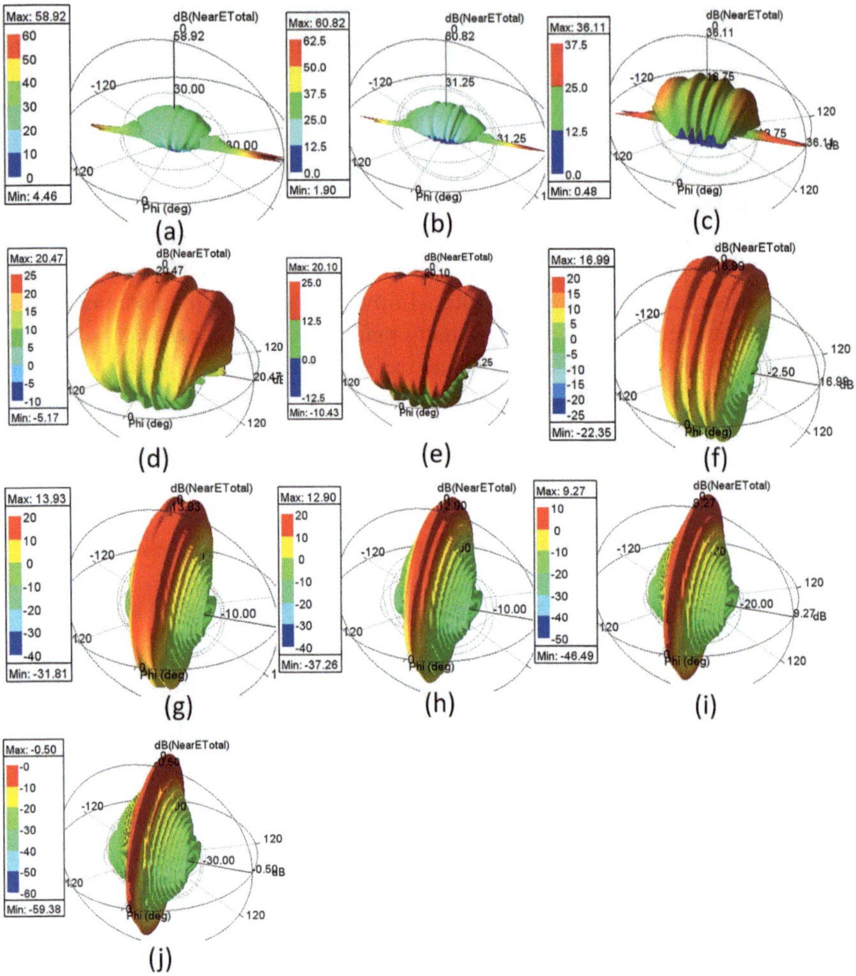


**Fig. 8.3** 3D-radiation patterns of $1 \times 24$ array operating at 3.5 GHz with field observation distance at $\tilde{r} =$ (**a**) 24 cm, (**b**) 40 cm, (**c**) 48 cm, (**d**) 52 cm, (**e**) 56 cm (**f**) 100 cm, (**g**) 200 cm, (**h**) 300 cm, (**i**) 600 cm, (**j**) 20 m

distance $\tilde{r}$ ranging from approximately $3\lambda$ to the Rayleigh distance. All antenna elements are configured in the setup with uniform feeding phases and uniform input power.[2] The results illustrate the variation of the beam pattern at various distances and the impact of phase differences across the array elements on wave summation.

[2] In the simulation, an element feeding power of 7 dBm is used, which corresponds to the maximum power our lab's power amplifier can deliver while operating within its linear region.

As shown in Fig. 8.3a, b, and c, when the observation distance is less than $6\lambda$, the antenna array does not radiate effectively, with most power confined as reactive fields near the array. Significant wave leakage occurs close to the conductor surface or within the dielectric materials along the axis aligned with ULA, making effective and radiative beamfocusing infeasible.

At a distance of $6.5\lambda$ or greater, the antenna array enters its radiative region, with the main radiation direction established in the forward or broadside direction, perpendicular to the array surface. This is indicated by a strong wave propagating away from the array. However, at these relatively short distances, large phase differences between elements result in a broader beamwidth due to the out-of-phase summation of waves. The rapid phase variation across array elements also offers an opportunity to create distance-variant beam radiation patterns. As the distance from the antenna array increases, the phase differences between elements decrease, leading to a narrower beamwidth when uniform phases are applied. Beyond $75\lambda$, the diminishing phase differences cause the radiation pattern to increasingly resemble that of the far-field radiation patterns. At such large distances, distance-dependent beamfocusing becomes challenging, as the beam patterns at different distances become less distinguishable. Beyond the Rayleigh distance, the beam pattern converges to that of the far-field, making distance-dependent beamfocusing ineffective.

### 8.2.2 Fresnel Region and Rayleigh Distance

In practice, the Fresnel distance, which marks the boundary between the reactive and radiative near-field regions, and the Rayleigh distance, which separates the near-field from the far-field regions, are determined by evaluating phase errors that occur when approximating spherical wavefronts with parabolic and linear models, respectively. As illustrated in Fig. 8.4, for an antenna array with an aperture size (or longest dimension) $D$ consisting of $M$ antenna elements, the greatest phase difference between the center and edge elements occurs when the field observation point impinges perpendicularly on the receiver. If the propagation distance to the antenna's center is $r_k$, then the distance to the edge element, which has the largest propagation path difference, can be calculated as

$$r_k^{\text{edge}} = \sqrt{(r_k)^2 + \left(\frac{D}{2}\right)^2}. \tag{8.1}$$

Compared to the path to the center element, the wave travels an additional distance of $r_k^{\text{edge}} - r_k$ to reach the edge element, resulting in a phase shift relative to the center. This phase difference can be approximated as $\beta\left(r_k^{\text{edge}} - r_k\right)$, where $\beta$ is the wave propagation constant, given by $\beta = \frac{2\pi}{\lambda}$. Consequently, the maximum

**Fig. 8.4** Wavefronts (**a**) Near-field spherical and (**b**) far-field planar

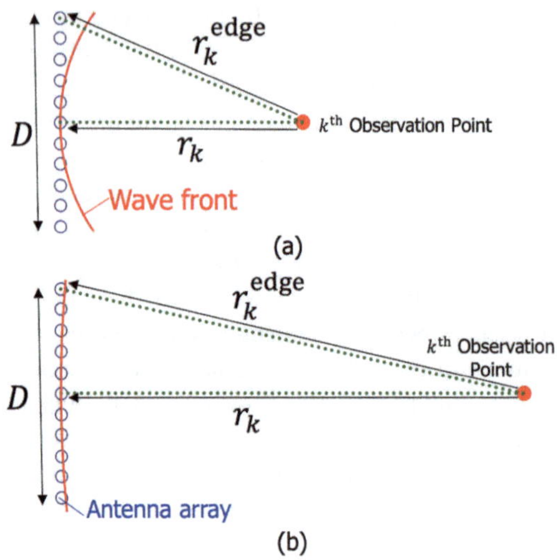


phase shift across the array elements can be derived as

$$\beta\left(r_k^{\text{edge}} - r_k\right) = \frac{2\pi}{\lambda}\left(\sqrt{(r_k)^2 + \left(\frac{D}{2}\right)^2} - r_k\right) \approx \frac{2\pi}{\lambda}\frac{D^2}{8r_k}, \tag{8.2}$$

where the first-order Taylor approximation $\sqrt{1+x} \approx 1 + \frac{x}{2}$ is used. In EM wave studies, a maximum phase shift of $\frac{\pi}{8}$ is conventionally considered an acceptable criterion between the spherical and planar wavefront assumptions [4]. Based on this threshold, the Rayleigh distance is defined as

$$r_{\text{FF}} = \frac{2D^2}{\lambda}, \tag{8.3}$$

meaning that when the distance between the $k$-th observation point and the array center, $r_k$, satisfies the condition $r_k > r_{\text{FF}}$, the combined wavefront is assumed to appear planar, as shown in Fig. 8.4b. This region is known as the far-field. In the far-field region, when uniform phase feedings are utilized, the electric fields generated or received by multiple antenna elements constructively interfere to form a concentrated beam, resulting in a field strength that diminishes inversely with the propagation distance. The beam radiation pattern depends solely on the angular location.

### 8.2.3 Far-Field and Near-Field Channel Model Comparison

The received signal of an antenna array consisting of $M$ antenna elements can be characterized by the weighted steering vector $\mathbf{h}(\boldsymbol{p}_k) \in \mathbb{C}^{M\times 1}$, which represents the signal response from the antenna array to the $k$th user (field observation point) located at position $\boldsymbol{p}_k$. The position vector $\boldsymbol{p}_k$ is defined in spherical coordinates $(\theta_k, \phi_k, r_k)$ relative to the array center. The array response vector of the $M$-element array can be expressed as

$$\mathbf{h}(\boldsymbol{p}_k) = \left[g_1 e^{j\beta\rho_1},\ g_2 e^{j\beta\rho_2},\ \ldots,\ g_M e^{j\beta\rho_M}\right]^T, \tag{8.4}$$

where the term $e^{j\beta\rho_m}$ for $m \in \{1, \ldots, M\}$ represents the phase difference attributable to the varying wave paths traveled by the $m$th element in the array. In near-field scenarios, the proximity between the antenna array and the $k$th user at position $\boldsymbol{p}_k$ causes significant relative distance differences that contribute to phase shifts across the elements. The phase difference in the near-field can be estimated as

$$\rho_m = r_k^m - r_k, \tag{8.5}$$

where $r_k$ represents the distance between the array center (or phase reference point) and the $k$th field observation point, and $r_k^m$ denotes the distance between the $m$th antenna element and the $k$th observation point. In spherical coordinates, $r_k^m$ can be calculated as

$$\begin{aligned} r_k^m &= \| r_k \boldsymbol{v}\left(\theta_k, \phi_k\right) - r_m \boldsymbol{v}\left(\theta_m, \phi_m\right) \| \\ &= \sqrt{(r_k)^2 + (r_m)^2 - 2 r_k r_m \boldsymbol{v}\left(\theta_k, \phi_k\right) \boldsymbol{v}\left(\theta_m, \phi_m\right)^T}, \end{aligned} \tag{8.6}$$

where $r_k\ \boldsymbol{v}(\theta_k, \phi_k)$ represents the position vector pointing from the reference point to the $k$th observation point, and $r_m\ \boldsymbol{v}(\theta_m, \phi_m)$ is the position vector pointing from the reference point to the $m$th antenna element. $r_m$ is the distance of the $m$th element relative to the array center. The unit direction vector $\boldsymbol{v}(\theta, \phi)$ can be expressed in terms of its projection components as

$$\boldsymbol{v}(\theta, \phi) = [\cos(\theta)\cos(\phi), \sin(\theta)\cos(\phi), \sin(\phi)], \tag{8.7}$$

where the $m$th antenna element is located at angles $(\theta_m, \phi_m)$, while the $k$th field observation point is in the direction $(\theta_k, \phi_k)$ relative to the array center.

In the array response vector, the weighting term $g_m$ for $m \in \{1, \ldots, M\}$ encapsulates the radiation characteristics of the $m$th antenna element, characterizing its directional magnitude response adjusted by the distance. Therefore, $g_m$ can be expressed as

$$g_m = \sqrt{G_{e,m}(\theta_k, \phi_k) G_r\left(r_k^m, f\right)}, \tag{8.8}$$

where $G_{e,m}(\theta_k, \phi_k)$ is the radiation power gain of the $m$th element towards the $k$th observation direction at $(\theta_k, \phi_k)$. The term $G_r\left(r_k^m, f\right)$ represents the free-space power path loss due to the wave traveling a distance $r_k^m$ at the operating frequency $f$, and can be expressed as

$$G_r\left(r_k^m, f\right) = \left(\frac{\lambda}{4\pi r_k^m}\right)^2 = \left(\frac{\lambda}{4\pi r_k}\frac{r_k}{r_k^m}\right)^2, \tag{8.9}$$

where the factor $\frac{r_k}{r_k^m}$ represents the normalized distance ratio from the $m$th element to the $k$th observation point relative to the reference distance from the array center to the $k$th observation point.

In far-field scenarios, where $r_k \gg r_m$, the approximation $r_k^m \approx r_k$ can be made. This simplifies the expression as

$$r_k^m \approx r_k - r_m \boldsymbol{v}\left(\theta_k, \phi_k\right) \boldsymbol{v}\left(\theta_m, \phi_m\right)^T. \tag{8.10}$$

Therefore, for a given $r_k$ and $r_m$, the phase difference across the elements in the array depends only on the directional component $\boldsymbol{v}\left(\theta_k, \phi_k\right) \boldsymbol{v}\left(\theta_m, \phi_m\right)^T$. Additionally, in the wave propagation path loss estimation, since $r_k^m \approx r_k$, the far-field path loss can be approximated as

$$G_r^{\text{far}}\left(r_k^m, f\right) = \left(\frac{\lambda}{4\pi r_k^m}\right)^2 \approx \left(\frac{\lambda}{4\pi r_k}\right)^2. \tag{8.11}$$

### 8.2.4 Near-Field Maximum Directivity Beamfocusing

The feeding weights across the antenna elements play a pivotal role in managing the gain of signals transmitted or received at various locations. The MDB vector is designed to align the phases across all elements, creating a constructive summation of signals, thereby concentrating radiation at the target location and resulting in a maximum beamfocusing gain. The MDB vector $\boldsymbol{w}_0\left(\boldsymbol{p}_i\right) \in \mathbb{C}^{M\times 1}$ for beamfocusing at location $\boldsymbol{p}_i$ can be expressed as

$$\boldsymbol{w}_0\left(\boldsymbol{p}_i\right) = \left[e^{j\beta\rho_1}, e^{j\beta\rho_2}, \ldots, e^{j\beta\rho_M}\right]^T. \tag{8.12}$$

The array beamfocusing gain at field observation location $\boldsymbol{p}_k$ with beamfocusing vector $\boldsymbol{w}\left(\boldsymbol{p}_i\right)$ is calculated by multiplying the beamfocusing vector with the normalized steering vector and can be expressed as

$$G_{\text{BF}}\left(\boldsymbol{w}\left(\boldsymbol{p}_i\right), \boldsymbol{p}_k\right) = ||\boldsymbol{w}^H\left(\boldsymbol{p}_i\right)\mathbf{h}'\left(\boldsymbol{p}_k\right)||^2, \tag{8.13}$$

where $\mathbf{h}'\left(\boldsymbol{p}_k\right) \in \mathbb{C}^M$ represents the steering vector of the $M$-element antenna array normalized by the distance-related path loss $G_r\left(r_k^m, f\right)$ and the element gain $G_{e,m}(\theta_k, \phi_k)$. It is defined as

$$\mathbf{h}'\left(\boldsymbol{p}_k\right) = \left[\frac{r_k}{r_k^1}e^{j\beta\rho_1}, \frac{r_k}{r_k^2}e^{j\beta\rho_2}, \ldots, \frac{r_k}{r_k^M}e^{j\beta\rho_M}\right]^T. \tag{8.14}$$

For an ULA, which facilitates beam steering in the $\theta$ angle, the phase shift values $\rho_m$ for the $m$th element can be estimated as

$$\rho_m = \sqrt{r_k^2 + r_m^2 - 2r_k r_m \sin\left(\theta_k\right)} - r_k, m \in \{1, \ldots, M\}. \tag{8.15}$$

The $1 \times 24$ patch antenna array was used to simulate and investigate the radiation pattern variations with the MDB vector as shown in Eq. (8.12). The simulation results at eight sample field observation distances are illustrated in Fig. 8.5. It can be observed that the alignment of the phases across the element effectively creates a concentrated radiation pattern. However, as depicted in the beam radiation patterns, when the observation distance is less than $6\lambda$, the antenna array does not operate effectively within its radiation domain. Although a narrow and concentrated beam can be achieved by aligning the element phases, most of the power remains confined as reactive fields near the antenna array. The primary leakage waves occur along the axis aligned with the ULA, within the dielectric substrate, or near the conducting patch surface. Intense fields are observed at $\theta = \pm 90°$, indicating very low radiation efficiency in the broadside direction and making effective radiative beamfocusing infeasible.

Starting at $6.5\lambda$, the antenna array enters its effective radiative region as shown in Fig. 8.5d. A uniform main-lobe beamwidth, with 3dB-beamwidth around 5°, can be maintained at various distances using the MDB vector, indicating consistent and concentrated radiation in the broadside direction. Due to the proximity of the field observation distance, leakage fields still exist within the dielectric substrates and near the array surface. As the observation distance increases, the sidelobe levels and resonant leakage waves along the axis aligned with the ULA decrease. Beyond the Rayleigh distance, the resonant leakage fields at $\theta = \pm 90°$ disappear, as shown in Fig. 8.5g. Moreover, due to increased path loss, the achievable peak power of the main lobe decreases inversely proportional to $(r_k)^2$ as the distance $r_k$ increases, which is consistent with theoretical free-space path loss model as shown in (8.9).

## 8.3 MU-mMIMO System Model

In the radiative near-field region, the close proximity between the antenna array and users results in spherical wavefronts. As illustrated by Eqs. (8.5) and (8.6), this proximity causes rapid phase variations across the elements as the distance

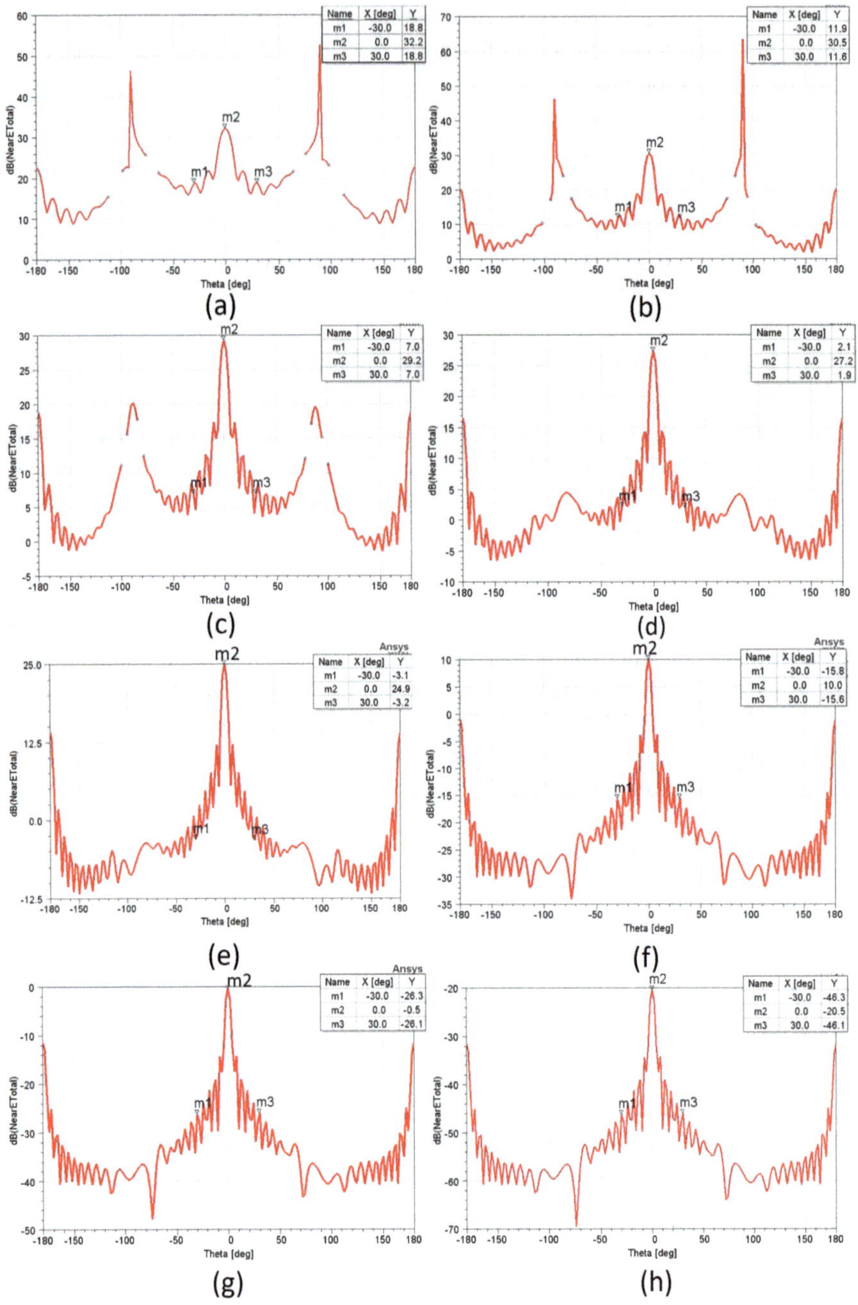


**Fig. 8.5** 2D radiation pattern $\theta$ cuts using the MDB vector, plotted with the field observation distance $\tilde{r} = r_i$. Beamfocusing location at $\theta_i = 0°$ and $\tilde{r} =$ (**a**) 24 cm, (**b**) 40 cm, (**c**) 52 cm, (**d**) 72 cm, (**e**) 100 cm, (**f**) 600 cm, (**g**) 20 m, (**h**) 200 m

changes. Consequently, distance-dependent beam radiation patterns and signal gains can be achieved. Unlike the far-field region, where beamforming primarily relies on direction due to planar wavefronts, the near-field region allows for the differentiation of beamfocusing gains for colinear users—a capability not feasible in far-field scheme.

In the near-field MU-mMIMO system setup, a partially connected HBF scheme is adopted, which boosts hardware efficiency by reducing the number of connections between RF chains and antenna elements [35]. In this architecture, each RF chain is connected to a subset of antenna elements, forming sub-arrays. This configuration simplifies hardware requirements while preserving effective beamforming capabilities. Each antenna sub-array features its own beam pattern, controlled by a BFN. In the problem formulation, we focus on scenarios involving the receiving (uplink) array. The BS operates in an MU-mMIMO setup, equipped with $M_s$ antenna sub-arrays, and serves $K$ users concurrently, where $M_s \geq K$. All users are located in the radiative near-field region of the antenna array, meaning that for all the users, the distance between the user and the array is less than the Rayleigh distance. The signal received by the $i$th sub-array consisting of $M$ elements can be expressed as

$$y_i(\boldsymbol{p}_1, \ldots, \boldsymbol{p}_K) = \underbrace{\boldsymbol{w}_i^H \mathbf{h}_i(\boldsymbol{p}_i) s_i}_{\text{Desired Signal}} + \underbrace{\sum_{k \neq i}^{k=1,\ldots,K} \boldsymbol{w}_i^H \mathbf{h}_i(\boldsymbol{p}_k) s_k}_{\text{Multi-user Interference}} + \underbrace{\boldsymbol{w}_i^H \boldsymbol{n}_i}_{\text{Noise}}, \tag{8.16}$$

where $\boldsymbol{n}_i \in \mathbb{C}^{M\times 1}$ is the Gaussian thermal noise vector at receiver sub-array $i$. The first term denotes the desired received signal from user $i$, located at $\boldsymbol{p}_i = (\theta_i, \phi_i, r_i)$. The transmitted signal symbol from user $i$ is $s_i$, adhering to the power constraint $E\{|s_i|^2\} = 1$. The combining vector $\boldsymbol{w}_i \in \mathbb{C}^{M\times 1}$ manipulates the phase and amplitude of the received signal at the $i$th sub-array, controlling the constructive and destructive summation across the antenna elements. The channel vector $\mathbf{h}_i(\boldsymbol{p}_i) \in \mathbb{C}^{M\times 1}$ characterizes the channel between user $i$ and sub-array $i$, encapsulating the antenna array's configuration, element patterns, and the user's spatial location as shown in expression (8.4).

The second term in Eq. (8.16) highlights the challenge posed by MUI when $K$ users are served concurrently. The signals from the other $K - 1$ users act as interference, significantly impacting the performance of the $i$th sub-array and the overall MU-mMIMO system. For each interfering user $k$ (where $k = 1, \ldots, K$ and $k \neq i$), the transmitted symbol is denoted as $s_k$, satisfying $E\{|s_k|^2\} = 1$. The interference channel vector $\mathbf{h}_i(\boldsymbol{p}_k) \in \mathbb{C}^{M\times 1}$ represents the channel from interfering user $k$ to receiving sub-array $i$. Each interfering user $k$ is located at position $\boldsymbol{p}_k = (\theta_k, \phi_k, r_k)$.

The combining vector $\boldsymbol{w}_i$ plays a pivotal role in managing this beamfocusing gain. In conventional near-field beamfocusing, the MDB vector, as expressed in Eq. (8.12), is used to optimize radiation concentration towards the target location, maximizing the power of the desired signal in Eq. (8.16). NCB serves as an effective strategy to further enhance the SIR in MU-mMIMO systems by increasing spatial discrimination among users. This method modulates the beam's radiation intensity across different locations, creating variable gains for users depending on their direction and distance from the antenna array. The radiation gain at the target user's location is maintained compared to the MDB scheme, while the beamfocusing gain at interference users is minimized through effective nulling processes. In the following sections, we present two distinct near-field NCB optimization algorithms that significantly mitigate the MUI.

## 8.4 LCMV-Based Nulling Control Beamfocusing

### 8.4.1 LCMV Algorithm

The LCMV beamforming technique is a widely used method in linear optimization for calculating array weighting coefficients [19]. It minimizes the variance of the beamformer's output while satisfying specified linear constraints. This approach is effective in spatial filtering, as it preserves desired signals and reduces interference from other sources. By handling multiple constraints simultaneously, it allows flexible control over the beam pattern, ensuring the main lobe is optimally shaped without significant distortion. To maximize the desired signal strength at location $\boldsymbol{p}_i$ while reducing interference from other locations $\boldsymbol{p}_k$ for $k = 1, \ldots, K$ and $k \neq i$, the LCMV algorithm computes the beamforming weights $\boldsymbol{w}_i$. This ensures that the $i$th sub-array's response adheres to specified constraints expressed as

$$\boldsymbol{C}_i^H \boldsymbol{w}_i = \boldsymbol{d}, \tag{8.17}$$

where the response matrix $\boldsymbol{C}_i \in \mathbb{C}^{M \times K}$ for the $i$th sub-array includes array response vectors for both the desired and interfering locations, represented as

$$\boldsymbol{C}_i(\boldsymbol{p}_1, \ldots, \boldsymbol{p}_K) = [\mathbf{h}_i(\boldsymbol{p}_1), \ldots, \mathbf{h}_i(\boldsymbol{p}_K)], \tag{8.18}$$

and the target response constraint vector $\boldsymbol{d} \in \mathbb{R}^{K \times 1}$ specifies the sub-array response at each user location with expression

$$\boldsymbol{d} = [d_1, \ldots, d_K]^T. \tag{8.19}$$

In this context, $d_i = 1$ imposes a unit gain constraint at the target user's location $\boldsymbol{p}_i$ for the $i$th sub-array, ensuring $\boldsymbol{w}_i^H \mathbf{h}_i(\boldsymbol{p}_i) = 1$. To minimize interference at the locations of other users, all other constraint elements $d_k = 0$, enforcing $\boldsymbol{w}_i^H \mathbf{h}_i(\boldsymbol{p}_k) = 0$ for $k = 1, \ldots, K$ and $k \neq i$. The closed-form solution can be derived using the Lagrange multiplier method with multiplier vector $\boldsymbol{\lambda} \in \mathbb{C}^{K \times 1}$. The Lagrangian function, incorporating the constraints, can be expressed as

$$\mathcal{L}(\boldsymbol{w}_i, \boldsymbol{\lambda}) = \boldsymbol{w}_i^H \mathbf{R}_i \boldsymbol{w}_i - 2\,\mathrm{Re}\{\boldsymbol{\lambda}^H (\boldsymbol{C}_i^H \boldsymbol{w}_i - \boldsymbol{d})\}, \tag{8.20}$$

where $\mathbf{R}_i \in \mathbb{C}^{M\times M}$ denotes the covariance matrix of the received signal for the $i$th sub-array before combining with the beamforming vector $\boldsymbol{w}_i$. The covariance matrix encapsulates the effects of the desired signal, interference, and noise prior to beamforming. The solution to the optimization problem can be found by taking the gradient of the Lagrangian with respect to $\boldsymbol{w}_i$ and setting it to zero, while also considering the constraints.

The LCMV beamformer weights for the $i$th can be calculated as

$$\boldsymbol{w}_{i,\mathrm{lcmv}}(\boldsymbol{p}_1, \ldots, \boldsymbol{p}_K) = \mathbf{R}_i^{-1}\boldsymbol{C}_i(\boldsymbol{C}_i^H\mathbf{R}_i^{-1}\boldsymbol{C}_i)^{-1}\boldsymbol{d}. \tag{8.21}$$

For an $M$-element antenna array, the LCMV-based nulling process involves inverting and solving a system with an $M \times M$ covariance matrix, resulting in a computational complexity on the order of $\mathcal{O}(M^3)$.

### 8.4.2 LCMV Near-Field Nulling Beamfocusing E-Field Radiation Patterns

The illustrative results of the full-wave simulation for the LCMV-based NCB E-field radiation patterns, using the sample 1 × 24 patch array, are presented in Fig. 8.6. For comparison, the E-field radiation patterns obtained using the MDB scheme

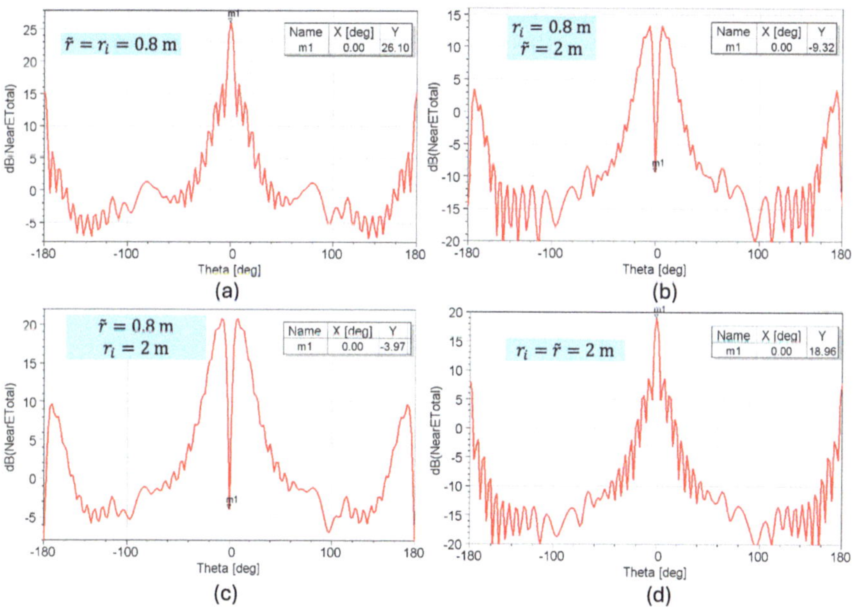


**Fig. 8.6** Two colinear users $(r_1, r_2) = (0.8\,\mathrm{m},\ 2\,\mathrm{m})$ with the LCMV NCB vector. $r_i = 0.8$ m E-field patterns with (**a**) $\tilde{r} = 0.8$ m and (**b**) $\tilde{r} = 2$ m; $r_i = 2$ m E-field patterns with (**c**) $\tilde{r} = 0.8$ m and (**d**) $\tilde{r} = 2$ m

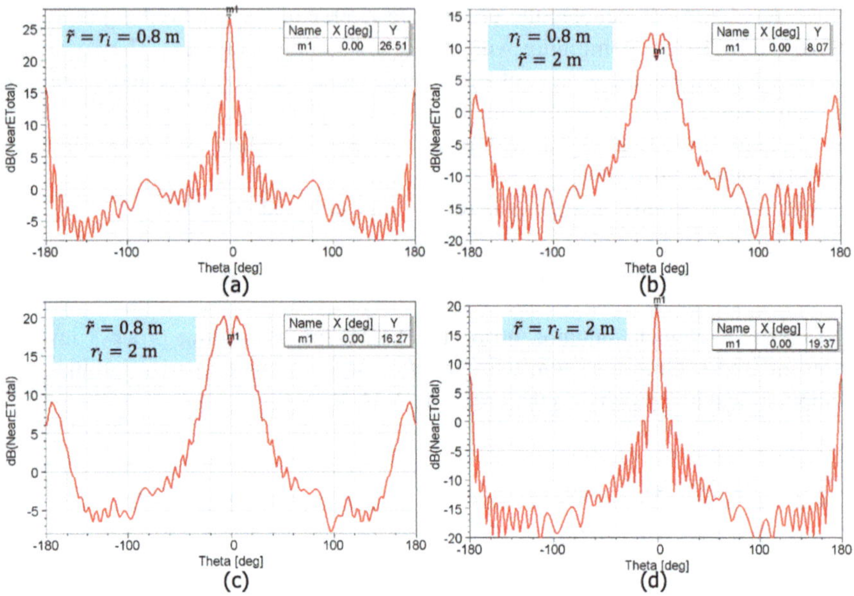


**Fig. 8.7** Two colinear users $(r_1, r_2) = (0.8\,\text{m},\ 2\,\text{m})$ with MDB vector. $r_i = 0.8$ m E-field patterns with (**a**) $\tilde{r} = 0.8$ m and (**b**) $\tilde{r} = 2$ m; $r_i = 2$ m E-field patterns with (**c**) $\tilde{r} = 0.8$ m and (**d**) $\tilde{r} = 2$ m

are shown in Fig. 8.7 and serve as the baseline. Two colinear-user scenarios are considered, where $(r_1, r_2) = (0.8\,\text{m},\ 2\,\text{m})$. The steering angle is set to $\theta_1 = \theta_2 = 0°$ and $\phi_1 = \phi_2 = 0°$. One of the two users is designated as the target user.

An average power difference enhancement of 18.4 dB between the desired and undesired users is observed with the LCMV scheme. This improvement is attributed to the greater beamforming gain difference $G_{\text{BF}}$ at the two locations when using $\boldsymbol{w}_{i,\text{lcmv}}$ compared to $\boldsymbol{w}_0(\boldsymbol{p}_i)$, increasing from an average of 10.8–29.2 dB. At the same time, the main lobe gain at the desired user is well preserved, with minimal gain degradation of only 0.4 dB. Deep and accurately positioned radiation nulls appear at the target nulling locations, demonstrating a considerable and effective enhancement in the MUI mitigation capability.

When the desired user is closer to the array (i.e., $r_i = 0.8$ m), the LCMV-based NCB scheme improves the power difference between the desired and interference signal locations to 35.4 dB, whereas the MDB approach achieves only an 18.4 dB power difference. Moreover, when the desired user is farther from the array than the interference user, the increased distance results in greater path loss for the desired signal. For instance, in the case $r_i = 2$ m, the desired signal power experiences approximately 8.0 dB more attenuation. This extra path loss causes the observed power levels at the interference user to approach those of the desired user. When the MDB vector is employed, only a moderate radiation E-field strength difference is achieved between the desired and interference users. As shown in Fig. 8.7c and d, the MDB approach yields a power difference of only 3.1 dB. In contrast, the LCMV scheme significantly improves the power difference to 23.0 dB, indicating a 19.9 dB enhancement in beamfocusing gain difference and demonstrating considerable near-field MUI suppression.

### 8.4.3 LCMV Scheme Achievable $G_{BF}$ with Different User Separation

To assess the performance of the LCMV algorithm across different colinear intended and nulling user separation, the simulation setup considers a two-user colinear scenario where $\theta_1 = \theta_2 = 0°$ and $\phi_1 = \phi_2 = 0°$. Three intended user locations are examined: (i) $r_i = 0.8$ m, (ii) $r_i = 2$ m (approximately $0.1 r_{\mathrm{FF}}$), and (iii) $r_i = 4$ m (approximately $0.2 r_{\mathrm{FF}}$). The interference user's distance to the array is iteratively varied from 0.8 to 20 m (approximately $r_{\mathrm{FF}}$), excluding $r_i$. Figure 8.8 presents the full-wave simulation results, showing the beamfocusing gain $G_{\mathrm{BF}}$ difference between the intended and interference users. To investigate the impact of the nulling process on the beamfocusing performance at the desired user, the E-field strength at the desired user location $r_i$ with different nulling location is also illustrated in Fig. 8.9.

With the LCMV-based NCB scheme, a high power difference is achieved when the two users are well separated. However, when the two users are closely located, only a moderate $G_{\mathrm{BF}}$ difference can be achieved, accompanied by significant degradation in the beamfocusing gain at the target location, due to the high correlation between the channel vectors at the user locations. When one user is located within 1 m (approximately $0.05 r_{\mathrm{FF}}$), and the other location beyond 2 m (approximately $0.1 r_{\mathrm{FF}}$), a substantial power difference is maintained, with the $G_{\mathrm{BF}}$

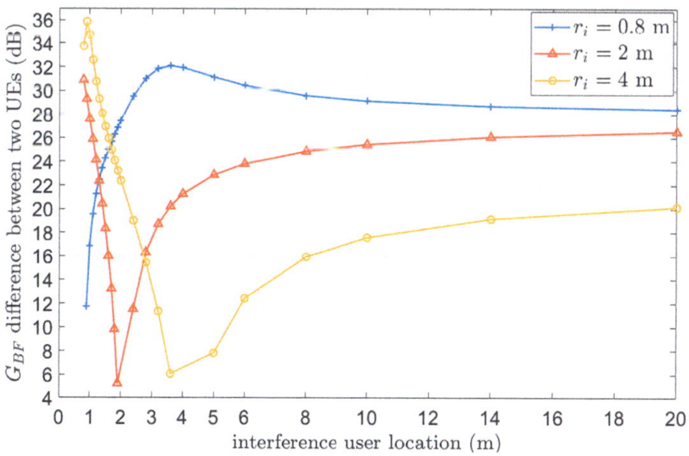


**Fig. 8.8** Achievable $G_{\mathrm{BF}}$ difference (in dB) employing LCMV vector versus user separation

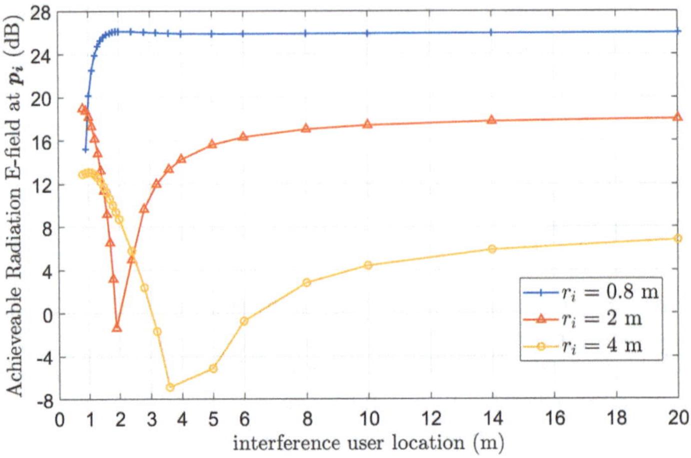


**Fig. 8.9** E-field radiation (in dB) at desired user employing LCMV vector with different user separation

difference exceeding 28 dB and maximum achievable $G_{\mathrm{BF}}$ difference better than 32 dB. The wide spatial separation also minimizes the nulling process's impact on the target user's gain, with a main lobe peak level degradation of less than 1 dB observed during the nulling process, as shown in Fig. 8.9.

In cases where one user is positioned at 2 m (approximately $0.1r_{\mathrm{FF}}$), the other user needs to be located either closer than 1.3 m or farther than 5 m (approximately $0.25r_{\mathrm{FF}}$) to achieve a $G_{\mathrm{BF}}$ difference greater than 22 dB. Furthermore, when both users are situated at distances greater than 2 m (approximately $0.1r_{\mathrm{FF}}$), the rapid degradation in beam resolution makes achieving a large differential gain between the two user locations more challenging, resulting in a smaller $G_{\mathrm{BF}}$ difference. When the target user is located at 4 m (approximately $0.2r_{\mathrm{FF}}$), the other user should be kept closer than 2.4 m or farther than 10 m to maintain a $G_{\mathrm{BF}}$ difference better than 18 dB. Moreover, when both users are situated at distances greater than 2 m, significant gain degradation can be observed at the desired user location, with a $G_{\mathrm{BF}}$ degradation exceeding 6 dB.

## 8.5 Near-Field Perturbation-Based Nulling Control Beamfocusing

While the LCMV method can generate the NCB vector using a closed-form equation based on accurate estimation of the array response vectors at desired and non-desired user locations, it requires the use of both attenuators and power amplifiers in the BFN to modify the beamforming weights. Moreover, the LCMV method can substantially diminish the desired user's signal strength when users are closely positioned, owing to the high correlation between the array response vectors at their respective locations. In scenarios where maintaining the desired signal strength is critical, a more adaptive approach that limits the maximum allowable gain degradation of the target signal is needed. In this section, we propose a constant-modulus beamforming technique employing an iterative perturbation-based NCB approach, which accurately controls the beamfocusing and nulling process using only phase shifters in the BFN.

### 8.5.1 Near-Field PNCB Algorithm

Near-field beamfocusing with large-scale antenna arrays potentially introduces excessive computational complexity. For instance, determining the optimal result from a set of $\gamma$-level uniformly quantized phase perturbations for an array with $M$ elements using exhaustive search requires $\gamma^M$ iterations. To circumvent this prohibitive computational burden, PSO offers an effective solution. PSO utilizes a swarm of intelligent particles, each representing a candidate beamforming vector, to collectively explore the search domain within predefined boundaries and optimize a given objective function. The movements of the particles are guided by their personal best-known positions in the search space and the swarm's global best-known position. The stochastic nature of PSO is particularly suited for non-convex optimization challenges, making it a robust tool for complex problem-solving scenarios.

During the optimization process, a maximum allowable degradation threshold of $\epsilon$ dB in the target user's beamfocusing gain is permitted as a trade-off to enhance MUI suppression capability. The optimization problem is formulated as follows

$$\min_{\boldsymbol{w}} \max_{k\in\{1,\ldots,K\},k\neq i} \left(G_{\mathrm{BF}}(\boldsymbol{w}, \boldsymbol{p}_k)\right), \tag{8.22}$$

$$\text{subject to } G_{\mathrm{BF}}(\boldsymbol{w}, \boldsymbol{p}_i) \geq G_{\mathrm{BF}}(\boldsymbol{w}_0(\boldsymbol{p}_i), \boldsymbol{p}_i) - \epsilon. \tag{8.23}$$

where $G_{\mathrm{BF}}(\boldsymbol{w}, \boldsymbol{p}_k)$ represents the beamfocusing gain at the location of the interfering user $k$ for $k = 1, \ldots, K$ and $k \neq i$ when using the beamforming vector $\boldsymbol{w}$, and $\boldsymbol{w}_0(\boldsymbol{p}_i)$ is the beamforming vector for the target user with MDB scheme. In this formulation, we aim to prevent the algorithm from focusing solely on optimizing the nulling depth for a single interfering user while neglecting the suppression of MUI at other interfering locations. By focusing on minimizing the worst-case scenario through the $\max(\cdot)$ function, which identifies the highest interference level among all interfering users, the algorithm ensures that interference is effectively reduced across all locations.

In developing the constant-modulus NCB vector, we consider only the iteration and perturbations of the phase values in the complex NCB vector. The iterative beamforming vector for the $p$th particle during the $n$th iteration, $\boldsymbol{w}_p^n \in \mathbb{C}^{M\times 1}$, is represented as

$$\boldsymbol{w}_p^n = c[e^{i\zeta_{p,1}^n}, e^{i\zeta_{p,2}^n}, \ldots, e^{i\zeta_{p,M}^n}]^T, \tag{8.24}$$

where $c$ is the constant magnitude of the beamforming vector determined by the power constraint, and $\zeta_{p,m}^n$ represents the phase of the $m$th element for particle $p$ at iteration $n$. The phase vector $\boldsymbol{\zeta}_p^n \in \mathbb{R}^{1\times M}$ containing the phase information of the beamforming vector is expressed as

$$\boldsymbol{\zeta}_p^n = [\text{ang}(\boldsymbol{w}_p^n)] = [\zeta_{p,1}^n, \zeta_{p,2}^n, \ldots, \zeta_{p,M}^n]. \tag{8.25}$$

To monitor performance during the optimization process, the cost function for $\boldsymbol{\zeta}_p^n$ at the $p$th particle during the $n$th iteration in the PSO is defined as

$$\Omega(\boldsymbol{\zeta}_p^n) = \alpha\left(\max_{k\in\{1,\ldots,K\},k\neq i}\left(G_{\text{BF}}(\boldsymbol{w}_p^n, \boldsymbol{p}_k)\right)\right) + Q, \tag{8.26}$$

where $\alpha > 0$ is an arbitrary scalar constant. The penalty factor $Q \geq 0$ is defined as

$$Q = \begin{cases} 0, & \text{if condition (8.23) is met,} \\ Q_1, & \text{if condition (8.23) is not met,} \end{cases} \tag{8.27}$$

with $Q_1$ being a large positive constant penalty to enforce the constraint. During the $n$th iteration, the personal best beamforming vector $\boldsymbol{\zeta}_{\text{best},p}^n$ for the $p$th particle and the current global best beamforming vector $\boldsymbol{\zeta}_{\text{best}}^n$ among all particles are updated as follows

$$\boldsymbol{\zeta}_{\text{best},p}^n = \mathop{\arg\min}_{\boldsymbol{\zeta}_p^{(n*)},\forall n*=0,1,\cdots,N_n} \Omega\left(\boldsymbol{\zeta}_p^{(n*)}\right), \tag{8.28}$$

$$\boldsymbol{\zeta}_{\text{best}}^n = \mathop{\arg\min}_{\boldsymbol{\zeta}_{\text{best},p}^n,\forall p=1,\cdots,N_p} \Omega\left(\boldsymbol{\zeta}_{\text{best},p}^n\right). \tag{8.29}$$

The shifting factor $\boldsymbol{\delta}_p^{(n+1)} \in \mathbb{R}^{1\times M}$ for the $p$th particle at the $(n+1)$th iteration is dynamically updated based on the values of $\boldsymbol{\zeta}_{\text{best},p}^n$ and $\boldsymbol{\zeta}_{\text{best}}^n$. This update is crucial for modulating the beamforming vector's rate of change, enhancing convergence probability. The shifting factor is initialized as a zero vector. The update can be expressed as

$$\boldsymbol{\delta}_p^{(n+1)} = \mathbf{J}\boldsymbol{\delta}_p^n + \mathbf{K}(\boldsymbol{\zeta}_{\text{best},p}^n - \boldsymbol{\zeta}_p^n) + \mathbf{L}(\boldsymbol{\zeta}_{\text{best}}^n - \boldsymbol{\zeta}_p^n), \tag{8.30}$$

where $\mathbf{J}$, $\mathbf{K}$, and $\mathbf{L}$ are diagonal matrices. $\mathbf{J}$ is the inertia weight matrix, essential for maintaining a balance between exploration and exploitation during the search for

**Algorithm 8.1:** Constant-Modulus PNCB Scheme

**Input**: $p_1, \ldots, p_K, i, \epsilon$.
**Output**: $\hat{\zeta} = \zeta^n_{\text{best}}$
**for** $n$ =0:$N_n$ **do**
  **for** $p$=1:$N_p$ **do**
    **if** $n$=0 **then**
      Initialize the shifting factor $\delta_p^{(0)} \leftarrow 0$.
      **if** $p$=1 **then**
        Initialize $w_p^n \leftarrow w_0(p_i)$ based on (8.12).
        Initialize $\zeta_p^n$ based on $w_p^n$ and (8.25).
      **else**
        Initialize $\zeta_p^n$ uniformly distributed in the search region $[\zeta_L, \zeta_U]$.
      **end**
    **else**
      Update shifting factor $\delta_p^n$ using (8.30). Update vector $\zeta_p^n$ using (8.31).
    **end**
    Update penalty factor $Q$ via (8.23) and (8.27)
    and evaluate cost $\Omega(\zeta_p^n)$ via (8.13) and (8.26).
    Find the personal best $\zeta^n_{\text{best},p}$ using (8.28).
  **end**
  Find the global best $\zeta^n_{\text{best}}$ using (8.29).
**end**

the optimal solution. **K** and **L** are positive diagonal matrices representing cognitive (self-adjustment) and social adjustment components, respectively. These matrices govern particle interactions and the tendency of each particle to move towards its personal best and the global best positions. Subsequently, the beamforming phase vector is updated by

$$\zeta_p^{(n+1)} = \text{clip}\left(\zeta_p^n + \delta_p^{(n+1)},\ \zeta_L,\ \zeta_U\right), \tag{8.31}$$

where $\zeta_L, \zeta_U \in \mathbb{R}^{1\times M}$ define the lower and upper bounds of the phase perturbations, respectively. In this optimization, $\zeta_L$ and $\zeta_U$ are set to $-\pi$ and $\pi$, respectively. The clip function enforces the specified bounds by constraining the phase values within the predefined search area. If the updated phase vector exceeds these limits, it is adjusted to remain within the feasible region. The detailed steps of the PNCB algorithm are outlined in Algorithm 8.1.

### *8.5.2 PNCB Near-Field Nulling Beamfocusing E-field Radiation Patterns*

Figure 8.10 presents the full-wave simulation results for E-field radiation pattern 2D cuts, utilizing sample $1\times 24$ patch antenna array with the proposed PNCB algorithm.

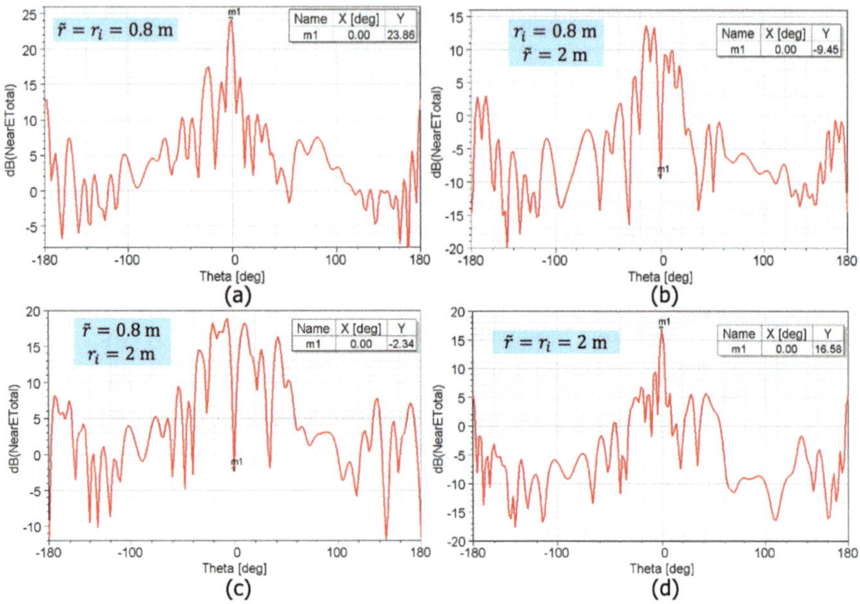


**Fig. 8.10** Two colinear users $(r_1, r_2) = (0.8\,\text{m},\ 2\,\text{m})$ with the PNCB vector. $r_i = 0.8$ m E-field patterns with (**a**) $\tilde{r} = 0.8$ m and (**b**) $\tilde{r} = 2$ m; $r_i = 2$ m E-field patterns with (**c**) $\tilde{r} = 0.8$ m and (**d**) $\tilde{r} = 2$ m

Two colinear-user scenarios are examined, where $(r_1, r_2) = (0.8\,\text{m},\ 2\,\text{m})$. The steering angle is set to $\theta_1 = \theta_2 = 0°$ and $\phi_1 = \phi_2 = 0°$. One of the two users is designated as the target user. The PNCB scheme demonstrates an average gain difference enhancement of 15.3 dB between the desired and undesired users compared with the MDB technique, where the average gain difference increases from 10.8 to 26.1 dB. Additionally, a desired user beamfocusing gain degradation of less than 2.8 dB is observed. The formation of deep radiation nulls at the interference user locations underscores the PNCB scheme's effectiveness in significantly mitigating MUI.

When the desired user is positioned closer to the array (i.e., $r_i = 0.8$ m), the PNCB scheme enhances the power difference between the desired and interference signal locations to 33.3 dB, compared to an 18.4 dB power difference achieved with the MDB approach. Moreover, when the desired user is located farther from the antenna array (i.e., $r_i = 2$ m), the PNCB scheme achieves a power difference greater than 18.9 dB, representing a 15.8 dB MUI mitigation capability improvement over the MDB scheme.

Furthermore, compared with the LCMV scheme in Sect. 8.4, the PNCB achieves a comparable level of differential gain between the two user locations, with only a 3.1 dB achieved beamfocusing gain difference degradation. The performance is attained without the need for attenuators or adjustments to the feeding weight magnitudes, highlighting the PNCB's hardware efficiency. The results demonstrate the PNCB algorithm's robust capability in optimizing desired user beamfocusing gain and effectively suppressing MUI in near-field environments.

### 8.5.3 PNCB Scheme Achievable $G_{BF}$ with Different User Separation

The $1\times24$ patch antenna was simulated to evaluate the performance of the PNCB algorithm across varying intended and nulling user separations. The same collinear two-user setup as Sect. 8.4.3 is considered. Full-wave simulation results, shown in Fig. 8.11, illustrate the $G_{\mathrm{BF}}$ difference between the intended and interference users. In the near-field region of the array, large beamfocusing gain differences are observed when the two colinear users are spatially separated. However, when the users are in close proximity, only a moderate $G_{\mathrm{BF}}$ difference can be achieved.

When one user is positioned very close to the array, within approximately $0.05r_{\mathrm{FF}}$, and the other is located beyond 2 m (approximately $0.1r_{\mathrm{FF}}$), a substantial power difference is maintained. The average difference in beamfocusing gain ($G_{\mathrm{BF}}$) of 26.4 dB can be achieved, with a minimum difference of better than 22.8 dB. In scenarios where one user is situated at 2 m (approximately $0.1r_{\mathrm{FF}}$), the second user must be either closer than 1.1 m or farther than 8 m (approximately $0.4r_{\mathrm{FF}}$) to achieve a $G_{\mathrm{BF}}$ difference exceeding 22.3 dB. When both users are positioned beyond 2 m (approximately $0.1r_{\mathrm{FF}}$), the rapid decrease in beam resolution makes it more challenging to attain a significant differential gain between their locations, thereby reducing the achievable $G_{\mathrm{BF}}$ difference. For instance, when the target user is located at 4 m (approximately $0.2r_{\mathrm{FF}}$), only a moderate $G_{\mathrm{BF}}$ difference of 4 dB can be achieved if the other user is kept farther than 10 m. The maximum allowable

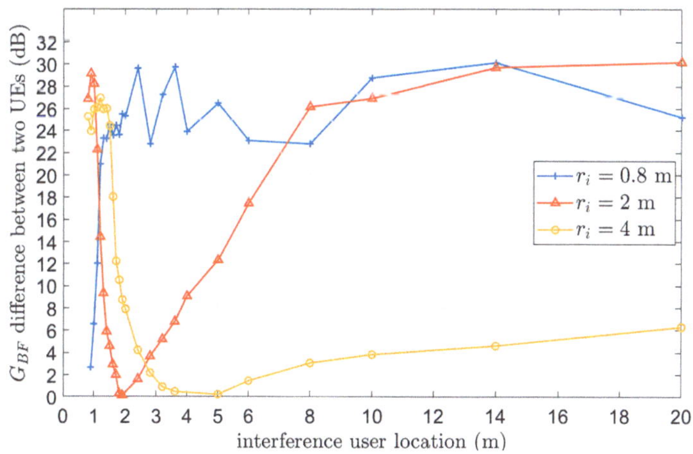


**Fig. 8.11** Achievable $G_{\mathrm{BF}}$ difference employing PNCB vector versus user separation

gain degradation threshold set during optimization ensures that the beamfocusing gain at the desired user's location is maintained. However, this constraint limits the flexibility of the NCB algorithm in achieving a larger gain difference, as it necessitates a trade-off with the degradation of the main user's signal.

### *8.5.4 Simulation Results for Both Colinear and Non-Colinear User Scenarios with PNCB*

To thoroughly evaluate the proposed PNCB scheme's capability for suppressing MUI across different user configurations, we simulated six-user scenarios incorporating both colinear and non-colinear user placements. User distances were varied within the range $r \in \{0.8\,\text{m}, 2.5\,\text{m}\}$, and user angular positions were set at $\theta \in \{-30°, 0°, 30°\}$ with a constant $\phi = 0°$, as illustrated in Fig. 8.12. Figure 8.13 displays the E-field radiation pattern $\theta$ cuts using the PNCB algorithm with a $1 \times 24$ patch antenna array. In these simulations, the PNCB vectors iteratively focus on one desired user while nulling interference at the other five user locations. The achievable power and beamfocusing gain ($G_{\text{BF}}$) differences are summarized in Table 8.1.

In six-user scenarios, the antenna array demonstrates a substantial $G_{\text{BF}}$ difference between the target user and both colinear and non-colinear interfering users. Specifically, an average $G_{\text{BF}}$ difference of 25.2 dB is achieved between colinear users, while a slightly higher average $G_{\text{BF}}$ difference of 26.5 dB is observed between non-colinear users. All simulation results exhibit a gain difference exceeding 20 dB, with the maximum gain difference reaching 34.1 dB. The consistent performance across both colinear and non-colinear user arrangements further emphasizes the robustness of the PNCB beamforming approach.

When the distance between the antenna array and the focal point increases from 0.8 to 2.5 m, there is an average power decrease of 9.4 dB. This corresponds to the average radiation power dropping from 23.5 dB at 0.8 m to 14.1 dB at 2.5 m.

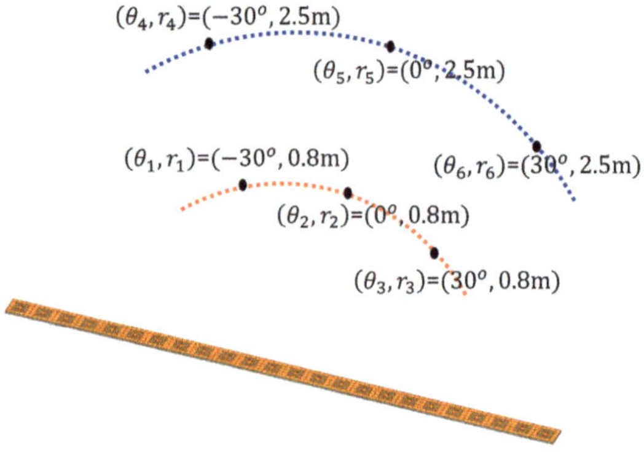


**Fig. 8.12** HFSS simulation setup with 6 users
$p_1 = (-30°, 0°, 0.8\,\text{m})$;
$p_2 = (0°, 0°, 0.8\,\text{m})$;
$p_3 = (30°, 0°, 0.8\,\text{m})$;
$p_4 = (-30°, 0°, 2.5\,\text{m})$;
$p_5 = (0°, 0°, 2.5\,\text{m})$;
$p_6 = (30°, 0°, 2.5\,\text{m})$

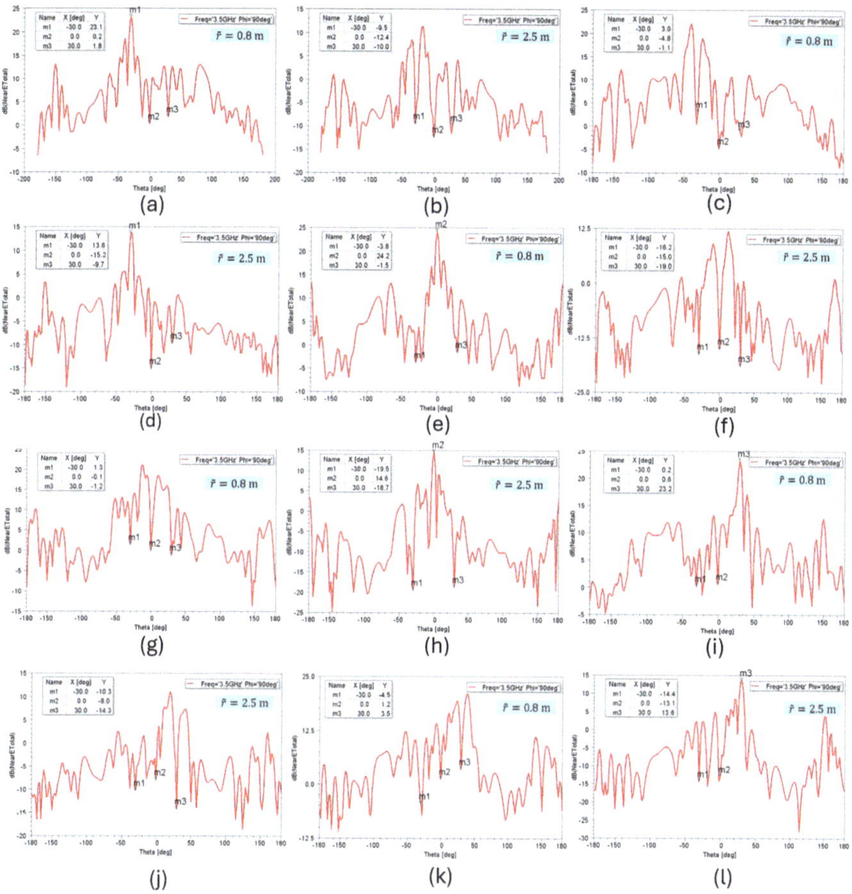


**Fig. 8.13** 2D radiation patterns for six users with the PNCB vector. E-field radiation pattern $\theta$ cuts with $\boldsymbol{p}_i = (-30°, 0°, 0.8\,\text{m})$ and $\tilde{r} =$ (**a**) 0.8 m, (**b**) 2.5 m; $\boldsymbol{p}_i = (-30°, 0°, 2.5\,\text{m})$ and $\tilde{r} =$ (**c**) 0.8 m, (**d**) 2.5 m; $\boldsymbol{p}_i = (0°, 0°, 0.8\,\text{m})$ and $\tilde{r} =$ (**e**) 0.8 m, (**f**) 2.5 m; $\boldsymbol{p}_i = (0°, 0°, 2.5\,\text{m})$ and $\tilde{r} =$ (**g**) 0.8 m, (**h**) 2.5 m; $\boldsymbol{p}_i = (30°, 0°, 0.8\,\text{m})$ and $\tilde{r} =$ (**i**) 0.8 m, (**j**) 2.5 m; $\boldsymbol{p}_i = (30°, 0°, 2.5\,\text{m})$ and $\tilde{r} =$ (**k**) 0.8 m, (**l**) 2.5 m

This observed power reduction aligns with the theoretical expectations of path loss in free-space propagation, where power density decreases with the square of the distance from the source.

At a focal distance of 0.8 meters, the system achieves an average radiation power difference of 23.2 dB and a minimum of 21.3 dB between the target user and interference users. When the focal distance is extended to 2.5 meters, the average and minimum radiation power differences are 12 and 10.1 dB, respectively. Despite the increased distance and associated path loss, the system maintains a significant power difference, ensuring reliable reception for the target user while effectively suppressing MUI.

**Table 8.1** Radiation E-field strength in dB (achievable normalized beamfocusing gain ($G_{BF}$) in dB) for six users using the PNCB scheme

| Desired | E-field strength (Normalized $G_{BF}$) in dB at | | | | | |
|---|---|---|---|---|---|---|
| user at | $p_1$ | $p_2$ | $p_3$ | $p_4$ | $p_5$ | $p_6$ |
| $p_1$ | 23.1 (0) | 0.2 (−22.9) | 1.8 (−22.9) | −9.5 (−22.7) | −12.4 (−25.6) | −10 (−23.2) |
| $p_2$ | −3.8 (−28) | 24.2 (0) | −1.5 (−25.7) | −16.2 (−30.5) | −15 (−29.3) | −19 (−33.3) |
| $p_3$ | 0.2 (−23) | 0.6 (−22.6) | 23.2 (0) | −10.3 (−23.6) | −8 (−21.3) | −14.3 (−27.6) |
| $p_4$ | 3 (−20.5) | −4.8 (−28.3) | −1.1 (−24.6) | 13.6(0) | −15.2(−28.8) | −9.7 (−23.3) |
| $p_5$ | 1.3 (−23.2) | −0.1 (−24.6) | −1.2 (−25.7) | −19.5 (−34.1) | 14.6 (0) | −18.7 (−33.3) |
| $p_6$ | −4.5 (−28) | 1.2 (−22.3) | 3.5 (−20) | −14.4 (−28) | −13.1 (−26.7) | 13.6 (0) |

## 8.6 Concluding Remarks

This chapter presents comprehensive full-wave simulation analyses of near-field beam radiation patterns for large-scale arrays, accounting for realistic EM wave characteristics, heterogeneous ERPs, and interactions among array elements. To leverage the advantages of distance-dependent radiation patterns in the near-field, we develop two NCB algorithms to effectively mitigate MUI in mMIMO systems by achieving significant focusing gain differences between desired and interfering locations. The first algorithm adjusts the array feeding magnitudes and phase shifters using the LCMV-based NCB scheme, achieving an average gain difference of 29.2 dB between desired and undesired users, with minimal gain degradation of 0.4 dB at the desired user compared to the MDB scheme. Later, a constant-modulus beamfocusing scheme based on a perturbation-based NCB algorithm employing PSO is developed. Using only phase shifters, an average gain difference enhancement of 26.1 dB between desired and undesired users can be achieved. Iterative full-wave simulations are conducted to investigate how the achievable beamfocusing gain difference varies with different desired and interfering user separations.

The optimized beamforming vectors generated by the LCMV and PNCB algorithms serve as a robust and practical dataset for training various neural network or learning-based models. For instance, in [J5], a deep neural network (DNN) model was trained using LCMV-generated beamfocusing vectors from this work for MUI suppression. The model achieved a phase error of less than 0.021 radians and a magnitude error of 0.17 dB in the predicted feeding weights. The near-field beam patterns produced using the LCMV-based vectors and the DNN-predicted vectors showed strong agreement, demonstrating the efficacy of the approach. Moreover, full-wave simulations provide valuable insights into the achievable gain differences for users in various collinear and non-collinear locations. These insights lay the groundwork for further experimental validation. The algorithms will be thoroughly evaluated in anechoic chamber measurements, utilizing real quantized beamforming networks (BFNs) and XL-antenna array prototypes, to assess their practical performance and refine their effectiveness in real-world scenarios.

## References


1. M. Cui, L. Dai, Channel estimation for extremely large-scale MIMO: far-field or near-field? IEEE Trans. Commun. **70**(4), 2663–2677 (2022)
2. H. Lei, J. Zhang, Z. Wang, B. Ai, D.W.K. Ng, Hybrid-field channel estimation for XL-MIMO systems with stochastic gradient pursuit algorithm. IEEE Trans. Signal Process. **72**, 2998–3012 (2024)
3. M. Cui, L. Dai, Near-field wideband beamforming for extremely large antenna arrays. IEEE Trans. Wireless Commun. **23**(10), 13110–13124 (2024)
4. K.T. Selvan, R. Janaswamy, Fraunhofer and Fresnel distances: Unified derivation for aperture antennas. IEEE Antennas Propag. Mag. **59**(4), 12–15 (2017)

5. M. Cui, Z. Wu, Y. Lu, X. Wei, L. Dai, Near-field MIMO communications for 6G: Fundamentals, challenges, potentials, and future directions. IEEE Commun. Mag. **61**(1), 40–46 (2022)
6. H. Zhang, N. Shlezinger, F. Guidi, D. Dardari, M.F. Imani, Y.C. Eldar, Beam focusing for near-field multiuser MIMO communications. IEEE Trans. Wirel. Commun. **21**(9), 7476–7490 (2022)
7. J. An, C. Yuen, L. Dai, M. Di Renzo, M. Debbah, L. Hanzo, Near-field communications: research advances, potential, and challenges. IEEE Wirel. Commun. **31**(3), 100–107 (2024)
8. Z. Wu, L. Dai, Multiple access for near-field communications: SDMA or LDMA? IEEE J. Sel. Areas Commun. **41**(6), 1918–1935 (2023)
9. P. Nepa, A. Buffi, Near-field-focused microwave antennas: Near-field shaping and implementation. IEEE Antennas Propag. Mag. **59**(3), 42–53 (2017)
10. J.D. Kraus, R.J. Marhefka, Antennas for all applications (2002)
11. H. Lu, Y. Zeng, Near-field modeling and performance analysis for multi-user extremely large-scale MIMO communication. IEEE Commun. Lett. **26**(2), 277–281 (2021)
12. Z. Wang, X. Mu, Y. Liu, Beamfocusing optimization for near-field wideband multi-user communications. IEEE Trans. Commun. **73**(1), 555–572 (2024)
13. A. Kosasih, E. Björnson, Finite beam depth analysis for large arrays. IEEE Trans. Wireless Commun. **23**(8), 10015–10029 (2024)
14. R. Qian, M. Sellathurai, D. Wilcox, A study on MVDR beamforming applied to an ESPAR antenna. IEEE Signal Process. Lett. **22**(1), 67–70 (2014)
15. J. Capon, High-resolution frequency-wavenumber spectrum analysis. Proc. IEEE **57**(8), 1408–1418 (1969)
16. L. Yang, M. R. McKay, R. Couillet, High-dimensional MVDR beamforming: optimized solutions based on spiked random matrix models. IEEE Trans. Signal Process. **66**(7), 1933–1947 (2018)
17. X. Wang, M. Amin, X. Cao, Analysis and design of optimum sparse array configurations for adaptive beamforming. IEEE Trans. Signal Process. **66**(2), 340–351 (2017)
18. Z.D. Zaharis, I.P. Gravas, P.I. Lazaridis, T.V. Yioultsis, T.D. Xenos, Improved beamforming in 3D space applied to realistic planar antenna arrays by using the embedded element patterns. IEEE Trans. Veh. Technol. **71**(6), 6145–6157 (2022)
19. O.L. Frost, An algorithm for linearly constrained adaptive array processing. Proc. IEEE **60**(8), 926–935 (1972)
20. B.R. Breed, J. Strauss, A short proof of the equivalence of LCMV and GSC beamforming. IEEE Signal Process. Lett. **9**(6), 168–169 (2002)
21. J. Xiong, H. Hong, H. Zhang, N. Wang, H. Chu, X. Zhu, Multitarget respiration detection with adaptive digital beamforming technique based on SIMO radar. IEEE Trans. Microw. Theory Technol. **68**(11), 4814–4824 (2020)
22. Z. Liu, S. Zhao, G. Zhang, B. Jiao, Robust adaptive beamforming for sidelobe canceller with null widening. IEEE Sensors J. **19**(23), 11213–11220 (2019)
23. Z. Liu, S. Zhao, C. Zhang, G. Zhang, Flexible robust adaptive beamforming method with null widening. IEEE Sensors J. **21**(9), 10579–10586 (2021)
24. J.A. Srar, K.-S. Chung, A. Mansour, Adaptive array beamforming using a combined LMS-LMS algorithm. IEEE Trans. Antennas Propag. **58**(11), 3545–3557 (2010)
25. Y. Chen, T. Le-Ngoc, B. Champagne, C. Xu, Recursive least squares constant modulus algorithm for blind adaptive array. IEEE Trans. Signal Process. **52**(5), 1452–1456 (2004)
26. R.C. de Lamare, M. Haardt, R. Sampaio-Neto, Blind adaptive constrained reduced-rank parameter estimation based on constant modulus design for CDMA interference suppression. IEEE Trans. Signal Process. **56**(6), 2470–2482 (2008)
27. Y. Cai, R.C. de Lamare, Low-complexity variable step-size mechanism for code-constrained constant modulus stochastic gradient algorithms applied to CDMA interference suppression. IEEE Trans. Signal Process. **57**(1), 313–323 (2008)
28. M.M. Khodier, C.G. Christodoulou, Linear array geometry synthesis with minimum sidelobe level and null control using particle swarm optimization. IEEE Trans. Antennas Propag. **53**(8), 2674–2679 (2005)

29. S.K. Goudos, V. Moysiadou, T. Samaras, K. Siakavara, J.N. Sahalos, Application of a comprehensive learning particle swarm optimizer to unequally spaced linear array synthesis with sidelobe level suppression and null control. IEEE Antennas Wirel. Propag. Lett. **9**, 125–129 (2010)
30. Y. Liao, J. Wang, Q.H. Liu, Transmit beampattern synthesis for frequency diverse array with particle swarm frequency offset optimization. IEEE Trans. Antennas Propag. **69**(2), 892–901 (2020)
31. Y. Gong, A. Yaseen, T. Zhang, R. Morawski, T. Le-Ngoc, Perturbation-based nulling control beamforming with measured element radiation patterns for MU-mMIMO. IEEE Open J. Veh. Technol. **5**, 1273–1293 (2024)
32. A. Hamza, H. Attia, Fast beam steering and null placement in an adaptive circular antenna array. IEEE Antennas Wirel. Propag. Lett. **19**(9), 1561–1565 (2020)
33. M.R.R. Khan, V. Tuzlukov, Null steering beamforming for wireless communication system using genetic algorithm, in *2011 IEEE International Conference on Microwave Technology & Computational Electromagnetics* (IEEE, 2011), pp. 289–292
34. W.L. Stutzman, G.A. Thiele, *Antenna theory and design* (John Wiley & Sons, Hoboken, 2012)
35. X. Gao, L. Dai, S. Han, I. Chih-Lin, R.W. Heath, Energy-efficient hybrid analog and digital precoding for mmWave MIMO systems with large antenna arrays. IEEE J. Sel. Areas Commun. **34**(4), 998–1009 (2016)

# Part II
# Full-Duplex Communications

# Chapter 9
# Large-Scale Antenna Arrays for In-Band Full-Duplex Massive MIMO

## 9.1 Introduction

In Part I, the role of metamaterial large-scale arrays in facilitating MU-mMIMO technology in HD communications was presented. Recognized as one of the most promising solutions for next-generation wireless systems, mMIMO leverages large-scale antenna arrays to generate highly directional and controllable beams with narrow beamwidths. This capability results in enhanced power efficiency, reduced spatial correlation, and significantly improved system capacity and spectral efficiency. Through configuring sub-arrays, mMIMO systems can form multiple beams to simultaneously serve multiple users from various directions, thereby greatly boosting system throughput [1–3].

Meanwhile, conventional HD systems, such as TDD and FDD, inherently constrain maximum data rates and spectral efficiency. In contrast, FD communication technology has the potential to double spectral efficiency and reduce latency by enabling simultaneous transmission and reception within the same frequency band [4–6]. Integrating FD with mMIMO offers a promising solution to meet the increasing throughput and latency demands of future wireless systems, particularly in spectrum-limited scenarios [7, 8]. A critical challenge in FD mMIMO systems is mitigating SI. Since the Tx and Rx operate concurrently on the same frequency, the transmitted signal can leak into the receiver, causing significant SI. The local Rx antenna is much closer to the Tx antenna than to remote users, leading to SI power levels that can far exceed the received signal power from intended users. This strong SI severely degrades the SINR, impacting the FD system performance [9]. As discussed in Chap. 2, various SI isolation and cancellation techniques have been proposed to suppress residual SI to or below the noise floor.

Tx/Rx isolation, which separates transmitted and received signals, is a crucial first step in mitigating Tx-to-Rx MC and preventing the overloading of front-end components such as ADCs or LNAs. One straightforward method to enhance isolation involves increasing the separation distance between the Tx and local



T. Le-Ngoc et al., *Half-Duplex/Full-Duplex MU-mMIMO*, Wireless Networks,
https://doi.org/10.1007/978-3-032-09195-6_9

Rx arrays, thereby introducing higher path loss for the SI signal. Experimental studies have demonstrated that separation distances of 3.5 to 5 meters can achieve attenuation levels of 65–85 dB [10, 11]. However, given practical system compactness requirements, it is essential to balance system dimensions with Tx-Rx isolation levels. Additionally, cross-polarized Tx/Rx designs can reduce SI by maximizing polarization mismatch between the local Tx and Rx antennas. Studies show that orthogonal polarization enhances isolation by 10–18 dB [12–16]. Other techniques, such as FSS [17, 18], DGS [19], and metamaterial absorbers [20–22], can further suppress, cancel, or dissipate leakage coupling waves, providing isolation improvements of 10–20 dB. However, most antenna isolation designs and investigations are limited to small-scale arrays due to challenges in simulation, manufacturing, and measurement [12, 15, 16, 18–21, 23, 24]. Experimental studies involving Tx-to-Rx elements in large-scale antenna arrays can provide realistic and critical SI channel information. These investigations facilitate the evaluation of antenna isolation techniques and offer opportunities for advancing the design and practical implementation of FD mMIMO systems.

Building on these observations, a cross-polarized 8×8Tx/8×8Rx large-scale array prototype has been developed and fabricated. The antenna element structure is based on the dual-layer metamaterial-based design introduced in Chap. 3, with the CP orientation mirrored to achieve cross-polarized Tx/Rx element pairs. HFSS-based full-wave simulation tools are employed to validate the radiation characteristics of the Tx/Rx antenna elements and to examine the effects of polarization, Tx/Rx layout, and separation distance on the achieved Tx-Rx isolation levels. Then, comprehensive experimental measurements of the Tx-Rx element-level MC provide a detailed evaluation of isolation performance and practical SI channel information. This array prototype will also be utilized in subsequent chapters for experimental studies on Tx-Rx beam-level isolation algorithm development. The main contributions presented in this chapter are summarized as follows:

- **Dual-layer metamaterial-based cross-polarized patch antenna element:** A pair of cross-polarized, dual-layer metamaterial-based patch antennas is developed for FD mMIMO array designs operating around 3.5 GHz. Orthogonal CP is achieved by mirroring the L-slotted radiating patch design, leading to a best AR of 1.9 dB for both the Tx and Rx elements. The Tx and Rx elements exhibit comparable radiation characteristics, with a gain exceeding 5.1 dB and an effective impedance bandwidth from 3.3 to 3.6 GHz. Full-wave simulations explore the effects of cross-polarization and separation distance on Tx-to-Rx antenna element isolation. The cross-polarization configuration yields an average MC level improvement of 9.7 dB. Insights from the simulations show that a Tx-Rx MC of −51.3 dB can be achieved with a separation distance of 20 cm (approximately 2.5 wavelength separation).
- **8×8Tx/8×8Rx large-scale array prototype for FD mMIMO:** A cross-polarized, dual-layer metamaterial-based 8×8Tx/8×8Rx large-scale array prototype for FD mMIMO is developed. Comprehensive experimental measurements of the SI channels between the 64 Tx and 64 Rx elements are

conducted in an anechoic chamber. A total of 64×64 = 4096 SI channels are measured with 1601 sampling points within a frequency range of 3–4 GHz for each channel. The experimental results illustrate that an average isolation of 59.1 dB across the 4096 Tx-Rx element pairs can be achieved over a bandwidth of 20 MHz (centered at 3.5 GHz) by the prototype of FD-mMIMO antenna arrays. The relationship between Tx-Rx element MC and the relative positions of Tx/Rx elements is thoroughly investigated, providing critical insights into the spatial isolation capabilities of large-scale arrays.

The rest of the chapter is structured as follows: Sect. 9.2 explores the design process of cross-polarized antenna elements tailored for large-scale FD mMIMO arrays. This section presents the Tx/Rx antenna element radiation patterns, circular polarization AR, and Tx/Rx impedance bandwidth. Additionally, the impact of Tx/Rx element separation and polarization status on MC performance is systematically investigated based on HFSS simulation results. Then, Sect. 9.3 introduces the development of an 8×8Tx/8×8Rx cross-polarized large-scale array prototype for FD mMIMO applications. This section details the experimental measurement setup used to evaluate the Tx-Rx SI channels. Comprehensive experimental SI measurement results between the 64 Tx and 64 Rx elements in the anechoic chamber are presented. Section 9.4 concludes the chapter by summarizing the findings and outlining potential directions for future work.

## 9.2 Cross-Polarized Metamaterial-Based Patch Antenna for FD Communications

### *9.2.1 Mirrored L-Slotted Radiating Patch*

As outlined in Chap. 3, a dual-layer EBG CP patch antenna is proposed for mMIMO systems operating at 3.5 GHz. The dual-layer slow-wave structure reduces the radiating patch size by 65%, enabling compact designs without compromising performance. An L-slot patch antenna is designed on the top side of the dual-layer metamaterial structure. The structure features a central square patch surrounded by copper strips, interconnected by two copper bridges. These bridges create two orthogonal current paths, generating the EM fields associated with the antenna's operation. The currents flowing through these paths excite EM waves in two directions. The resulting fields are perpendicular to each other, with the wave's polarization determined by the interaction and rotation of these field components as they propagate. The direction of CP is established by the rotation of these two field components during propagation. The initial phase and direction of the emitted fields are controlled by the feed's position relative to the patch's geometry. The orthogonal and equal-magnitude fields ensure the excitation of pure CP using a simple probe-feeding mechanism. Simultaneously, the simplicity of the CP excitation structure allows for easy modification of the CP direction, achieving a cross-polarized antenna configuration with identical radiation characteristics.

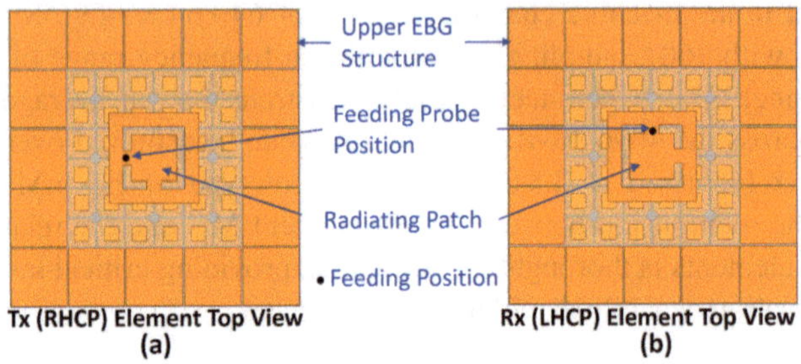


**Fig. 9.1** (**a**) Tx antenna top view; (**b**) Rx antenna top view

As illustrated in Fig. 9.1, when the CP antenna structure is mirrored, the handedness of the polarization is reversed. This reversal occurs because the EM fields are now oriented oppositely relative to the antenna's structure due to the mirrored direction of input current flow. Mirroring the antenna structure across a vertical axis reverses all asymmetric features that initially determined the polarization direction. As a result, the interaction between the electric and magnetic fields changes, altering the wave's rotational direction. In the original design, the electric field rotates clockwise to create RHCP. In contrast, in the mirrored design, it rotates counterclockwise, resulting in left-hand circular polarization (LHCP). When cross-polarized antenna pairs are used in FD operations, the polarization mismatch between the Tx and Rx elements is maximized. This configuration enhances the immunity of the local Rx antenna to leakage waves from the Tx, effectively preventing the orthogonal wave from being received and thereby improving Tx-Rx isolation in the FD system.

Additionally, the surface capacitance is controlled by the gaps between and within the top patch's metallic circuits, as well as by the dielectric material sandwiched between these circuits and the ground plane. The inductance is determined by the geometry of the conducting strips in the patch antenna, which governs the magnetic field linkage within the structure. The symmetric design of the cross-polarized antenna pairs also guarantees identical radiation characteristics in terms of impedance bandwidth, element gain, and polarization purity, eliminating the need for parametric optimization of the mirrored element structure. The performance of this design will be examined in the following sections.

### *9.2.2 Tx/Rx Antenna Element Impedance Bandwidth Simulation*

The HFSS simulated $S_{11}$-parameters, which illustrate the impedance matching bandwidth of the proposed Tx and Rx antenna elements, are presented in Fig. 9.2. The reflection characteristics of the Tx and Rx elements exhibit a consistent trend with respect to the operating frequency. Both elements achieve a return loss

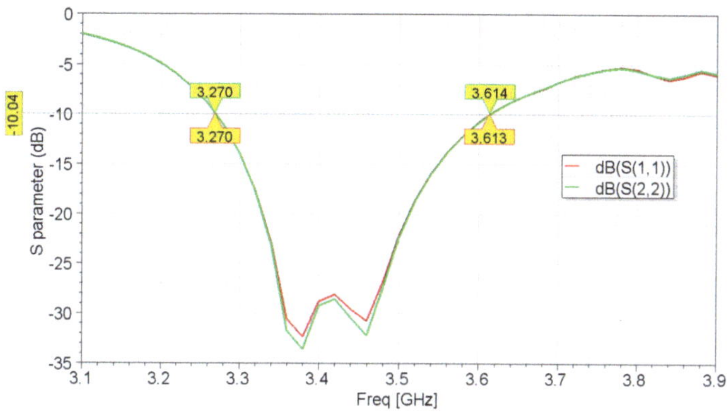


**Fig. 9.2** Return loss of Tx element (labelled by S(1,1)) and Rx element (labelled by S(2,2)) (simulation results)

better than 10 dB within the band from 3.27 to 3.61 GHz, indicating an effective impedance matching bandwidth of 340 MHz or 9.7% at 3.5 GHz. Furthermore, the return loss is better than 20 dB within the band from 3.33 to 3.51 GHz. A good return loss indicates well-matched antenna impedance, ensuring that the majority of input power is transmitted rather than reflected, thereby enhancing signal transmission efficiency and overall system performance.

### *9.2.3 Tx/Rx Antenna Element Radiation Pattern Simulation*

The simulated radiation patterns for the Tx and Rx antenna elements are depicted in Fig. 9.3. Both elements display symmetric radiation patterns, each with a directivity of 6.1 dB. The realized gains for the Tx and Rx elements stand at 5.1 and 5.2 dB, respectively, corresponding to radiation efficiencies of 79.4% for the Tx and 81.3% for the Rx. The 3dB-beamwidth is around 94° in both axis directions $\phi = 0^o$ and $\phi = 90^o$. The copper ground structure effectively eliminates back radiation; as a result, the normalized front-to-back ratios of the Tx and Rx elements, which quantify the power difference between forward and backward radiation, are $-11.8$ and $-12$ dB, respectively. Notably, the Rx element's radiation pattern can be derived by rotating the Tx element pattern by 90° in the $\phi$ plane.

### *9.2.4 Tx/Rx Antenna Element Circular-Polarization Axial Ratio Simulation*

High CP purity is essential for antennas to effectively transmit or receive the intended polarization, thereby minimizing signal degradation and improving com-

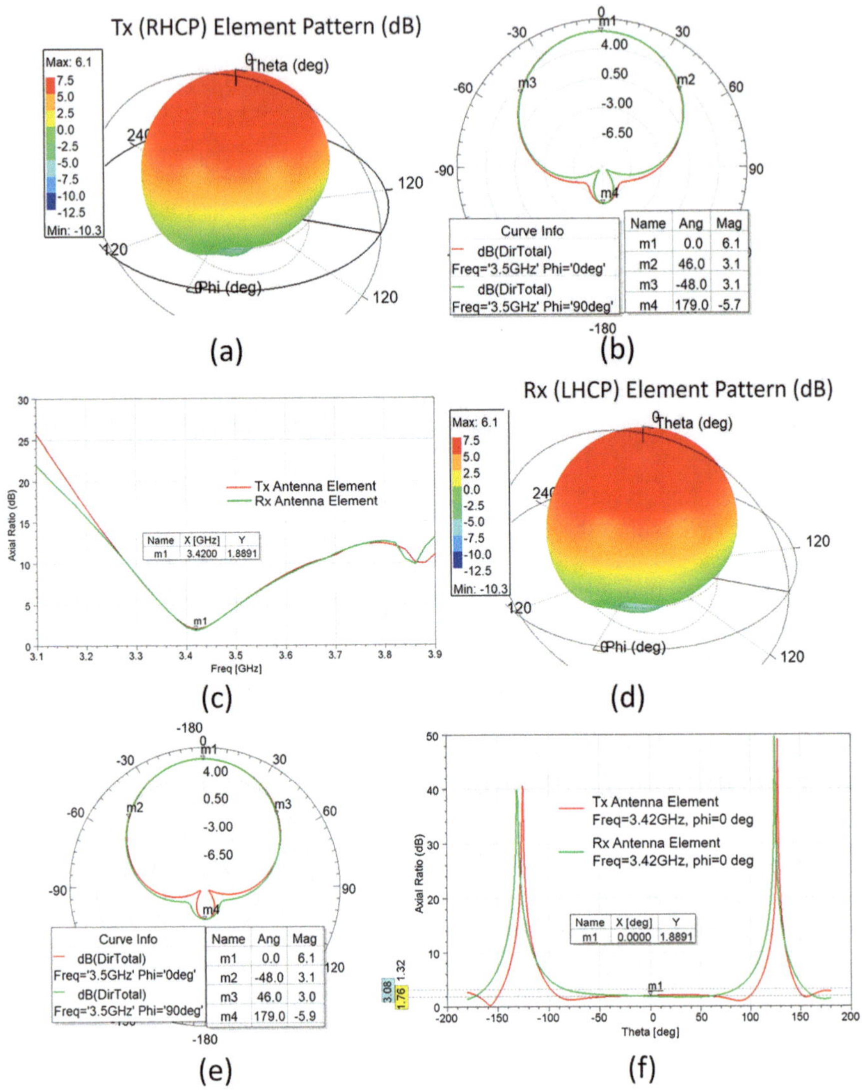


**Fig. 9.3** Element radiation patterns (directivity in dB) and element AR, (**a**) Tx element 3D at 3.5 GHz, (**b**) Tx 2D cut at $\phi = 0^o, 90^o$ at 3.5 GHz, (**c**) Tx/Rx AR (dB) versus operating frequency (GHz), (**d**) Rx element 3D at 3.5 GHz, (**e**) Rx 2D cut at $\phi = 0^o, 90^o$, (**f**) Tx/Rx AR at 3.42 GHz (simulation results)

munication performance. The AR, defined as the ratio of the magnitudes of the orthogonal electric field components of the radiated wave, serves as a key metric for assessing CP quality. A low AR indicates a high degree of CP purity.

The HFSS simulation AR results at different operating frequencies of the proposed Tx and Rx antenna elements are presented in Fig. 9.3c. It can be observed that the best AR of 1.9 dB can be achieved at 3.42 GHz. The proposed antenna structure maintains an AR below 3 dB across an operating bandwidth of approximately 80 MHz, from 3.38 to 3.46 GHz, ensuring high polarization purity. Across a 190 MHz bandwidth, from 3.34 to 3.53 GHz, the AR remains better (lower) than 6 dB. Beyond this frequency range, the AR rapidly increases, indicating a deterioration in CP purity due to the degradation of the antenna's radiation characteristics. The Tx/Rx antenna element AR in the $\theta$ angle range at 3.42 GHz is illustrated in Fig. 9.3f. The results show that the Tx/Rx elements achieve an AR better (lower) than 3 dB in the whole broadside radiation range, i.e., from $-90°$ to $90°$. In FD communications, CP purity directly influences the effectiveness of cross-polarization designs. High CP purity in both Tx and Rx antennas increases the Rx antenna's immunity to leakage signals from the Tx antenna, thereby improving Tx-Rx isolation and overall system performance.

### 9.2.5 *Tx-Rx Antenna-Element Mutual Coupling Simulation*

The MC between Tx and Rx antennas is influenced by both the separation distance ($d_e$) and the polarization efficiency. To understand the interplay between separation distance and MC, as well as the effects of cross-polarization, a series of simulations are conducted. As shown in Fig. 9.4, the simulations begin with both Tx and Rx antennas having identical polarization and being aligned in the same direction, both RHCP with an orientation angle of $\phi = 0°$. In subsequent trials, the orientation of the Rx antenna is rotated within the $\phi$ plane in increments of 90°. The edge-to-edge separation $d_e$ is increased in increments of 4 cm (approximately half-wavelength at 3.5 GHz), ranging from 0 to 20 cm. Additionally, the Rx antenna is also mirrored in the simulation, where the Rx antenna is adjusted to orthogonal polarization (OP) status. The HFSS simulation results are summarized in Table 9.1.

#### 9.2.5.1 Tx-to-Rx MC Versus Polarization States

It can be observed that the variations in MC levels between the Tx and Rx antennas in the first four configurations, where the elements are co-polarized and $\phi$ ranges

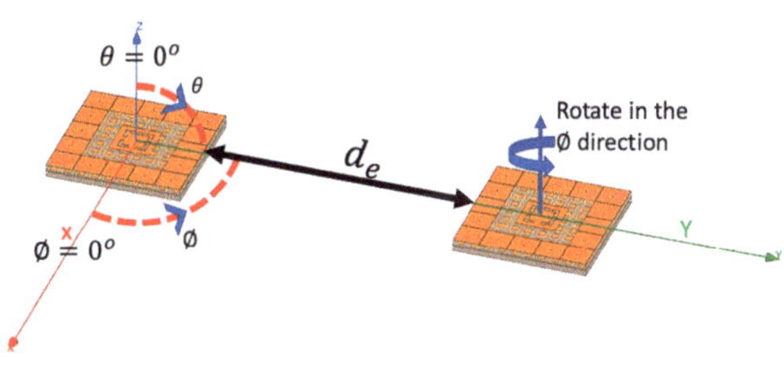


**Fig. 9.4** Locations of Tx and Rx elements in the MC simulation

**Table 9.1** MC between Tx and Rx antenna VS separation distance ($d_e$) at 3.5 GHz (dB) (simulation results)

| | Separation $d_e$ (cm) | 0 | 4 | 8 | 12 | 16 | 20 |
|---|---|---|---|---|---|---|---|
| 1 | $\phi = 0^o$ | −15.2 | −28.4 | −32.7 | −35.7 | −37.8 | −39.6 |
| 2 | $\phi = 90^o$ | −14.5 | −28.5 | −32.5 | −35.1 | −37.2 | −38.9 |
| 3 | $\phi = 180^o$ | −15.4 | −28.7 | −33.2 | −36.2 | −38.3 | −40.2 |
| 4 | $\phi = 270^o$ | −15.9 | −29.6 | −33.5 | −36.1 | −38.3 | −39.9 |
| 5 | Mirrored (OP) | −24.2 | −36.3 | −42.3 | −46.1 | −48.4 | −51.3 |

from 0° to 270°, are less than 1.4 dB. This indicates that rotating the Rx antenna does not effectively reduce the MC level when both antennas share the same polarization state. As the separation distance increases, a lower (better) Tx-Rx MC is achieved; however, the rate of improvement slows when the separation distance exceeds two wavelengths. The best Tx-Rx isolation levels of 33.5 and 40.2 dB are achieved at separations of 8 cm (approximately one wavelength) and 20 cm (approximately 2.5 wavelengths), respectively, in the first four Tx/Rx layout.

A considerable improvement in isolation is observed when the Tx and Rx antenna elements are placed in a mirrored configuration, with the Tx antenna being RHCP and the Rx antenna being LHCP. Compared to the average MC magnitude of the first four configurations, this orthogonal polarization configuration yields an average MC level that is 9.7 dB lower. The improvement in isolation varies, reaching a maximum reduction of 11.7 dB at a 20 cm separation and a minimum reduction of 7.5 dB at a 4 cm separation. Tx-Rx isolation of 42.3 and 51.3 dB are achieved at separations of 8 cm (approximately one wavelength) and 20 cm (approximately 2.5 wavelengths), respectively. As a result, Configuration 5 is utilized in our design for the subsequent simulations and measurements.

#### 9.2.5.2 Tx-to-Rx MC Versus Separation Distance

Figure 9.5 displays the HFSS simulation results of MC level versus element separation distance for the cross-polarized Tx/Rx setup over the entire target frequency range of 3.35–3.60 GHz. At an 8 cm (approximately 1 wavelength) and 20 cm (2.5 wavelengths) separation, the Tx-Rx MC remains lower (better) than −42 and −50 dB across the entire target bandwidth. All six curves exhibit a similar trend. Beyond a separation distance of 20 cm, the slope of the curves begins to flatten, indicating a reduced rate of decrease in MC magnitude. Further increasing the separation distance by approximately one wavelength (8 cm) results in a decrease in MC magnitude of less than 5 dB. These observations suggest that as the separation distance exceeds 20 cm, the impact of surface leakage wave components following the LoS path diminishes.

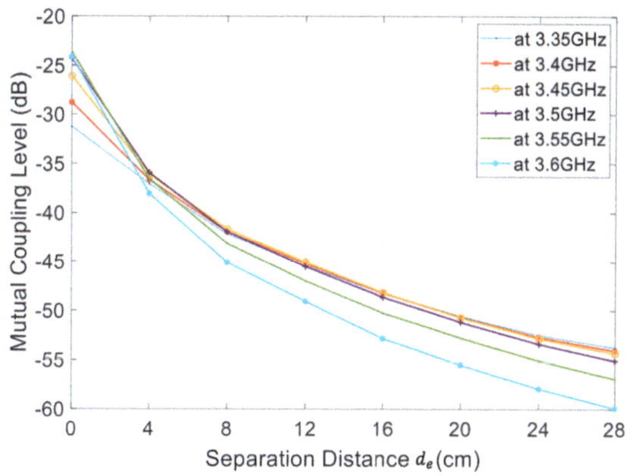


**Fig. 9.5** Tx-to-Rx MC magnitude vs separation distance (operating frequency from 3.35 to 3.6 GHz) (simulation results)

#### 9.2.5.3 Tx-to-Rx Element MC Between Sample Arrays

To simulate a realistic scenario for estimating the isolation levels between Tx-Rx elements when configured into arrays, two sample 4×4 Tx/Rx antenna arrays were constructed in HFSS using the proposed cross-polarized Tx/Rx element pairs. The arrays were placed 20 cm apart (edge-to-edge). Simulation results show an average MC of −48.4 dB across 256 Tx-Rx element pairs at 3.5 GHz, which is 2.9 dB higher than the level observed when only two elements are simulated. Increasing the array edge-to-edge separation to 24 cm reduces the average Tx-Rx MC to −49.5 dB, an improvement of only 1.1 dB compared to the 20 cm separation. Therefore, a distance of 20 cm was selected between the Tx and Rx arrays to balance antenna isolation performance with the overall structure dimensions.

## 9.3 Cross-Polarized 8×8Tx/8×8Rx Antenna Array Prototypes

### *9.3.1 Tx/Rx Large-Scale Array Prototype*

Figure 9.6 presents the fabricated 8×8Tx/8×8Rx array prototype, which incorporates the proposed cross-polarized Tx/Rx antenna elements along with their corresponding element indices. To avoid spatial aliasing, coherent plane-wave addition, and the formation of grating lobes, an inter-element spacing of 4.0 cm ($0.47\lambda_0$) center-to-center is employed in the Tx/Rx antenna arrays. All elements in each Tx or Rx array share the same FR-4 dielectric substrate and ground plane. Each array has dimensions of 32 cm in width and 32 cm in length, while the two arrays are spaced 20 cm apart. Accounting for this separation, the complete Tx-Rx antenna array prototype set measures 84 cm in length and 32 cm in width, accommodating a total of 64 transmitting and 64 receiving elements. On the back of the antenna

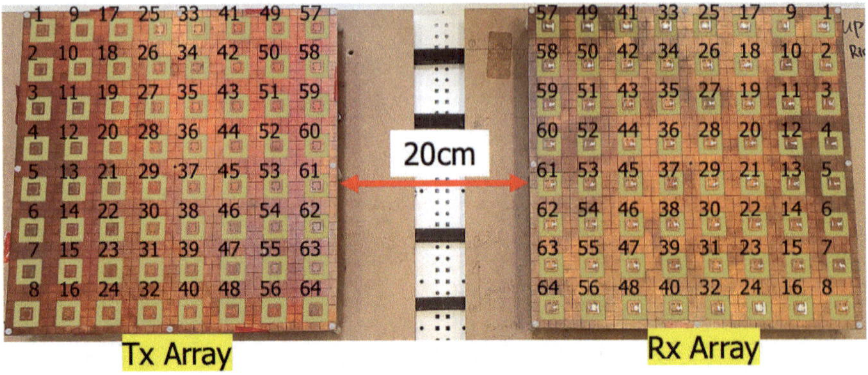


**Fig. 9.6** 8×8Tx/8×8Rx array prototype for FD mMIMO with element indices

array, each antenna element is soldered with a JSC coaxial connector, which is further connected to the BFN components or RF switches, depending on the type of measurement being performed.

### 9.3.2 Tx-Rx Element Mutual Coupling Measurement Experimental Setup

To comprehensively assess the MC between the Tx-Rx elements and obtain realistic SI channel data, measurements were conducted using an 8×8Tx/8×8Rx array prototype, as shown in Fig. 9.7. The over-the-air (OTA) FD measurement setup was performed in our anechoic chamber with internal dimensions of 20 ft (6.q m) in length, 8 ft (2.4 m) in width, and 8 ft (2.4 m) in height, eliminating external reflections. The chamber was large enough to accommodate the antenna array under test, measuring 84 × 32 cm (width × height), positioned approximately 4 ft away from the five walls, which were covered with C-RAM SFC-48 absorbers that provided reflections of less than −45 dB at 3.5 GHz. S-parameters were measured using a Keysight N5247A VNA with an output power of 10 dBm. To minimize noise impact on the high-isolation measurements, the intermediate frequency bandwidth of the PNA was set to 300 Hz (resulting in an integrated thermal noise power of approximately −149 dBm). The trace smoothing function was set to 1%, and averaging was disabled. The measurement bandwidth ranged from 3 to 4 GHz, with a step size of 625 kHz, yielding 1601 sample points. Each Tx/Rx antenna element pair was measured over a sweep time of approximately 10 seconds.

To streamline the iterative measurement process and protect the fragile JSC connectors on the antenna board, two 1-to-16 RF switches were employed, with MATLAB-based programs controlling the switching between different Tx/Rx antenna element pairs. Both the Tx and Rx antennas were connected to the VNA

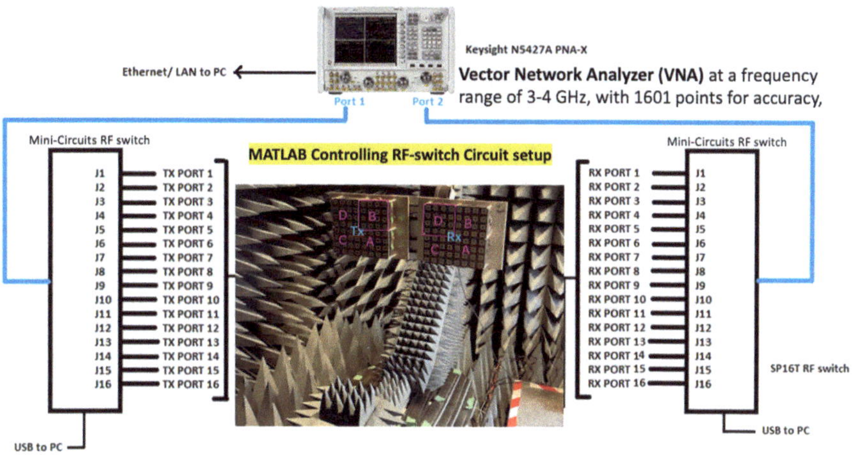


**Fig. 9.7** 8×8Tx/8×8Rx array prototype for FD mMIMO measurement setup

via 5-meter-long coaxial cables and Mini-Circuits USB-1SP16T-83H RF absorptive switches, resulting in an approximate total insertion loss of 12.5 dB on both sides. Measurements were conducted in sections of 16Tx/16Rx using SP16T RF switches to capture the full 64×64 S-parameter matrix, with the data transmitted to a PC for post-processing in MATLAB. Variations in the trace over frequency, for a fixed RF attenuator of 100 dB at the VNA ports, were approximately $\pm$ 0.75 dB in magnitude, $\pm$ 3° in phase, and $\pm$ 100 ps in group delay. The MC between individual Tx-to-Rx element pair ranges from −37 to −80 dB. Consequently, the minimum expected SNR at the VNA receiver was approximately 54 dB (calculated as 10 dBm—12.5 dB—80 dB—12.5 dB—(−149 dBm)).

The resulting S-parameters, saved as .s2p files, were imported into MATLAB for post-processing. The insertion losses from the measurement setup were de-embedded to account for losses in the connections. These de-embedded S-parameters were then combined into .s128p files to construct the 64×64 SI channel matrix, which represented the MC between all the Tx and Rx elements. The complete complex SI channel matrix, $\mathbf{H}_{SI}$, had dimensions of 64×64×1601, corresponding to the 1601 frequency points measured.

### 9.3.3 *Tx-Rx Element Mutual Coupling Versus Operating Frequency*

Given the extensive number of Tx-to-Rx element pairs, the magnitude plot of measured MC (in dB) versus the whole operating frequency bandwidth for three representative pairs: the closest (T52/T60 to R52/R60), the pairs located at the array center (T28/T36 to R28/R36), and the farthest (T4/T12 to R4/R12), are presented in

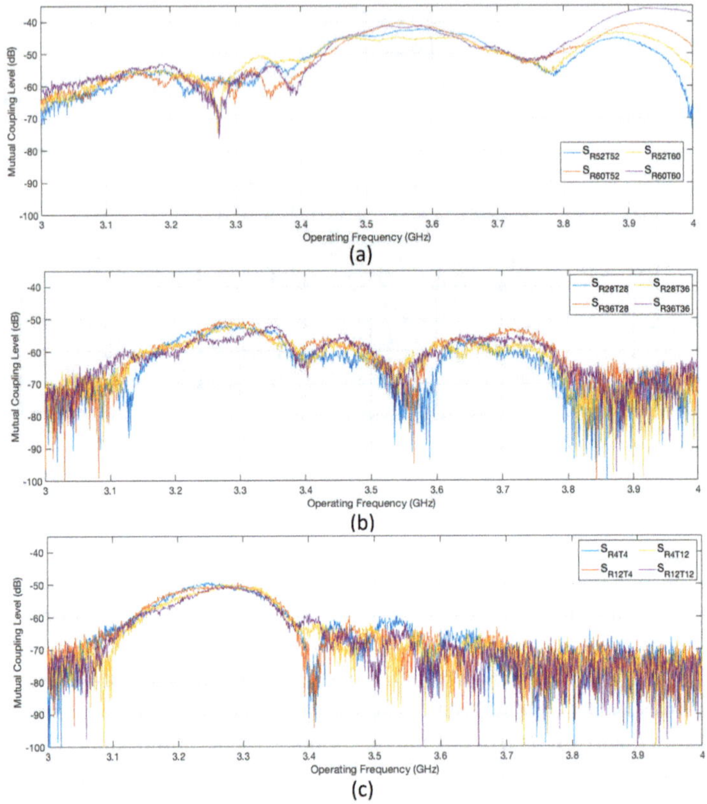


**Fig. 9.8** Measured S-parameter magnitude plots of sample Tx-to-Rx antenna element pairs: (**a**) closest: T52/T60 to R52/R60; (**b**) elements at the array centers: T28/T36 to R28/R36; (**c**) farthest: T4/T12 to R4/R12

Fig. 9.8. As the distance between the Tx and Rx antennas increases, along with the number of EBG elements placed between them increasing, a considerable improvement in Tx-Rx isolation can be observed. At 3.5 GHz, the MC values average at −44.2, −60.8, and −68.5 dB for the pairs shown in Fig. 9.8a, b, and c, respectively. Moreover, when analyzed over bandwidths of 20 and 100 MHz, the variation in average MC values within these bandwidths is less than 2.3 dB compared to the values at the center frequency of 3.5 GHz.

Considering the 200 MHz bandwidth from 3.4 to 3.6 GHz, the average MC levels are −44.4 dB for T52/T60 to R52/R60, −61.1 dB for T28/T36 to R28/R36, and −66.4 dB for T4/T12 to R4/R12, indicating consistent Tx-Rx element isolation performance at 3.5 GHz. Outside this bandwidth, a more rapid change in MC levels is observed due to degradation in antenna impedance matching and polarization purity.

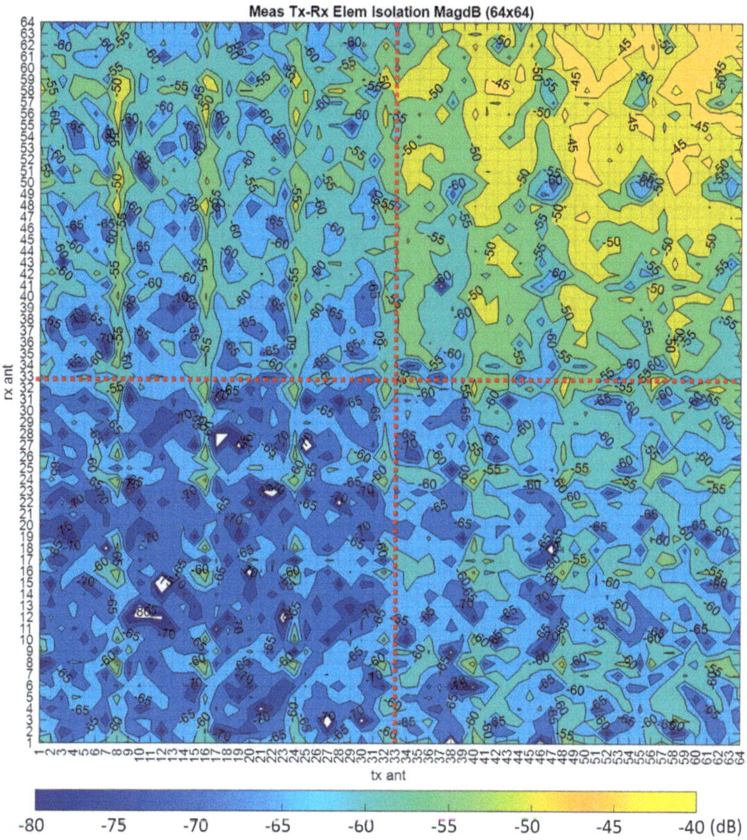


**Fig. 9.9** 64 Tx element to 64 Rx element mutual coupling magnitude in dB maps (at 3.5 GHz) (measurement results)

## *9.3.4 Tx-Rx Element Mutual Coupling*

### 9.3.4.1 Measured Tx-Rx MC Magnitude Results

To illustrate the SI channels between the 64 Tx and 64 Rx elements at various locations, the measured MC magnitudes (in dB) at 3.5 GHz are presented in Fig. 9.9. Across all 4096 SI channels, an average[1] MC level of −59.6 dB is achieved at 3.5 GHz. An average MC level of −59.1 dB is achieved over the 20 MHz range (3.49–3.51 GHz), and −58.4 dB over the 100 MHz range (3.45–3.55 GHz).

---

[1] This study utilizes geometric means to characterize the average mutual coupling or isolation level, as the mutual coupling value represents a ratio of the received power (Rx) to the transmitted power (Tx). For averaging MC power across a bandwidth, where it signifies the cumulative power within the operating frequencies, an arithmetic mean is employed to depict the average MC within a band.

At 3.5 GHz, the two Tx-Rx element pairs with the highest (worst) MC are: 1) Tx-64 and Rx-62, which has the Tx element at the array corner and the Rx element at the array edge, shows −37.8 dB MC; 2) Tx-58 and Rx-58, which have the smallest separation distance, shows MC level −38.5 dB. The lowest (best) MC value of −101.3 dB can be achieved. Across all 4096 Tx-Rx pairs, 99.9% (4092 pairs) achieve an MC level of better than −40 dB, 87.0% (3562 pairs) achieve an MC better than −50 dB or better, and 48.3% (1980 pairs) achieve an MC better than −60 dB. 8.6% (351 pairs) achieve an MC level of −70 dB or better.

A clear trend in MC behavior versus the separation distance is observed: Tx-Rx pairs with closer spacing, for instance, in the top-right quadrant of the MC heatmap, exhibit stronger SI levels. The average MC between Tx-33 to Tx-64 and Rx-33 to Rx-64 is −46.7 dB. In contrast, the average MC between Tx-1 to Tx-32 and Rx-1 to Rx-32 is significantly lower at −66.1 dB, as indicated in the bottom-left quadrant of the graph, a difference of approximately 19.4 dB. Among these 1024 Tx-Rx pairs, 99.4% Tx-Rx element pairs can achieve an MC better than −55 dB, and 85.7% can achieve an MC better than −60 dB. For Tx-Rx element pairs with similar relative locations, comparable MC levels are observed. For instance, the average MC for the Tx/Rx element pairs in the top-left quadrant and bottom-right quadrant of the Fig. 9.9 is −59.5 and −61.4 dB, respectively, indicating an average difference of approximately 1.9 dB.

This analysis demonstrates that distance and element surrounding environments play a crucial role in determining MC levels, with greater Tx-Rx separation leading to reduced coupling. Furthermore, the presence of additional EBG patches between Tx and Rx elements enhances isolation. As the distance increases, surface wave leakage between the Tx-Rx elements diminishes and eventually becomes negligible.

#### 9.3.4.2 Measured Tx-Rx MC Phase Results

The measured MC phase (in degrees) between the 64 Tx and 64 Rx antenna elements at 3.5 GHz is shown in Fig. 9.10. A distinct periodic pattern in the MC phase can be observed along the diagonal direction of the plot. This behavior stems from variations in coupling path lengths, which change with half-wavelength separation between the elements. Specifically, when the distance between Tx and Rx elements approaches a wavelength or its multiples, the coupled waves exhibit a consistent phase relationship, thus creating the periodic phase shifts across the Tx and Rx arrays.

Additionally, out-of-phase MC waves (where the phases of the coupled signals are shifted by 180°) can be advantageous for improving spatial isolation. Their inherent destructive interference reduces unwanted signal leakage among neighboring elements in the signal summation, a feature especially advantageous in antenna sub-array beamforming isolation design. By leveraging this out-of-phase SI signal cancellation, SI can be further mitigated, enhancing isolation between sub-arrays, which will be presented in the following chapter.

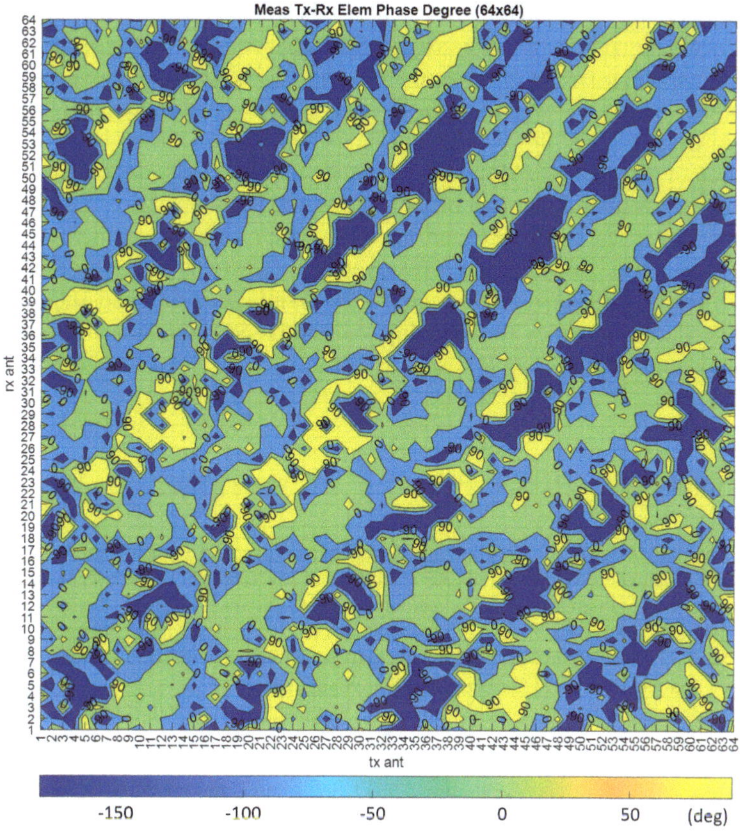


**Fig. 9.10** 64 Tx element to 64 Rx element mutual coupling phase in degree maps (at 3.5 GHz) (measurement results)

## 9.4 Concluding Remarks

In this chapter, a cross-polarized 8×8Tx/8×8Rx large-scale array prototype was designed, fabricated, and characterized to demonstrate its capability in Tx-Rx antenna isolation for FD mMIMO applications around 3.5 GHz. Leveraging the dual-layer metamaterial-based antenna design from Chap. 3, cross-polarized Tx/Rx pairs are achieved through mirroring the L-slotted radiating patch and achieve comparable radiation characteristics. Detailed full-wave simulations in HFSS confirmed the effectiveness of the cross-polarized configuration, showing that appropriate element separation and polarization matching are key to achieving high Tx-Rx isolation. An in-depth experimental study involving measurements of 4096 Tx–Rx SI channels over a 1 GHz bandwidth (3–4 GHz) further validated the prototype's isolation performance, with an average element isolation of 59.1 dB observed within a 20 MHz band centered at 3.5 GHz. In addition, the array's spatial isolation capabilities were thoroughly investigated based on the relative positioning of Tx and Rx elements, providing valuable insights for large-scale FD deployments.

The comprehensive set of measured SI channel data provides a realistic foundation for developing and testing beam-level isolation and cancellation algorithms. In the following chapters, this dataset will be used to examine beam-level isolation as a function of array length. Moreover, the fabricated $8\times8$Tx/$8\times8$Rx array prototype will serve as a testbed for further investigations into beam-level isolation algorithms, exploring the relationship between beam radiation patterns and Tx-Rx isolation levels, as well as assessing the effectiveness of metamaterial absorbers. Additionally, the illustrative results highlight the significance of multi-stage isolation and cancellation schemes for realizing substantial SI mitigation in practical FD communication systems.

## References

1. P. Xingdong, H. Wei, Y. Tianyang, L. Linsheng, Design and implementation of an active multibeam antenna system with 64 RF channels and 256 antenna elements for massive MIMO application in 5G wireless communications. China Commun. **11**(11), 16–23 (2014)
2. S. Malkowsky, J. Vieira, L. Liu, P. Harris, K. Nieman, N. Kundargi, I.C. Wong, F. Tufvesson, V. Öwall, O. Edfors, The world's first real-time testbed for massive MIMO: Design, implementation, and validation. IEEE Access **5**, 9073–9088 (2017)
3. E.G. Larsson, O. Edfors, F. Tufvesson, T.L. Marzetta, Massive MIMO for next generation wireless systems. IEEE Commun. Mag. **52**(2), 186–195 (2014)
4. A. Sabharwal, P. Schniter, D. Guo, D.W. Bliss, S. Rangarajan, R. Wichman, In-band full-duplex wireless: challenges and opportunities. IEEE J. Sel. Areas Commun. **32**(9), 1637–1652 (2014)
5. T. Le-Ngoc, A. Masmoudi, Full-duplex wireless communications systems. *Wirel. Netw.* (Springer, 2017)
6. Z. Zhang, K. Long, A.V. Vasilakos, L. Hanzo, Full-duplex wireless communications: challenges, solutions, and future research directions. Proc. IEEE **104**(7), 1369–1409 (2016)
7. G.C. Alexandropoulos, M.A. Islam, B. Smida, Full-duplex massive multiple-input, multiple-output architectures: Recent advances, applications, and future directions. IEEE Veh. Technol. Mag. **17**(4), 83–91 (2022)
8. A. Shojaeifard, K.-K. Wong, M. Di Renzo, G. Zheng, K.A. Hamdi, J. Tang, Massive MIMO-enabled full-duplex cellular networks. IEEE Trans. Commun. **65**(11), 4734–4750 (2017)
9. F. Chen, H.H. Lee, R. Morawski, T. Le-Ngoc, RF/Analog self-interference canceller for $2\times2$ MIMO full-duplex transceiver, in *2017 IEEE International Conference on Communications (ICC)* (IEEE, 2017), pp. 1–6
10. M. Duarte, A. Sabharwal, V. Aggarwal, R. Jana, K.K. Ramakrishnan, C.W. Rice, N. Shankaranarayanan, Design and characterization of a full-duplex multiantenna system for WiFi networks. IEEE Trans. Veh. Technol. **63**(3), 1160–1177 (2013)
11. W. Slingsby, J. McGeehan, Antenna isolation measurements for on-frequency radio repeaters, in *1995 Ninth International Conference on Antennas and Propagation, ICAP'95 (Conf. Publ. No. 407)*, vol. 1 (IET, 1995), pp. 239–243
12. B. Debaillie, D.-J. van den Broek, C. Lavin, B. van Liempd, E.A. Klumperink, C. Palacios, J. Craninckx, B. Nauta, A. Pärssinen, Analog/RF solutions enabling compact full-duplex radios. IEEE J. Sel. Areas Commun. **32**(9), 1662–1673 (2014)
13. H. Nawaz, I. Tekin, Dual-polarized, differential fed microstrip patch antennas with very high interport isolation for full-duplex communication, IEEE Trans. Antennas Propag. **65**(12), 7355–7360 (2017)

14. J. Ha, M.A. Elmansouri, P.V. Prasannakumar, D.S. Filipovic, Monostatic co-polarized full-duplex antenna with left-or right-hand circular polarization. IEEE Trans. Antennas Propag. **65**(10), 5103–5111 (2017)
15. F. Chen, R. Morawski, T. Le-Ngoc, Self-interference channel characterization for wideband 2× 2 MIMO full-duplex transceivers using dual-polarized antennas. IEEE Trans. Antennas Propag. **66**(4), 1967–1976 (2018)
16. D. Wu, Y.-X. Sun, B. Wang, R. Lian, A compact, monostatic, co-circularly polarized simultaneous transmit and receive (STAR) antenna with high isolation. IEEE Antennas Wirel. Propag. Lett. **19**(7), 1127–1131 (2020)
17. S. Ghosh, T.-N. Tran, T. Le-Ngoc, Dual-layer EBG-based miniaturized multi-element antenna for MIMO systems. IEEE Trans. Antennas Propag. **62**(8), 3985–3997 (2014)
18. H.S. Farahani, M. Veysi, M. Kamyab, A. Tadjalli, Mutual coupling reduction in patch antenna arrays using a UC-EBG superstrate. IEEE Antennas Wirel. Propag. Lett. **9**, 57–59 (2010)
19. D. Ahn, J.-S. Park, C.-S. Kim, J. Kim, Y. Qian, T. Itoh, A design of the low-pass filter using the novel microstrip defected ground structure. IEEE Trans. Microw. Theory Technol. **49**(1), 86–93 (2001)
20. G. Zhai, Z.N. Chen, X. Qing, Enhanced isolation of a closely spaced four-element mimo antenna system using metamaterial mushroom. IEEE Trans. Antennas Propag. **63**(8), 3362–3370 (2015)
21. A. Dadgarpour, B. Zarghooni, B.S. Virdee, T.A. Denidni, A.A. Kishk, Mutual coupling reduction in dielectric resonator antennas using metasurface shield for 60-GHz MIMO systems. IEEE Antennas Wirel. Propag. Lett. **16**, 477–480 (2016)
22. Y. Gong, R. Morawski, T. Le-Ngoc, Metamaterial absorber structure for tx-rx antenna isolation improvement in full-duplex massive MIMO. IEEE Access **12**, 64571–64588 (2024)
23. J. Wu, M. Li, N. Behdad, A wideband, unidirectional circularly polarized antenna for full-duplex applications. IEEE Trans. Antennas Propag. **66**(3), 1559–1563 (2018)
24. E. Everett, A. Sahai, A. Sabharwal, Passive self-interference suppression for full-duplex infrastructure nodes. IEEE Trans. Wirel. Commun. **13**(2), 680–694 (2014)

# Chapter 10
# Tx-Rx Beam Isolation Improvement for FD mMIMO

## 10.1 Introduction

FD mMIMO has emerged as a promising technique for boosting throughput and spectral efficiency in wireless communication systems, as it enables simultaneous transmission and reception over the same bandwidth. However, this concurrent operation introduces strong SI due to signal leakage from the nearby Tx to its own Rx. The resulting high SI power degrades the SINR and limits channel throughput [1]. As discussed in Chap. 2, a single stage of SI isolation or cancellation cannot effectively reduce SI near the noise floor and mitigate its adverse impact on FD performance, primarily because of: (1) Excessive hardware and computational complexity in a single-stage SIC process. (2) Nonlinear distortions at the front-end ADC or LNA due to the excessive SI power. (3) Inadequate RF isolation provided by single-stage antenna design or decoupling and isolation-enhancement structures.

To address these challenges, multi-stage approaches that combine Tx-Rx isolation techniques with SIC methods have gained traction, leveraging the additional degrees of freedom in large-scale antenna arrays to effectively mitigate SI and overcome the bottlenecks of single-stage solutions in FD mMIMO [2]. To begin with, Tx/Rx isolation denotes the ability to separate transmit and receive signals by minimizing RF leakage between the local Tx and Rx. Structural antenna array design serves as the first line of defence against SI. As detailed in Chap. 9, our proposed 8×8Tx/8×8Rx FD antenna array achieves an average Tx-Rx element isolation of 59.1 dB by combining cross-polarization with strategic spacing. In addition, large-scale arrays allow for enhanced spatial isolation through narrow, steerable beams, which focus signal energy in a confined region and thus minimize leakage in unintended directions. For instance, experimental results in [3] show that directional antennas in low-reflection environments can increase isolation by about 17 dB compared to omnidirectional antennas. Improved beam resolution further enlarges the set of uplink-downlink steering angle pairs with high isolation by adding more null angles to the beam radiation patterns. [4] demonstrates

 
T. Le-Ngoc et al., *Half-Duplex/Full-Duplex MU-mMIMO*, Wireless Networks,
https://doi.org/10.1007/978-3-032-09195-6_10

how electronically steering beams toward the nulls of interference sources can significantly suppress SI. Nevertheless, there is a lack of research examining how beam-level isolation changes with array length or beamwidth based on realistic measured SI channels.

To reduce hardware complexity while efficiently harnessing large-scale arrays for FD mMIMO SI management, HBF has become popular for joint uplink/downlink beamformer design [5]. HBF splits the beamforming procedure into RF beamforming and digital baseband precoding/combining, connecting them with fewer RF chains to reduce hardware cost and power consumption. The efficacy of HBF in SI suppression has been widely studied in theoretical SI models and simulations. For example, [6] proposes sequential convex programming for separate and joint FD beamforming that maximizes sum rates for SU and MU MIMO FD systems. An iterative FD HBF scheme in [7] jointly designs transmit and receive RF beamformers, precoders, and combiners, realizing SI suppression of up to 30 dB. Then, [8] presents an HBF design for mmWave FD mMIMO that exploits angle-of-departure and angle-of-arrival information, reducing SI power by 81.5 dB in far-field channels and 44.5 dB in near-field channels, respectively. Building on these innovations, [4] introduces an FD HBF framework for MU-mMIMO that uses non-orthogonal beams to simultaneously serve multiple users while suppressing near-field SI and maximizing sum-rate capacity, achieving 78.1 dB of SI isolation. By incorporating instantaneous SI channel state information into HBF design, FD MU-mMIMO systems can accommodate a single downlink UE alongside a single uplink UE on the same frequency band [9] or multiple downlink and uplink UEs [10, 11].

Although existing multi-stage SI suppression techniques in FD mMIMO systems have shown efficacy in mitigating strong SI from Tx to Rx and an isolation level of 78 to 81 dB can be achieved [4, 8], their performance has primarily been evaluated based on theoretical SI channel models and simulations [4, 6, 8, 10–13]. For instance, [4, 6, 8, 12, 13] base their assumptions and analyses on simulated near-field SI channels over LoS paths and far-field SI channels via reflected non-line-of-sight (NLoS) paths. Moreover, there is a gap in experimental investigations of how Tx/Rx beam radiation patterns correlate with beam-level isolation, a vital consideration in understanding the effectiveness of beamformer-based SI enhancement. Hence, a more realistic evaluation of beamforming-based SI suppression, which considers both quantized BFNs and real EM interactions within and between arrays, is needed to experimentally validate the SI suppression capabilities of FD mMIMO system implementations.

Moreover, rather than solely adjusting the uplink and downlink beam directions [4, 8], which can compromise the quality of the intended signal, null steering beamforming (NSB) techniques, such as LCMV [14, 15], MVDR [16], offer a viable solution for effectively creating radiation nulls in the Tx and Rx beams along directions of potential strong SI. NSB operates by manipulating the feeding magnitude and the phase shift, enabling dynamic steering of the array's main lobe towards the intended signal's direction of arrival while concurrently establishing a radiation null in the direction of unwanted interference [17]. Nonetheless, only a very limited number of studies investigate the NSB within FD system design, particularly involving realistic SI channels and experimental validation.

After combining antenna and beamforming isolation, the residual SI can be lowered sufficiently to avoid saturating critical components in the Rx RF chains. Then, SIC techniques can be employed to further mitigate or eliminate the remaining SI, which arises from local transmission signals leaking into the receive path. These methods use advanced algorithms and signal processing strategies to estimate, reconstruct, and subtract residual interference [18, 19]. However, the accuracy of SI channel estimation is crucial for effective SIC, highlighting the need for beamformer designs that provide adequate margins to suppress SI while maintaining the FD system's overall performance.

Building on these observations, this chapter delves into beamforming isolation techniques for FD mMIMO systems. It begins by introducing the system model underlying the HBF scheme for FD mMIMO. Then, the relationship between array length, which dictates beam resolution, and Tx-Rx beam-level isolation is investigated using both simulated and experimentally measured SI channels. Finally, two beam-level isolation enhancement algorithms are proposed and validated, with their performance investigated by experimental measurements in an anechoic chamber. The main contributions of this chapter are summarized as follows:

- **Tx-Rx beam-level isolation versus Tx/Rx beamwidth:** The Tx-Rx beam-level isolation varies with the Tx/Rx array length across different uplink and downlink steering angles is investigated. Illustrative results, derived from both simulated and experimentally measured SI channels, demonstrate that changes in the average beam MC follow a trend similar to that of the sub-array beamwidth. Both the average and maximum MC values confirm that confining radiated power to a smaller scattering area significantly boosts spatial isolation. Specifically, when the sub-array length increases from $1\times3$ to $1\times8$, the average beam MC decreases by 9 dB and the maximum (worst-case) beam MC decreases by 9.9 dB.
- **Experimental studies of angular-perturbation beamforming isolation:** To further enhance the beam isolation, two RF beamformer design schemes are investigated, based on the angular locations of the users: 1) MDB-based RF beamformer, which maximizes the power pattern peaks in the desired user directions; and 2) beam angle-perturbation-based RF beamformer, which introduces controlled perturbations within the 3dB-beamwidth of the desired beam angle to further enhance isolation. Experimental measurements are conducted to evaluate the beam isolation performance. Average beam isolation of 70.2 dB and the highest isolation level of 79.6 dB can be achieved between the sample $1\times4$ Tx/Rx sub-arrays with the angular-perturbation scheme. Experimental measurements of far-field beam patterns have been conducted to elucidate their influence on beam-level isolation. The findings indicate that substantial beam level isolation can be attained when the Rx beams are aligned with the radiation nulls of the Tx beams. The results can offer valuable insights for angular-based or arbitrary radiation-nulling-based FD beamforming isolation design.

- **Experimental study of nulling steering beamforming isolation:** A NSB-based isolation optimization scheme to mitigate Tx-Rx SI in FD mMIMO systems is proposed. By searching for optimal nulling angles in both the uplink and downlink beamforming processes, experimental measurements demonstrate considerable improvements in beam-level isolation while maintaining the integrity of the original beamforming performance. Illustrative results using a 20 MHz modulated signal and the 8×8 Tx/8×8 Rx full-duplex array prototype in an anechoic chamber show an average Tx-Rx isolation improvement of 10.2 dB compared to the MDB-scheme, resulting in an average Tx-Rx MC level of −75.7 dB. Among the 25 measured Tx/Rx beam pairs, the most substantial isolation improvement of 22.2 dB is observed, achieving the best beam-level isolation of 92.7 dB.

The rest of the chapter is structured as follows: Sect. 10.2 illustrates the HBF-FD-mMIMO system models used in this work, emphasizing the necessity of multi-stage isolation and cancellation designs in FD mMIMO systems. Section 10.3 presents studies on Tx-Rx beam-level isolation in relation to Tx/Rx beamwidth variations. The results demonstrate that as the Tx/Rx array length increases, more confined beams and more densely distributed radiation nulls emerge as key mechanisms contributing to improved beam-level isolation. Then, Sect. 10.4 proposes two beam-level isolation optimization algorithms: one based on joint uplink/downlink beam angle perturbation and the other on the NSB algorithm. Section 10.5 investigates these algorithms through experimental measurements, analyzing the correlation between Tx-Rx beam-level isolation and radiation patterns. Finally, Sect. 10.6 concludes the chapter by summarizing the findings and outlining potential directions for future research.

## 10.2 System Model

In this study, an FD-HBF-mMIMO BS transceiver is considered, equipped with $M_{\text{Tx}}$ transmit and $M_{\text{Rx}}$ receive antenna elements. As illustrated in Fig. 10.1, the BS can concurrently serve up to $K_D$ downlink and $K_U$ uplink users in FD mode. The users operate in HD mode due to the hardware/software constraints (e.g., low power consumption, limited signal processing, and active/passive SIC capability).

For the downlink, the transmitted signal undergoes baseband precoding represented by $\mathbf{B}_D \in \mathbb{C}^{N_D \times K_D}$ and RF beamforming denoted by $\mathbf{F}_D \in \mathbb{C}^{M_{\text{Tx}} \times N_D}$. The number of transmit RF chains is $N_D$, satisfying $K_D \leq N_D \ll M_{\text{Tx}}$. In the uplink, the received signal at the BS is processed through RF beamforming $\mathbf{F}_U \in \mathbb{C}^{M_{\text{Rx}} \times N_U}$ and a baseband combiner $\mathbf{B}_U \in \mathbb{C}^{N_U \times K_U}$, utilizing $N_U$ receive RF chains where $K_U \leq N_U \ll M_{\text{Rx}}$. Both the uplink and downlink RF beamforming stages consist of phase shifters and attenuators.

For downlink transmission, the BS transmits the signal vector $\mathbf{x}_D = \mathbf{F}_D \mathbf{B}_D \mathbf{s}_D \in \mathbb{C}^{M_{\text{Tx}} \times 1}$, where $\mathbf{s}_D \in \mathbb{C}^{K_D \times 1}$ is the downlink data signal vector, adhering to signal

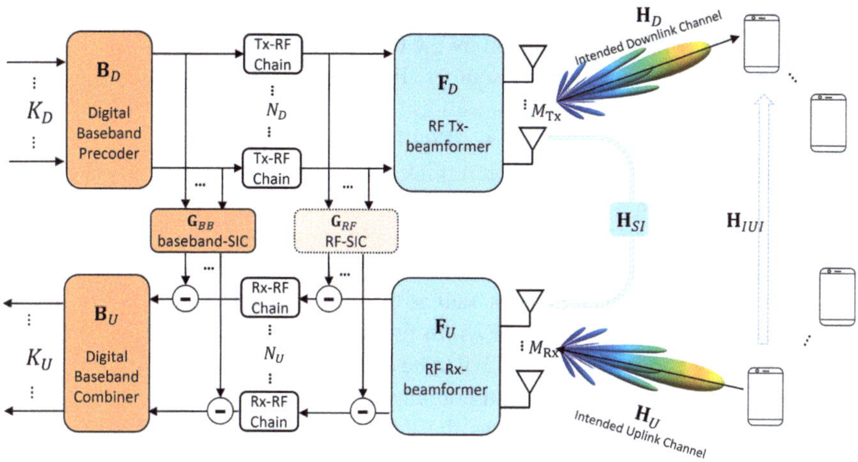


**Fig. 10.1** FD-HBF-mMIMO system with multiple SI isolation and cancellation

power constraints $\mathbb{E}[\mathbf{s}_D\mathbf{s}_D^H] = \mathbf{I}_{K_D}$. The received downlink signal vector at the downlink users is expressed as

$$\mathbf{r}_D = \underbrace{\mathbf{H}_D\mathbf{F}_D\mathbf{B}_D\mathbf{s}_D}_{\text{Desired Signal}} + \underbrace{\mathbf{H}_{\text{IUI}}\mathbf{s}_U}_{\text{IUI}} + \underbrace{\mathbf{n}_D}_{\text{Noise}}, \tag{10.1}$$

where $\mathbf{H}_D \in \mathbb{C}^{K_D \times M_{\text{Tx}}}$ is the downlink channel matrix from the BS to the downlink users. $\mathbf{s}_U \in \mathbb{C}^{K_U \times 1}$ following power constraints $\mathbb{E}[\mathbf{s}_U\mathbf{s}_U^H] = \mathbf{I}_{K_U}$ represents the uplink data signal vector transmitted by the uplink users. $\mathbf{n}_D \sim \mathcal{CN}(0, \sigma_n^2\mathbf{I}_{K_D})$ is the complex Gaussian noise vector at the downlink users. $\mathbf{H}_{\text{IUI}} \in \mathbb{C}^{K_D \times K_U}$ denotes the IUI channel, representing the interference from uplink users to downlink users due to simultaneous use of the same time and frequency resources.

The received uplink signal vector $\mathbf{r}_U \in \mathbb{C}^{K_U \times 1}$, after being processed by the RF beamformer and baseband combiner at the BS, can be represented as

$$\mathbf{r}_U = \underbrace{\mathbf{B}_U^T\mathbf{F}_U^T\mathbf{H}_U\mathbf{s}_U}_{\text{Desired Signal}} + \underbrace{\mathbf{B}_U^T\mathbf{F}_U^T\mathbf{n}_U}_{\text{Modified Noise}} + \underbrace{\mathbf{B}_U^T\mathbf{F}_U^T\mathbf{H}_{\text{SI}}\mathbf{F}_D\mathbf{B}_D\mathbf{s}_D}_{\text{Self-interference (SI)}}, \tag{10.2}$$

where $\mathbf{H}_U \in \mathbb{C}^{M_{\text{Rx}} \times K_U}$ denotes the uplink channel matrix from the uplink users to the BS. $\mathbf{n}_U \sim \mathcal{CN}(0, \sigma_n^2\mathbf{I}_{M_{\text{Rx}}})$ is the complex Gaussian thermal noise at the BS receive antennas. $\mathbf{H}_{\text{SI}} \in \mathbb{C}^{M_{\text{Rx}} \times M_{\text{Tx}}}$ is the SI channel matrix, characterizing the interference between the transmit and receive antennas at the BS due to simultaneous transmission and reception.

In HD communication, the IUI term in (10.1) and the SI term in (10.2) are absent because downlink transmission and uplink reception occur in separate time or frequency slots. As a result, HD system performance is primarily determined by thermal Gaussian noise and dictated by the operating SNR. In contrast, FD systems encounter additional interference sources. For downlink users, these include MUI caused by other downlink users, as well as IUI arising from concurrent uplink user transmissions, represented by $\mathbf{H}_{\text{IUI}}\mathbf{s}_U$. Similarly, in the uplink, the key interference contributors are MUI from multiple uplink users and the strong SI at the BS. To mitigate the effects of MUI, beamforming strategies can be employed. As discussed in Chaps. 7 and 8, NCB techniques effectively reduce MUI. Moreover, the BS can jointly design the digital baseband combiner ($\mathbf{B}_U$) and analog RF receive beamformer ($\mathbf{F}_U$) for the uplink, as well as the digital baseband precoder ($\mathbf{B}_D$) and analog RF transmit beamformer ($\mathbf{F}_D$), to further suppress MUI and drive it to the noise floor. These joint designs, shown to be effective in recent work [4], ensure that MUI does not become the dominant bottleneck in FD communications.

Moreover, the SI term, shown as the third component in (10.2), substantially impacts the detection performance of FD communication systems. To match the detection performance of HD systems, it is crucial to reduce this SI term below the thermal noise level. As presented in Chap. 2, achieving FD operation over a 20 MHz bandwidth with a transmitted power of 30 dBm and a noise floor at $-174$ dBm/Hz requires an SI suppression of approximately 131 dB relative to the transmitted power, that poses a significant design challenge for FD systems. In the initial phase of the design process, it is crucial to ensure that SI mitigation is achieved by ensuring $\mathbf{F}_U^T\mathbf{H}_{\text{SI}}\mathbf{F}_D \approx 0$. This can be accomplished through the following approaches

1. **Antenna design and isolation enhancement structure design**: Implementing antenna configurations that reduce the coupling between Tx and Rx antennas, effectively minimizing $\mathbf{H}_{\text{SI}}$;
2. **Beamformer design**: Designing the RF transmit and receive beamformers $\mathbf{F}_D$ and $\mathbf{F}_U$ to ensure their interaction with the SI channel approaches zero, that is, $\mathbf{F}_U^T\mathbf{H}_{\text{SI}}\mathbf{F}_D \approx 0$.

After the initial stage of suppression, additional SIC techniques can be employed at either the RF-SIC or the baseband-SIC to further reduce SI. SIC algorithms adaptively estimate the residual effective SI channel, $\mathbf{F}_U^T\mathbf{H}_{\text{SI}}\mathbf{F}_D$, through time-domain methods such as transversal filters and then subtract it from the received signal, as illustrated in Fig. 10.1. Conventional RF-SIC leverages adaptive tapped-delay-line (TDL) transversal filters in the RF analog domain. However, these filters tend to be both complex and limited in tap length. An alternative approach to RF-SIC employs adaptive finite-impulse-response (FIR) filters in the digital baseband domain, which are subsequently upconverted to the RF domain via separate auxiliary transmit RF chains. While this approach benefits from the simplicity of digital FIR filtering, the added auxiliary chains increase system complexity, hardware cost, and overall size and power consumption. Therefore, baseband-SIC, which operates in the digital domain using simple adaptive FIR filters, is generally more favored over RF-SIC due to its simpler implementation, lower hardware complexity, and reduced cost.

Since the Rx RF chains include LNA and ADCs with limited effective dynamic range, the residual SI at their input must be sufficiently suppressed to prevent

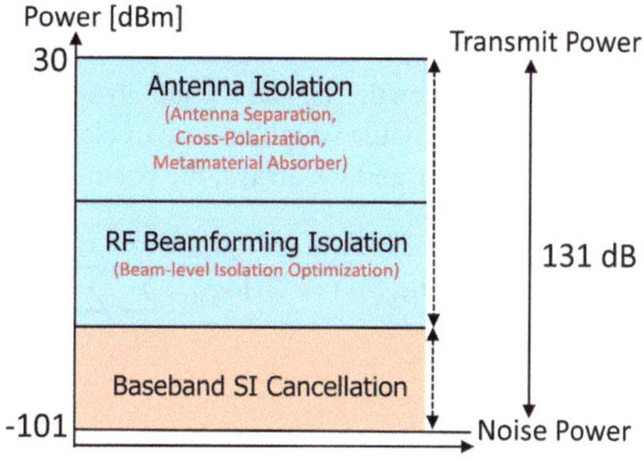


**Fig. 10.2** An example of performance breakdown of SI isolation and cancellation techniques

saturation and overloading, which could lead to nonlinear signal distortion. If the RF beamformers are designed to ensure that $\mathbf{F}_U^T \mathbf{H}_{\mathrm{SI}} \mathbf{F}_D$ is sufficiently small, only baseband-SIC is required, simplifying the implementation. Otherwise, RF-SIC is necessary to further reduce the residual SI before it reaches the receive RF chains, ensuring they operate within their linear region without increasing the noise floor.

Moreover, from the perspective of hardware and computational complexity, if FD mMIMO is implemented using single-stage DBF, only the baseband precoder $\mathbf{B}_D$ and combiner $\mathbf{B}_U$ are employed, with $\mathbf{F}_D$ and $\mathbf{F}_U$ being identity matrices. The dimensions of $\mathbf{B}_D$ and $\mathbf{B}_U$ are $M_{\mathrm{Tx}} \times K_D$ and $M_{\mathrm{Rx}} \times K_U$, respectively, where $M_{\mathrm{Tx}} \gg N_D$ transmit and $M_{\mathrm{Rx}} \gg N_U$ receive RF chains. Therefore, in the absence of Tx/Rx beamforming for SI isolation, an impractically large $M_{\mathrm{Rx}} \times M_{\mathrm{Tx}}$ RF-SIC would be required to maintain the $M_{\mathrm{Rx}}$ receive RF chains within their normal linear operating regions. Additionally, an excessively large $M_{\mathrm{Rx}} \times M_{\mathrm{Tx}}$ baseband-SIC would also be needed. Thus, these requirements make single-stage DBF unsuitable for FD-mMIMO systems. In contrast, HBF is not only a feasible solution but also a key enabler for achieving efficient operation.

An example of the achievable SI isolation and cancellation in the FD design is illustrated in Fig. 10.2. As presented in Chap. 9, we employ strategic separation and cross-polarization to reduce Tx-to-Rx antenna MC. In this chapter, various beam-level isolation optimization approaches are proposed and investigated to further exploit the benefits of antenna arrays. Following these stages, a baseband precoding technique, as presented in [2], is applied, achieving a SIC level exceeding 67 dB. The integration of these diverse techniques achieves over 131 dB of SI attenuation, effectively reducing the SI to the noise level.

## 10.3 Tx-Rx Array Beamwidth and Isolation

Antenna arrays play a pivotal role in enhancing spatial isolation through two principal mechanisms: (1) Generation of directional beams: These beams concentrate radiated power toward the intended directions, significantly reducing power leakage into potential interference paths. (2) Interference nulling: Beams are steered towards the radiation nulls of interference, thereby minimizing the receiver's sensitivity to interference power.

The MC between the Tx and Rx sub-arrays, as a function of beam steering angles, can be estimated by the weighted sum of element-wise MC contributions. For a pair of $M-$element Tx and Rx arrays, the beam-level MC in dB can be expressed as

$$S_b(M, \theta_D, \theta_U) = 20\log_{10} |\sum_{j=1}^{M}\sum_{i=1}^{M} w_i(\theta_D) w_j(\theta_U) S_{ji}|, \tag{10.3}$$

where $S_{ji}$ is the complex S-parameter that represents the coupling between Tx antenna element $i$ and Rx antenna element $j$ in the sub-arrays. For simplicity, both Tx and Rx are assumed to have ULA configurations. The terms $w_i(\theta_D)$ and $w_j(\theta_U)$ are the complex steering weights for Tx element $i$ and Rx element $j$ at the downlink beamforming angle $\theta_D$ and uplink beamforming angle $\theta_U$ in the $\theta$ plane.

### *10.3.1 Beamwidth and Isolation with HFSS Simulated SI Channel*

Experimental measurements on length-adjustable Tx/Rx arrays are inherently challenging due to limitations in hardware flexibility and scalability. To overcome these difficulties and obtain realistic estimates, Ansys HFSS full-wave simulation in this section is utilized to analyze the relationship between Tx/Rx beamwidth and beam-level isolation, while maintaining a fixed Tx-Rx array separation. Figure 10.3 illustrates the simulation setup, featuring two orthogonally polarized $1 \times M$ ULAs for the Tx and Rx, based on the proposed antenna element structure detailed in Chap. 9. An edge-to-edge separation of 20 cm is maintained between the arrays, consistent with the spacing used in the FD antenna array prototype. The simulation results for MC $S_{ji}$ between Tx element $i$ and Rx element $j$ at 3.5 GHz are used to estimate the beam-level MC between the Tx and Rx $1 \times M$ arrays at various steering angles, with Eq. (10.3).

The relationship between the average beam-level MC and the 3dB-beamwidth as a function of the number of elements $M$ is illustrated in Fig. 10.4. The Tx/Rx array length $M$ ranges from 3 to 16, with uplink and downlink beam steering angles spanning $\theta_U, \theta_D \in [-45° : 0.1° : 45°]$. As $M$ increases from 3 to 16, a considerable

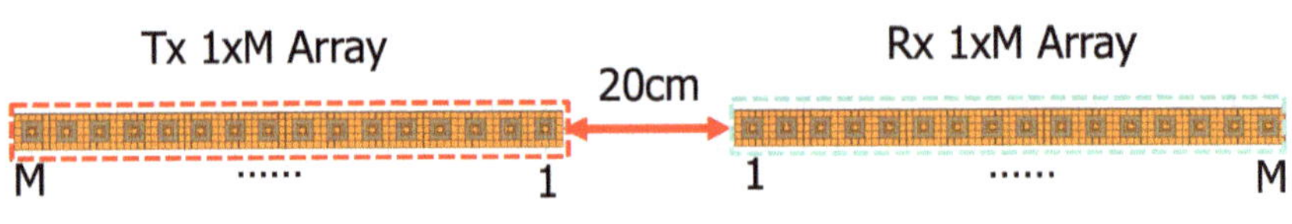


**Fig. 10.3** HFSS Simulation setup for array length versus beam mutual coupling

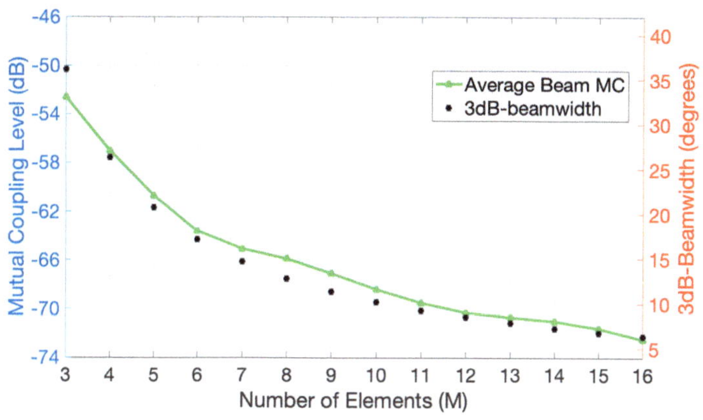


**Fig. 10.4** 3dB-beamwidth and average beam mutual coupling versus the number of elements in two Tx/Rx orthogonally-polarized $1 \times M$ ULAs (simulation results)

improvement in beam-level isolation is observed, evidenced by a 19.9 dB reduction in the average MC. Narrower beamwidths are associated with enhanced beam-level isolation as the array length increases. Specifically, the average beam MC decreases from −52.6 to −72.5 dB, accompanied by a reduction in the 3dB-beamwidth from 36° to 6°. The decreasing trend of both the beam-level MC and beamwidth becomes less pronounced for array lengths greater than 10, where further increases in $M$ yield only marginal reductions in the average beam MC, with an improvement of less than 4 dB as $M$ increases to 16.

To illustrate the Tx-Rx beam-level MC at various uplink and downlink steering angles, the relationship between beam MC and Tx/Rx steering angles for six sample Tx/Rx array lengths, with $\{\theta_U, \theta_D\} \in [-90°, 90°]$, is depicted in Fig. 10.5. It is evident that as $M$ increases, larger and denser regions of Tx-to-Rx isolation angle pairs (with MC $\leq -70$ dB) can be achieved. For a $1 \times M$ array, $M - 1$ radiation nulls can be created in the beamforming process through the destructive interference and summation of signals across the array elements. This mechanism reduces the angular spacing between adjacent radiation nulls. This decreased spacing between high-isolation Tx/Rx beam angle pairs results in a higher proportion of high-isolation $(\theta_D, \theta_U)$ pairs within the steering range. Consequently, this phenomenon is the primary driver of the observed improvement in average isolation.

### *10.3.2 Beamwidth and Isolation with Measured SI Channel*

The measured SI channel data at 3.5 GHz, presented in Chap. 9, is employed as the $S_{ji}$ parameter in Eq. (10.3) to estimate and analyze the variation in beam-level MC as a function of sub-array length. The Tx and Rx sub-arrays, situated in the fourth row of the 8×8Tx/8×8Rx array prototypes, are separated by 20 cm edge-to-edge

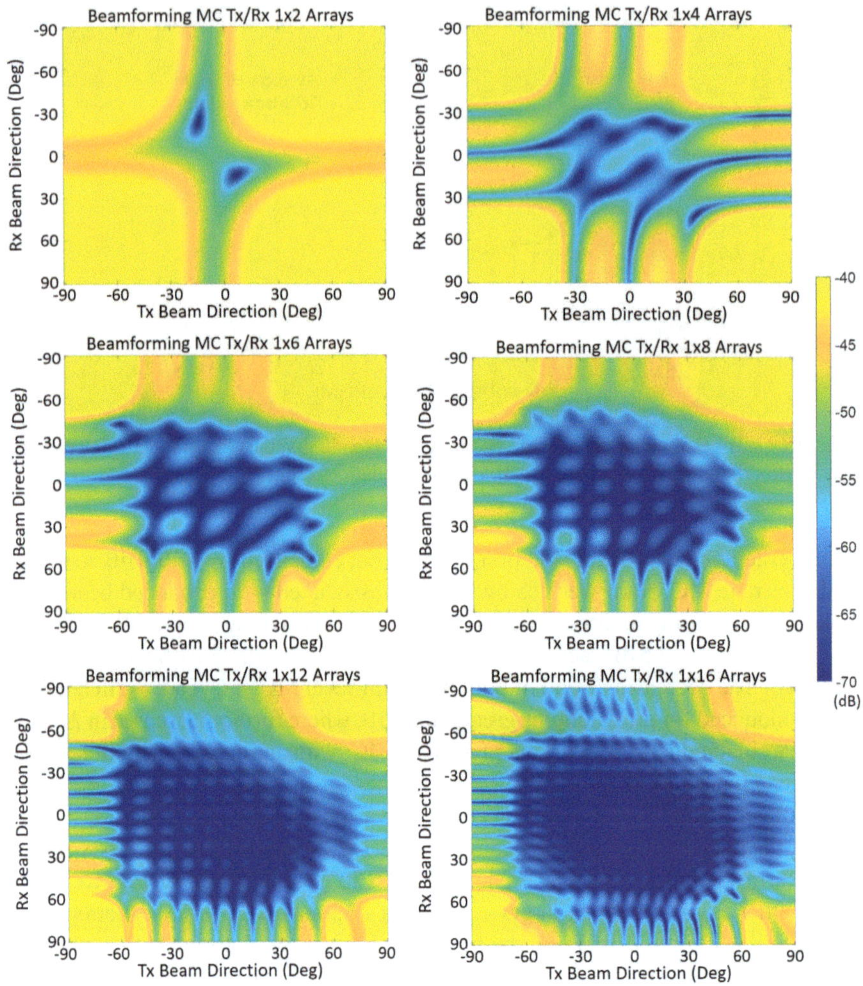


**Fig. 10.5** Tx/Rx beam mutual coupling within steering range $-90°$ to $90°$ of $1{\times}2$, $1{\times}4$, $1{\times}6$, $1{\times}8$, $1{\times}12$, and $1{\times}16$ arrays at 3.5 GHz (simulation results)

and are steered within the angular range $\theta_D, \theta_U \in [-45° : 0.1° : 45°]$. The Tx/Rx $1 \times M$ sub-arrays in the fourth row are composed of elements with indices ranging from T/R 4 to T/R $(4 + 8M)$, as labeled in Fig. 9.7.

The sub-array beamwidths at $\theta = 0°$ and the achieved average beam-level MC, as a function of the ULA length $M$, are presented in Fig. 10.6. The variations in average beam MC (in dB) closely align with the changes in sub-array beamwidth. Both the average and maximum MC values indicate that concentrating radiation power to a limited scattering area significantly enhances spatial isolation. Specifically, as the sub-array length increases from $1 \times 3$ to $1 \times 8$, the average Tx-Rx beam-level MC decreases by 9 dB, while the maximum (worst-case) beam MC decreases by 9.9 dB.

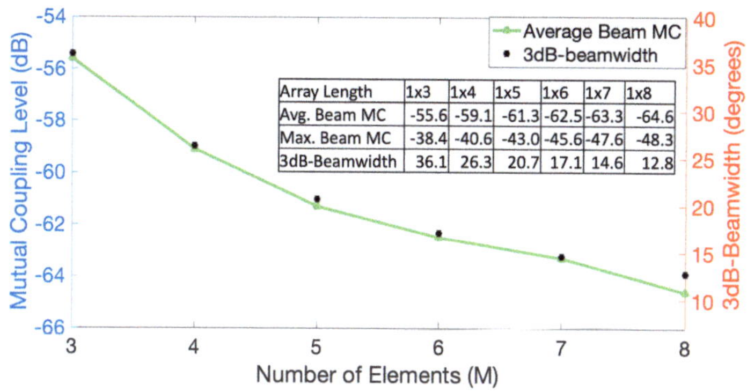


**Fig. 10.6** 3dB-beamwidth and average beam mutual coupling versus number of elements in $1 \times M$ ULAs (measurement results)

These results, derived from the measured SI channel, are in good agreement with the findings from the HFSS simulations.

## 10.4 Beamforming Optimization Algorithm for Tx-Rx Isolation Enhancement

### *10.4.1 Angular-Perturbation-Based Tx-Rx Beam-Level Isolation Enhancement*

As demonstrated in the previous section, increasing the concentration of the array's radiation pattern narrows the beamwidth and enhances the density of high-isolation downlink and uplink angle pairs. The conventional MDB scheme designs the uplink and downlink beamformer to maximize signal gain at the uplink and downlink user locations $(\theta_D, \theta_U)$ by aligning phase shifts to create constructive summation across the antenna elements. However, achieving good Tx-Rx isolation is limited to specific uplink and downlink user location pairs, as illustrated in Fig. 10.5. To optimize beam-level isolation while maintaining the desired uplink and downlink signal gains, beam angle perturbation techniques present a promising approach by allowing a controllable trade-off in desired beam gain at the downlink/uplink users. Instead of precisely aligning the Tx and Rx beam steering angles at the target downlink and uplink user angles ($\theta_D$ and $\theta_U$), introducing minor variations in the beam steering angles can enhance isolation and align the Rx beams with the local Tx nulls. This approach searches for new steering angles $(\hat{\theta}_D, \hat{\theta}_U)$ that prioritize isolation with a minor compromise in directivity.

**Lemma 10.1** *If* $\mathbf{F}_D$ *and* $\mathbf{F}_U$ *are the DL and UL beamforming stages, respectively. Then, for a given* $(i, j)^{th}$ *DL-UL pair* $\{\theta_{D_i}, \theta_{U_j}\}$ *in large array structures, the* *following holds:*

$$\mathbf{F}_U \mathbf{H}_{\text{SI}} \mathbf{F}_D \approx \mathbf{0}. \tag{10.4}$$

***Proof*** See Appendix. □

For an FD mMIMO system, given the downlink and uplink users located at angles $\theta_D$ and $\theta_U$, equipped with downlink and uplink RF beamformers $\mathbf{F}_D$ and $\mathbf{F}_U$, the optimization problem for Tx-Rx isolation can be formulated as

$$\begin{aligned} \min_{\{\hat{\theta}_D, \hat{\theta}_U\}} \quad & \left|\mathbf{F}_U^T(\hat{\theta}_U)\mathbf{H}_{SI}\mathbf{F}_D(\hat{\theta}_D)\right|^2, \\ \text{s.t.} \quad & C_1: \ |\hat{\theta}_D - \theta_D| \leq \Delta\theta_D, \\ & C_2: \ |\hat{\theta}_U - \theta_U| \leq \Delta\theta_U, \end{aligned} \tag{10.5}$$

where $C_1$ and $C_2$ represent the constraints on beam angle perturbation in the desired downlink and uplink beam directions. The ranges $\hat{\theta}_D \in [\theta_D - \Delta\theta_D, \theta_D + \Delta\theta_D]$ and $\hat{\theta}_U \in [\theta_U - \Delta\theta_U, \theta_U + \Delta\theta_U]$ help identify deep radiation nulls and enhance isolation. When maximum directivity beamformers are used, the maximum allowable angle perturbation ranges $\Delta\theta_D$ and $\Delta\theta_U$ implicitly indicate the maximum acceptable degradation in beam gain; a maximum $\epsilon$-dB beamwidth angle perturbation range will limit the desired gain degradation within $\epsilon$ dB.

In this study, we simplify our model to serve a single uplink user and a single downlink user, i.e., $K_D = K_U = 1$, using two ULA with length $M_D$ and $M_U$. The downlink and uplink beamformers $\mathbf{F}_D$ and $\mathbf{F}_U$ are constructed with phase shifters and uniform magnitudes, focusing on beam steering in azimuth directions $\theta_D$ and $\theta_U$ with constant elevation angles $\phi_D = \phi_U = 90°$. Then, the Tx and Rx RF beam steering vectors can be expressed as follows

$$\mathbf{F}_D(\theta_D) = \frac{1}{\sqrt{M_D}}\left[1, e^{j\beta d \sin(\theta_D)}, \ldots, e^{j\beta(M_D-1)d \sin(\theta_D)}\right]^T, \tag{10.6}$$

$$\mathbf{F}_U(\theta_U) = \frac{1}{\sqrt{M_U}}\left[1, e^{-j\beta d \sin(\theta_U)}, \ldots, e^{-j\beta(M_U-1)d \sin(\theta_U)}\right]^T, \tag{10.7}$$

where $\beta = 2\pi/\lambda_0$ is the propagation phase constant and $d$ is the separation distance between two neighboring antenna elements.

The pseudo-code of the proposed angular perturbation-based RF beamformer design is outlined in Algorithm 10.1. The algorithm follows four key steps:

- Define the search space: Specify the angular range of possible perturbations $\Delta\theta_{\max}$, based on the maximum $\epsilon$-dB beamwidth that satisfies the design constraint. For example, the average 3dB-beamwidth of a $1 \times 4$ array is approximately 30°. Accordingly, $\Delta\theta_D$ and $\Delta\theta_U$ are set to half this value, i.e., 15°,

**Algorithm 10.1:** Beam Angle Perturbation-Based RF Beamformer Design Algorithm

**Input**: Maximum angle perturbations $\Delta\theta_D$, $\Delta\theta_U$; SI channel matrix $\mathbf{H}_{\text{SI}}$; desired downlink angle $\theta_D$; desired uplink angle $\theta_U$.
**Output**: Optimal angles $(\hat{\theta}_D, \hat{\theta}_U)$ and the corresponding beamformers $\mathbf{F}_D(\hat{\theta}_D)$, $\mathbf{F}_U(\hat{\theta}_U)$.
Discretize the angular search space within the perturbation limits $(\theta_D \pm \Delta\theta_D,\ \theta_U \pm \Delta\theta_U)$. Initialize $\text{SI}_{\text{min}} \leftarrow \infty$.
**for** *each* $\tilde{\theta}_D$ *from* $\theta_D - \Delta\theta_D$ *to* $\theta_D + \Delta\theta_D$ **do**
  **for** *each* $\tilde{\theta}_U$ *from* $\theta_U - \Delta\theta_U$ *to* $\theta_U + \Delta\theta_U$ **do**
    Compute $\mathbf{F}_D(\tilde{\theta}_D)$ using Eq. (10.6).
    Compute $\mathbf{F}_U(\tilde{\theta}_U)$ using Eq. (10.7).
    Evaluate SI level: $SI = \left|\mathbf{F}_U^{\text{T}}(\tilde{\theta}_U)\,\mathbf{H}_{\text{SI}}\,\mathbf{F}_D(\tilde{\theta}_D)\right|^2$.
    **if** $SI < SI_{\text{min}}$ **then**
      Update optimal angles: $\hat{\theta}_D \leftarrow \tilde{\theta}_D$, $\hat{\theta}_U \leftarrow \tilde{\theta}_U$.
      Update $SI_{\text{min}} \leftarrow SI$.
    **end**
  **end**
**end**
Compute the final beamformers: $\mathbf{F}_D(\hat{\theta}_D)$ and $\mathbf{F}_U(\hat{\theta}_U)$.

allowing a maximum half-power desired beam gain reduction in exchange for isolation improvement.

- Discretize the search space: Divide the search space into discrete steps or intervals. For instance, the search space can use increments of 1° or larger. However, finer resolutions increase computational complexity and the number of iterations required during measurements.
- Evaluate the objective function: Calculate or measure the objective function for each uplink-downlink perturbed-angle pair $(\hat{\theta}_D, \hat{\theta}_U)$ to minimize SI. Iterate through all possible values of $(\hat{\theta}_D, \hat{\theta}_U)$ within the defined search space.
- Select optimal angles: Choose the perturbed angles that yield the minimum value of the objective function. These angles offer the best trade-off between beam gain and isolation while satisfying the 3dB-beamwidth constraint. Finally, construct the downlink and uplink RF beamforming stages using Eqs. (10.6) and (10.7), respectively.

### *10.4.2 Nulling Steering Beamforming-Based Tx-Rx Beam-Level Isolation Enhancement*

Although angular perturbation-based SI suppression techniques, as expressed in Sect. 10.4.1, offer the benefit of isolation enhancement by searching for adjacent high-isolation angles $\hat{\theta}_D$ and $\hat{\theta}_U$ and with a simple beamformer design, the tilting in the uplink and downlink steering angles often comes with trade-offs in the desired signal gain. The following experiments demonstrate a strong correlation between the Tx/Rx far-field patterns and beam MC. Significant improvements in beam-level isolation are observed when the Rx sub-array's main lobe direction aligns with the deep radiation nulls of the Tx beam. Therefore, instead of solely adjusting the uplink and downlink beam directions, which could compromise the quality of the intended signal gain, NSB provides an effective alternative. By controlling the locations of radiation nulls in the Tx and Rx beams, NSB enhances isolation while maintaining the integrity of the desired signal.

By employing the LCMV algorithm, the optimal weights are determined to minimize array radiation in a designated nulling direction ($\theta_N$), while maintaining the original beam directivity toward the desired user direction ($\theta_i$) as achieved with the MDB scheme. As derived in Chap. 7, the closed-form expression for the LCMV beamformer weights can then be obtained as

$$\mathbf{F}_{\text{lcmv}}(\theta_i, \theta_N) = \mathbf{R}^{-1}\boldsymbol{C}(\boldsymbol{C}^H\mathbf{R}^{-1}\boldsymbol{C})^{-1}\boldsymbol{d}, \tag{10.8}$$

where $\mathbf{R}$ denotes the covariance matrix of the received signal before being modified and combined by the beamformer. The response matrix $\boldsymbol{C}$ includes the sub-array response vectors in both the desired and interfering directions, represented as $\boldsymbol{C}(\theta_i, \theta_N) = [\mathbf{h}(\theta_i), \mathbf{h}(\theta_N)]$. The target response constraint vector $\boldsymbol{d}$ specifies the sub-array response at each user location, given by $\boldsymbol{d} = [1, 0]^T$.

The LCMV beamformer is employed for both uplink and downlink operations. Radiation nulls are created in the directions $\theta_{UN}$ and $\theta_{DN}$ to enhance Tx-Rx beam-level isolation. This approach also maintains directivity towards the targeted uplink and downlink user directions, $\theta_U$ and $\theta_D$, respectively. The uplink and downlink beamformers using the LCMV approach can be represented as

$$\mathbf{F}_U(\theta_U, \theta_{UN}) = \mathbf{F}_{\text{lcmv}}(\theta_U, \theta_{UN}), \tag{10.9}$$

$$\mathbf{F}_D(\theta_D, \theta_{DN}) = \mathbf{F}_{\text{lcmv}}(\theta_D, \theta_{DN}). \tag{10.10}$$

To identify the optimal nulling angles $\hat{\theta}_{DN}$ and $\hat{\theta}_{UN}$ that minimize the SI level, a PSO-based nulling-angle search scheme. Instead of employing exhaustive search methods, PSO leverages a swarm of intelligently guided particles with dynamic searching speeds that explore the search space within predefined constraints, aiming to minimize the objective function. Due to its stochastic characteristics, PSO is particularly adept at navigating non-convex optimization landscapes. The SI power modified by the uplink and downlink beamformers can be expressed as

$$A_{\text{SI},\theta_D,\theta_U}(\theta^n_{DN,p}, \theta^n_{UN,p}) = 10\log_{10}\left(\left|\mathbf{F}_U^T(\theta_U, \theta^n_{UN,p})\mathbf{H}_{\text{SI}}\mathbf{F}_D(\theta_D, \theta^n_{DN,p})\right|^2\right), \tag{10.11}$$

where $\theta^n_{UN,p}$ and $\theta^n_{DN,p}$ represents the iterative uplink and downlink nulling angles at the $p$th particle during the $n$th iteration. In the search process, $N_p$ particles are used in each iteration and total $N_n$ iterations are considered. Therefore, $p \in \{1, 2, \ldots, N_p\}$ and $n \in \{1, 2, \ldots, N_n\}$.

To avoid the nulling process degrading the original mainlobe beam pattern, the intended nulling locations are kept half mainlobe beamwidth[1] ($BW_{FN}$) away from the desired user direction. Additionally, the nulling angle search space focuses on the whole front-radiation angular span, i.e., $-90^\circ$ to $90^\circ$. Therefore, the optimization problem can be formulated as

$$\begin{aligned} \min_{\{\hat{\theta}_{DN},\hat{\theta}_{UN}\}} \quad & A_{\mathrm{SI},\theta_D,\theta_U}(\hat{\theta}_{DN},\hat{\theta}_{UN}), \\ \text{s.t.} \quad C_1: \; & |\hat{\theta}_{DN}-\theta_D| \geq \frac{BW_{FN}}{2}, \quad -90^\circ \leq \hat{\theta}_{DN} \leq 90^\circ \\ C_2: \; & |\hat{\theta}_{UN}-\theta_U| \geq \frac{BW_{FN}}{2}, \quad -90^\circ \leq \hat{\theta}_{UN} \leq 90^\circ \end{aligned} \tag{10.12}$$

Let define $\boldsymbol{\Theta}_p^n$ as the searching vector which includes the target nulling angles $\boldsymbol{\Theta}_p^n = [\theta_{DN,p}^n, \theta_{UN,p}^n]$ at the $n$th iteration and $p$th particle. The personal best for the $p$th particle at the $n$th iteration and the current global best among all particles are respectively found as

$$\boldsymbol{\Theta}_{\text{best},p}^n = \underset{\boldsymbol{\Theta}_p^{(n^*)},\forall n^*=1,\cdots,N_n}{\arg\min} \; A_{\mathrm{SI},\theta_D,\theta_U}(\boldsymbol{\Theta}_p^{(n^*)}), \tag{10.13}$$

$$\boldsymbol{\Theta}_{\text{best}}^n = \underset{\boldsymbol{\Theta}_{\text{best},p}^n,\forall p=1,\cdots,N_p}{\arg\min} \; A_{\mathrm{SI},\theta_D,\theta_U}(\boldsymbol{\Theta}_{\text{best},p}^n). \tag{10.14}$$

The effectiveness of the PSO-based perturbation scheme for SI suppression improvement relies on the searching velocity vector $\mathbf{v}_p$. This vector influences the optimization process toward both the personal best $\boldsymbol{\Theta}_{\text{best},p}^n$ and global best $\boldsymbol{\Theta}_{\text{best}}^n$ solutions and can be defined as

$$\mathbf{v}_p^{(n+1)} = \mathbf{a}_1(\boldsymbol{\Theta}_{\text{best}}^n - \boldsymbol{\Theta}_p^n) + \mathbf{a}_2(\boldsymbol{\Theta}_{\text{best},p}^n - \boldsymbol{\Theta}_p^n) + \mathbf{a}_3^n \mathbf{v}_p^n, \tag{10.15}$$

where $\mathbf{v}_p^n$ is the shifting factor of the $p$th particle at the $n$th iteration. $\mathbf{a}_1$ and $\mathbf{a}_2$ are the random diagonal matrices of self-adjustment weights and social-adjustment weights, reflecting the social interactions among particles and each particle's inclination to move towards its global best and personal best, respectively. The $\mathbf{a}_3^n$ denotes the diagonal inertia weight matrix, which is crucial for balancing exploration and exploitation in the search for the optimal solution within the search space. With this dynamically updated shift factor, the $n$th iterative particle is updated as

$$\boldsymbol{\Theta}_p^{(n+1)} = \mathbf{clip}(\boldsymbol{\Theta}_p^n + \mathbf{v}_p^{(n+1)}, \boldsymbol{\Theta}_L, \boldsymbol{\Theta}_U), \tag{10.16}$$

[1] In the simulations and experimental measurements of this study, we utilized $1\times 8$ arrays, which exhibited a mainlobe beamwidth $BW_{FN} = 32^\circ$.

**Algorithm 10.2:** NSB-Based Isolation Optimization Algorithm

**Input**: SI channel matrix $\mathbf{H}_{\mathrm{SI}}$, desired downlink angle $\theta_D$, desired uplink angle $\theta_U$.
**Output**: Optimal angles $[\hat{\theta}_{DN}, \hat{\theta}_{UN}] = \boldsymbol{\Theta}_{\mathrm{best}}^{(n)}$.
Initialize parameters: Number of iterations $N_n$, Number of particles $N_p$. Set initial velocities: $\mathbf{v}_p^{(1)} \leftarrow 0$ for all particles.
**for** $n = 1$ **to** $N_n$ **do**
  **for** $p = 1$ **to** $N_p$ **do**
    **if** $n = 1$ **then**
      Initialize particle positions: $\boldsymbol{\Theta}_p^{(1)} = [\theta_{DN,p}^{(1)}, \theta_{UN,p}^{(1)}]$ uniformly distributed within constraints $C_1$ and $C_2$ of Eq. (10.12).
    **else**
      Update velocity $\mathbf{v}_p^{(n)}$ via Eq. (10.15).
      Update position $\boldsymbol{\Theta}_p^{(n)}$ via Eq. (10.16).
    **end**
    Evaluate cost function $A_{\mathrm{SI},\theta_D,\theta_U}(\boldsymbol{\Theta}_p^{(n)})$ via (10.11).
    Update personal best $\boldsymbol{\Theta}_{\mathrm{best},p}^{(n)}$ via (10.13).
  **end**
  Update global best $\boldsymbol{\Theta}_{\mathrm{best}}^{(n)}$ via (10.14).
**end**

with $\boldsymbol{\Theta}_L$, $\boldsymbol{\Theta}_U$ defining the lower-bound and upper-bounds of perturbations based on $C_1$ and $C_2$ in expression (10.12), respectively. The clip function rigorously enforces the specified boundaries by adjusting the nulling angles to their nearest boundary values whenever they exceed the predefined search area. The pseudo-code for the proposed NSB algorithm is presented in Algorithm 10.2. Upon determining the optimal downlink and uplink nulling angles $[\hat{\theta}_{DN}, \hat{\theta}_{UN}]$, the corresponding downlink beamformer $\mathbf{F}_D(\theta_D, \hat{\theta}_{DN})$ and uplink beamformer $\mathbf{F}_U(\theta_U, \hat{\theta}_{UN})$ are computed using Eqs. (10.9) and (10.10).

## 10.5 Experimental Studies of the Tx-Rx Beam-Level Isolation

### *10.5.1 Measurement Setup*

The $8\times8$Tx/$8\times8$Rx cross-polarized antenna array prototype, detailed in Chap. 9, was used for beam radiation pattern measurements and Tx-Rx beam-level isolation assessments. The placement of the antenna array within the anechoic chamber is depicted in Fig. 10.7. This study focuses on a scenario with a single uplink user and a single downlink user, i.e., $K_D = K_U = 1$, served by two ULAs. Two sample sub-arrays, Tx $1 \times M_D$ and Rx $1 \times M_U$, located on the same rows of the Tx and Rx antenna arrays with an edge-to-edge separation of 20 cm (approximately $2.5\lambda_0$), were employed. In the first part of the experimental measurements, Algorithm 10.1

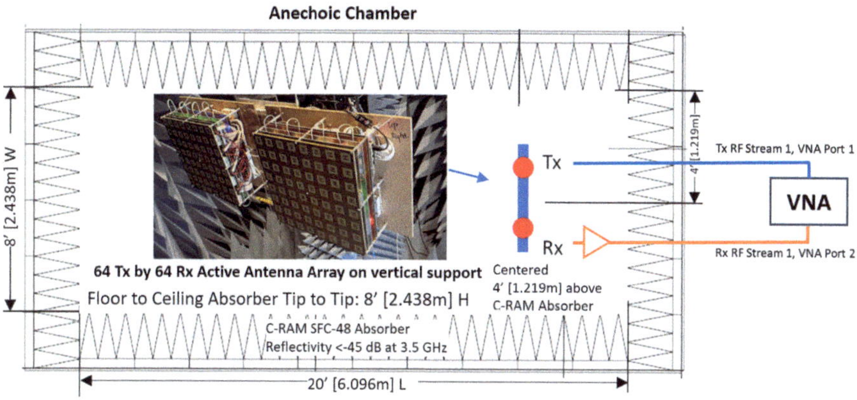


**Fig. 10.7** FD massive MIMO prototype setup in the anechoic chamber

was evaluated with $M_D = M_U = 4$, focusing on the angular-perturbation-based Tx-Rx isolation enhancement and the correlation between far-field beam radiation patterns and beam-level MC. In the second part, Algorithm 10.2 was tested with $M_D = M_U = 8$, focusing on the NSB-based isolation improvement method.

On the Tx side, a VSG was used to generate a modulated New Radio (NR) signal with a bandwidth of 20 MHz (3.49–3.51 GHz) and an output power of 10 dBm. The signal was equally divided into $M_D$ paths via a $1 \times M_D$ power splitter. Within the BFN, each signal path was modulated by a digitally controlled attenuator with seven-bit resolution, providing a 32 dB dynamic range in 0.25 dB steps. This was followed by a digitally controlled phase shifter with six-bit resolution, 360° coverage in steps of 5.625°. To compensate for system insertion losses, final-stage RF amplifiers were integrated before delivering the signal to the antenna elements.

On the Rx side, the received signal from each antenna element was first passed through LNAs to enhance measurement sensitivity by improving the SNR. Each receiving path was then processed by the BFN, which consisted of attenuators and phase shifters identical to those used in the Tx BFN, but operating in the reverse direction. The signals from these $M_U$ paths were subsequently combined using an $M_U$-to-1 power combiner and fed into a spectrum analyzer or VNA for measurement and analysis.

Short coaxial cables with JSC connectors were used to interconnect the components of both the Tx and Rx sub-arrays. Two customized SPI modules were employed for beamforming control: one for the Tx beamformer and the other for the Rx beamformer, both integrated with a Raspberry Pi 4. The beamforming algorithms were developed and executed in MATLAB on a local PC, which interfaced with the spectrum analyzer to monitor in-band Tx-Rx MC levels. The PC also communicated with the SPI modules via USB to update the uplink and downlink beamformer configurations. The Tx-Rx beam-level MC with MDB at the exact target steering angles corresponding to the downlink and uplink user locations served as the baseline for evaluating the proposed algorithms. Prior to the experiments, all phase shifters and attenuators were calibrated and characterized to ensure uniform reference phase and attenuation levels across all elements, effectively mitigating potential phase drift effects in the phase shifters.

### 10.5.2 Tx/Rx 1×4 Sub-array Radiation Characteristics

The measured 3D radiation patterns of the sample Tx 1 × 4 and Rx 1 × 4 sub-arrays with uniform power feeding and zero phase shift in the anechoic chamber are shown in Fig. 10.8. Both sub-arrays produce disk-shaped beams with two sidelobes flanking the main beam. The directivity of the Tx 1 × 4 sub-array is 11.3 dB, while the Rx 1 × 4 sub-array directivity is 11.9 dB. The realized gains are 9.1 and 9.4 dB, respectively, corresponding to radiation efficiencies of 61.0 and 55.6%. The Tx and Rx sub-arrays exhibit 3dB-beamwidths of 27.3° and 28.7° in the $\theta$ plane, respectively. Both sub-arrays demonstrate normalized sidelobe levels of − 14.1 dB and first null beamwidths of 66°. The concentrated beam radiation patterns validate the effectiveness and accuracy of the phase reference characterization process for the phase shifters, enabling coherent summation of signals across the antenna elements.

Figure 10.9 illustrates the AR measurement results of the Tx 1 × 4 and Rx 1 × 4 sub-arrays at a steering angle of $\theta_D = \theta_U = 0°$. Within the main lobe range, the minimum AR values achieved by the Tx and Rx 1 × 4 sub-arrays are 2.3 and 2.8 dB, respectively. The AR remains below 6 dB across the $\theta$ angle range of −32° to 28° for the Tx sub-array and − 28° to 23° for the Rx sub-array, which approximately aligns with the main lobe beamwidth. Beyond these ranges, larger fluctuations in AR are observed, which may reduce the cross-polarization ratio and adversely affect beam isolation performance.

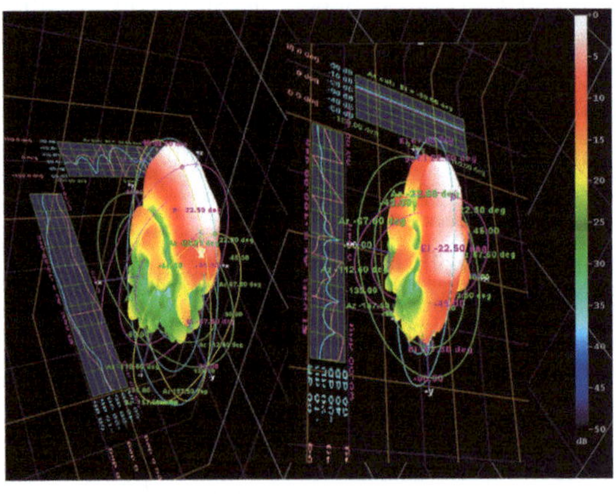

**Fig. 10.8** Tx(left)/Rx(right) 1×4 sub-array beams at 3.5 GHz (measurement results)

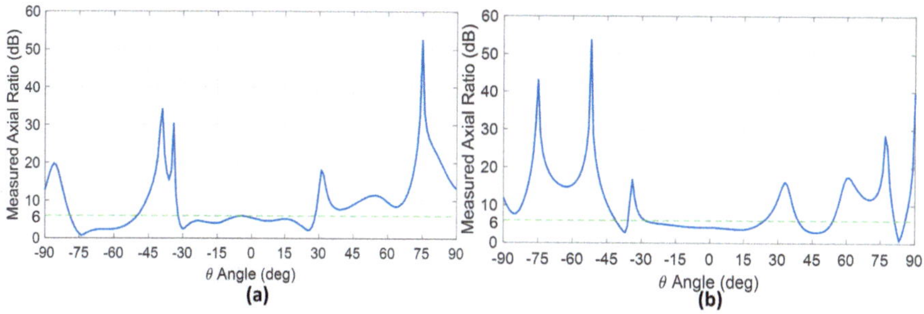


**Fig. 10.9** Beam axial ratio cut at $\phi$=0° (**a**) Tx 1×4 sub-array beam; (**b**) Rx 1×4 sub-array beam (measurement results)

### *10.5.3 Tx-Rx Mutual Coupling Level with Max-Directivity-Based Beamformer*

Beam isolation was evaluated using the sample Tx and Rx 1 × 4 sub-arrays with a 20 MHz modulated signal in the frequency range of 3.49–3.51 GHz. Given that the first null beamwidth of both 1 × 4 sub-arrays is approximately 66°, corresponding to 33° from the beam peak direction, four steering angles with separations of 30° to 33° were selected for the measurements: $(\theta_D, \theta_U) \in \{-48°, -15°, 15°, 48°\}$. This resulted in a total of 16 Tx and Rx beam angle combinations being tested.

The measured Tx-Rx beam MC levels, using the MDB-based beamformer at steering angles corresponding to the downlink and uplink user locations, are illustrated in Fig. 10.10a. Across the 16 beamforming combinations, an average MC level of − 58.1 dB is observed, with a minimum (best) MC level of − 69.1 dB. When compared to the average MC of 16 Tx-Rx element pairs over the 20 MHz bandwidth with the same relative locations, an average reduction of 7.6 dB in MC is observed. Among the tested cases, 11 Tx-Rx angle pairs (69%) achieve beam-level MC below (better than) − 57 dB.

Notable impacts of the beam steering angles on the beam-level MC can be observed. Steering the Tx and Rx beams toward each other results in an average of 12.6 dB lower (better) MC (as shown in the right top quadrant of the graph) compared to the scenarios when steering them away from each other (as shown in the left bottom quadrant of the graph). The average Tx-Rx beam-level isolation is enhanced from 51.2 to 63.8 dB. Additionally, it can be seen that an average MC of − 61.1 dB is attained when the Rx beam steers to the Tx beam's null location-that is, approximately 33° away from the Tx peak direction.

### *10.5.4 Far-Field Beam Pattern Influence on the Beam MC*

Plots of four sample pairs of measured Tx/Rx far-field beam patterns are presented. The Tx/Rx beams corresponding to the two highest MC levels, highlighted by blue

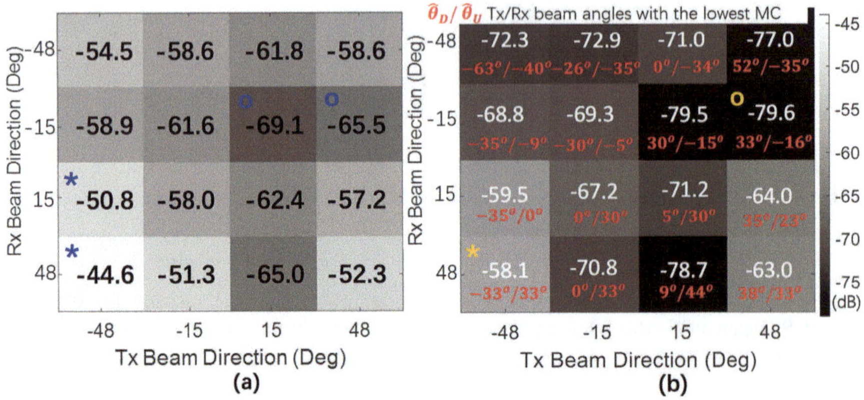


**Fig. 10.10** Tx-to-Rx beam-level MC level of 1×4 sub-arrays with (**a**) scheme 1: max-directivity-based beamformer; (**b**) scheme 2: angle-perturbation-based beamformer (measurement results)

stars in Fig. 10.10a, are shown in Fig. 10.11a and b. Meanwhile, the Tx/Rx beams with the two lowest (best) MC values, highlighted by blue circles in Fig. 10.10a, are displayed in Fig. 10.11c and d. The Tx beam is plotted in the cross-polarization (X-pol) state (i.e., Left-Hand Circular Polarization, LHCP) to showcase the far-field radiation intensity of the Tx beam in the co-polarization (Co-pol) state with the Rx beam.

A strong correlation between the Tx/Rx far-field patterns and the beam MC can be observed. Substantial beam-level isolation can be achieved by aligning the Rx beams with the deep radiation nulls of the Tx beams. On average, the two beam pairs with the highest MC levels are 19.6 dB higher than the two beam pairs with the lowest MC levels. Correspondingly, within the 3dB-beamwidth range of the Rx main lobe, the Tx far-field radiation intensity for the two highest MC beam pairs is, on average, 9.5 dB higher. These empirical findings demonstrate that the far-field radiation component significantly contributes to the total beam MC.

When the Tx beams have steering angle $\theta_D < 0°$, the Tx radiation intensity in the Rx main lobe direction is relatively stronger than when the beams are steered at positive angles. For example, at a Tx beam steering angle of $-48°$, relatively high Tx radiation power with shallow nulls (higher than $-18$ dB) is observed in the Rx main lobe direction. In contrast, when the Tx beam is steered at 15° and 48°, deeper nulls (lower than 25 dB) are present in the Rx main lobe range. Additionally, a discrepancy of more than 10 dB is observed between the average beam isolation level and the corresponding difference in far-field radiation intensity. This difference is mostly attributed to near-field leakage waves, indicating that near-field coupling remains a significant contributor to the total MC. The joint near-field and far-field MC designs will be discussed in the next Chapter.

### 10.5.5 Tx-Rx Mutual Coupling Level with Angular Perturbation-Based Beamforming Optimization

Beam isolation using Algorithm 10.1 is evaluated with the sample Tx and Rx 1 × 4 sub-arrays, transmitting a 20 MHz modulated signal within the frequency range of 3.49–3.51 GHz. The evaluation incorporated a maximum 3dB-beamwidth angle variation tolerance, centered on the target uplink and downlink user directions. An iterative measurement approach was employed, with an uplink and downlink angle search step of 1°. The achieved minimum (best) Tx-Rx beam-level MC values, along with the corresponding beam steering angles, are shown in Fig. 10.10b. An average MC level of − 70.2 dB was observed across the 16 beamforming combinations, which indicates an average MC reduction (improvement) of 12.0 dB compared to the results with the MDB scheme shown in Fig. 10.10a. When compared to the average Tx-Rx element MC over the 20 MHz bandwidth with the same relative locations, an average reduction of 19.7 dB in MC is observed. Additionally, the MC level is below − 58.1 dB in all 16 beam angle pairs, indicating a 13.5 dB isolation improvement compared to the highest (worst) beam-level MC in the maximum directivity scheme.

When the Tx and Rx beams are steered towards each other (top right quadrant of Fig. 10.10b), an average MC of − 76.8 dB is attained, which is 12.9 dB lower than the average MC when steering the two beams away from each other (bottom left quadrant of Fig. 10.10b). A best beam-level isolation of 79.6 dB can be achieved, which indicates a 10.5 dB improvement in the highest achievable isolation compared to the scenario without angular-perturbation optimization. The largest isolation improvement of 19.5 dB occurs at a Tx beam of − 15° and an Rx beam of 48°. Additionally, it is observed that the best isolation typically occurs at 15° away from the intended direction, resulting in approximately 3 dB lower power than the beam peak. Consequently, there is a trade-off between a maximum loss of 3 dB in desired signal strength and an average improvement of 12.0 dB in isolation level during beamforming.

The two samples of measured far-field beam pattern 2D cuts after beam steering angle perturbation, i.e., the highest and the lowest beam MC cases, are plotted in Fig. 10.12. When comparing Figs. 10.11a and 10.12a, which represent the beam pair with the highest MC in the maximum-directivity scheme and angular-perturbation scheme, it is observed that the Tx radiation power in the Rx 3dB-beamwidth range direction is lowered by an average of 6.2 dB. This reduction is one of the main contributors to the 13.8 dB beam isolation enhancement. Comparing the Tx/Rx beam patterns of the lowest MC pair (Tx steering angle at 33° and Rx steering angle at − 16°), the Tx beam exhibits very deep nulls (lower than − 30 dB) and an average normalized power level of − 26.5 dB in the Rx beam's 3dB-beamwidth angle range. This observation is consistent with the previous section.

These findings suggest that methods such as directing the Rx beam towards the radiation nulling range of the Tx beam or creating arbitrary deep nulls via adaptive beamforming techniques in the beamformer design can effectively enhance the Tx-to-Rx beam isolation in FD communications.

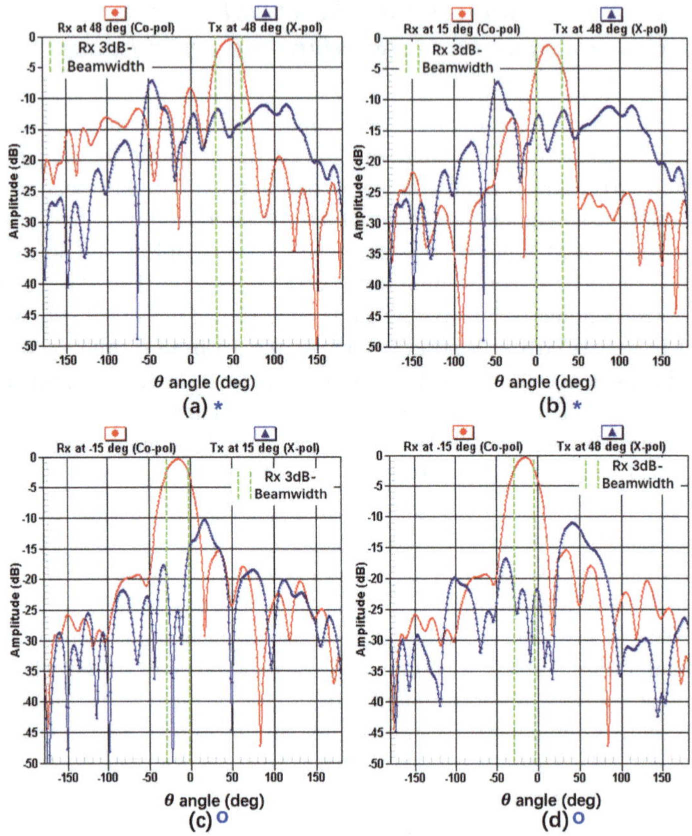


**Fig. 10.11** Measured Tx/Rx 1×4 sub-array beam radiation pattern 2D cuts Scheme 1 at 3.5 GHz. The Tx/Rx beam combinations with the highest MC (**a**) and (**b**). The Tx/Rx beam combinations with the lowest MC (**c**) and (**d**)

### *10.5.6 Nulling Control Beamforming-Based Tx-Rx Isolation Enhancement*

Algorithm 10.2 is assessed using Tx $1 \times 8$ and Rx $1 \times 8$ sub-arrays with a 20 MHz modulated signal. Given that the first null beamwidth of both $1 \times 8$ sub-arrays is approximately 32°, corresponding to 16° from the beam peak direction, five steering angles with separations of 16° were selected for the measurements: $(\theta_D, \theta_U) \in \{-32^\circ, -16^\circ, 0^\circ, 16^\circ, 32^\circ\}$. Totally, 25 downlink and uplink angle pairs are measured. The measurement results of the Tx-Rx beam-level MC with the MDB-scheme (i.e., no NSB consideration) and the proposed NSB-based beam-level isolation optimization are compared in Fig. 10.13a and b. The effectiveness of the proposed beam isolation enhancement is detailed in Fig. 10.13c.

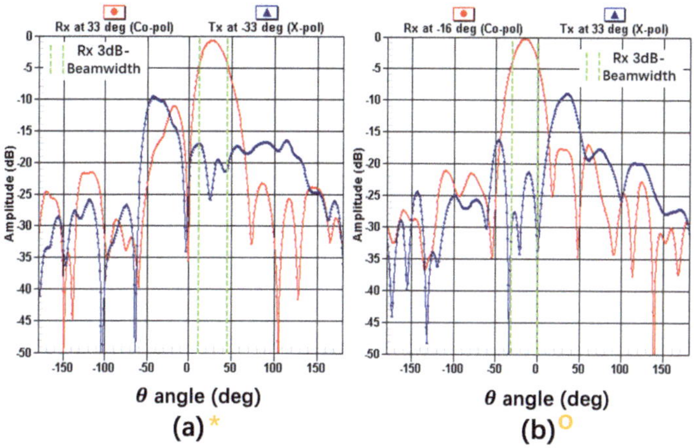


**Fig. 10.12** Tx/Rx 1×4 sub-array beam radiation pattern 2D cuts Scheme 2 at 3.5 GHz. The Tx/Rx beam combinations with the highest MC (**a**). The Tx/Rx beam combinations with the lowest MC (**b**) (measurement results)

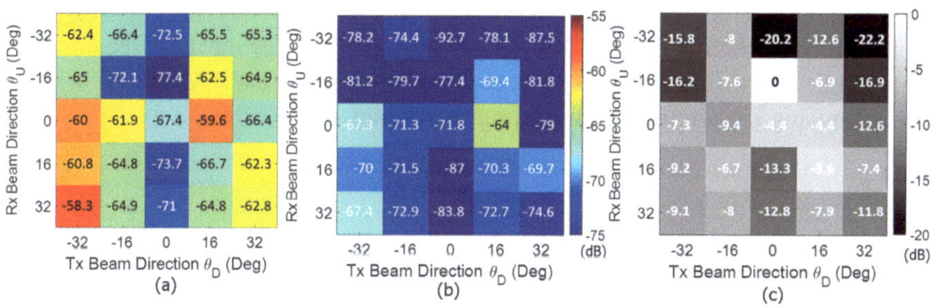


**Fig. 10.13** Tx-Rx beam-level mutual coupling (**a**) with Tx/Rx maximum-directivity beamformer; (**b**) with the proposed NSB-based isolation optimization; (**c**) beam-level mutual coupling improvement: $MC_{NSB} - MC_{MDB}$ (measurement results)

It can be observed that the proposed isolation optimization scheme leads to a considerable improvement in beam-level isolation. Specifically, an average isolation enhancement of 10.2 dB was achieved, reducing the average beam-level MC from − 65.5 to − 75.7 dB.[2] The lowest (best) beam-level MC of − 92.7 dB can be achieved, indicating the best achievable beam-level isolation improvement of 15.3 dB compared with the MDB scheme.

Furthermore, after applying the NSB isolation optimization, the highest (worst) MC is mitigated to − 64 dB, indicating a 5.7 dB improvement in the worst beam-level isolation case. Notably, except for three beam pairs, all other tested cases

[2] Since mutual coupling represents the ratio of the received power with respect to the transmitted power, geometric means are used to characterize the average mutual coupling levels.

exhibit isolation levels better than $-69.4\,\text{dB}$. Out of 25 measured beam pairs, 19 (76%) showed improvements exceeding 7.3 dB, with 10 pairs (40%) demonstrating enhancements above 11.8 dB. The most substantial isolation improvement of 22.2 dB is observed when $(\theta_D, \theta_U) = (32^\circ, -32^\circ)$.

Additionally, higher isolation enhancements are observed when either the Tx or Rx beams are steered at large angles toward the Rx or Tx sub-array, facilitated by the creation of radiation nulls. For instance, at $\theta_U = -32^\circ$, an average isolation improvement of 15.8 dB was recorded, attributed to the intense interaction between the main lobes of Tx and Rx at these beam pairs.

To evaluate the impact of the proposed NSB scheme isolation-optimization technique on the desired beam patterns, we employed MATLAB Phased Array Toolbox to analyze the radiation patterns of the Tx $1\times8$ array. Specifically, we compared the radiation patterns generated with $\mathbf{F}_D(\theta_D, \hat{\theta}_{DN})$ using the NSB against the patterns produced by the MDB. The results illustrate that the NSB scheme has a minimal impact on the main lobe radiation patterns. When comparing the beam patterns of the NSB scheme, which gives the best beam-level isolation at $(\theta_D, \theta_U)$, and the MDB scheme, a marginal average directivity difference of 0.2 dB is noted, with the maximum directivity degradation capped at 0.3 dB. The 3dB- and 10dB-beamwidth variations average around $0.8^\circ$ and $1.6^\circ$, respectively, with discrepancies consistently below $1.4^\circ$ for the 3dB-beamwidth and $3.0^\circ$ for the 10dB-beamwidth. Therefore, the NSB isolation-optimization scheme can greatly reduce SI without compromising the original beamforming performance.

## 10.6 Concluding Remarks

FD systems require SI levels that remain below the receiver noise floor to match the detection performance of HD systems. Achieving such stringent SI suppression is a central challenge in FD mMIMO system design. By leveraging the inherent flexibility of large-scale arrays, a joint uplink/downlink beamforming design can enhance gain toward desired directions while maintaining low SI levels to avoid overloading front-end components.

Using the $8\times8$Tx/$8\times8$Rx FD mMIMO antenna array prototype developed in previous chapters, we examined the relationship between array size, which determines beamwidth, and the achievable beam-level MC. Both full-wave simulated and measured SI channels confirm that concentrating the Tx/Rx beam radiation patterns substantially increases Tx-Rx isolation. Building on this insight, two beam-level isolation optimization algorithms are proposed. The first uses an angular perturbation search to find the best uplink and downlink steering angles within an acceptable desired power degradation limit. The second integrates an NSB approach with a PSO algorithm to find optimal uplink and downlink nulling angles for improved isolation. In the experiments using $1\times4$ sub-arrays with a 20 MHz modulated signal and allowing a 3dB-beamwidth steering variation, the average beam MC level can be reduced to $-70.2\,\text{dB}$, which is 19.7 dB lower (better) than the average element-level MC at the same location. The highest recorded beam isolation reaches 79.6 dB. The experiments also demonstrate that relatively high beam isolation can be achieved by steering Rx beams in the far-field null directions of the Tx beam.

Next, the joint uplink/downlink NSB algorithm is evaluated experimentally with $1\times 8$ sub-arrays. Measurements with a 20 MHz signal show an average isolation improvement of 10.2 dB, producing an average beam-level MC level of −75.7 dB. The largest observed isolation enhancement is 22.2 dB, with the highest beam-level isolation measured at 92.7 dB.

Although this chapter focuses on single-user uplink/downlink scenarios, the proposed algorithms can be adaptable to multi-user cases. Furthermore, the optimized datasets and experimental findings reported here support intelligent NN or other learning-based model development. For instance, in [20], an NN-based NSB scheme trained on a PSO-generated dataset achieves a 9.3 dB average beam-level MC reduction and over 74.7 dB of Tx-Rx isolation. More than 92.6% of tested downlink and uplink user locations show improved beam-level isolation, and 46.5% of cases exhibit enhancement above 10 dB. Additionally, these measurement-based insights into sub-array radiation patterns and achievable MC can inform other joint uplink/downlink beamforming methods aimed at further enhancing isolation.

## Appendix: Proof of Lemma 10.1

Let $\mathbf{F}_U$ and $\mathbf{F}_D$ be the UL and DL RF beamformers for $i$th DL UE and $j$th UL UE, respectively. Then, we can express the RF beamformers as

$$\mathbf{F}_D(\theta_D) = \frac{1}{\sqrt{M_D}}\left[1, e^{j\beta d\sin(\theta_D)}, \ldots, e^{j\beta(M_D-1)d\sin(\theta_D)}\right]^T, \tag{10.17}$$

$$\mathbf{F}_U(\theta_U) = \frac{1}{\sqrt{M_U}}\left[1, e^{-j\beta d\sin(\theta_U)}, \ldots, e^{-j\beta(M_U-1)d\sin(\theta_U)}\right]^T. \tag{10.18}$$

Then, we can rewrite the expression for $\mathbf{F}_U^T\mathbf{H}_{SI}\mathbf{F}_D$ as:

$$\begin{aligned}
&\mathbf{F}_U^T\mathbf{H}_{SI}\mathbf{F}_D \\
&= \frac{1}{\sqrt{M_D}}\mathbf{F}_U^T\mathbf{H}_{SI}[1, e^{j\beta d\sin(\theta_D)}, \cdots, e^{j\beta d(M_D-1)\sin(\theta_D)}]^T \\
&= \frac{1}{\sqrt{M_U}}[1, e^{-j\beta d\sin(\theta_U)}, \cdots, e^{-j\beta d(M_U-1)\sin(\theta_U)}]\mathbf{H}_{SI} \\
&\quad \frac{1}{\sqrt{M_D}}[1, e^{j\beta d\sin(\theta_D)}, \cdots, e^{j\beta d(M_D-1)\sin(\theta_D)}]^T
\end{aligned}$$

$$= \frac{1}{\sqrt{M_D M_U}} \sum_{n_u=1}^{M_U} \sum_{n_d=1}^{M_D} h_{n_u,n_d} e^{-j\beta d(n_u \sin(\theta_U) - n_d \sin(\theta_D))}, \quad (10.19)$$

where $h_{n_u,n_d}$ is the $(n_u, n_d)$-th element of $\mathbf{H}_{SI}$. Since $|\sin(\theta_{U_j})|, |\sin(\theta_{D_i})| \leq 1$, the term inside the exponent of (10.19) is bounded as follows:

$$|n_u \sin(\theta_U) - n_d \sin(\theta_D)| \leq n_u |\sin(\theta_U)| + n_d |\sin(\theta_D)|$$
$$\leq n_u + n_d \leq N_U + N_D - 2, \quad (10.20)$$

where the last inequality follows because $n_u$ and $n_d$ both range from 1 to $M_U$ and $M_D$, respectively. Now, let us consider the term $h_{n_u,n_d}$. Considering the large array structures, the path loss between the transmit and receive antennas within the same device is generally low due to the short distance between them. Thus, the SI channel, which is the channel from the transmit antenna to the receive antenna in the same device, is expected to be very strong in magnitude. Hence, we can assume that $h_{n_u,n_d}$ has non-negligible values. Next, we analyze the magnitude of the expression $\mathbf{F}_U^T \mathbf{H}_{SI} \mathbf{F}_D$. Applying the triangle inequality, we can write as follows:

$$|\mathbf{F}_U^T \mathbf{H}_{SI} \mathbf{F}_D|$$
$$= \frac{1}{\sqrt{M_D M_U}} \left| \sum_{n_u=1}^{M_U} \sum_{n_d=1}^{M_D} h_{n_u,n_d} e^{-j\beta d(n_u \sin(\theta_U) - n_d \sin(\theta_D))} \right|$$
$$\leq \frac{1}{\sqrt{M_D M_U}} \sum_{n_u=1}^{M_U} \sum_{n_d=1}^{M_D} |h_{n_u,n_d}|$$
$$\leq \frac{1}{\sqrt{M_D M_U}} (M_U + M_D - 2) \max_{n_u,n_d} |h_{n_u,n_d}|, \quad (10.21)$$

where the last inequality holds due to the bounds on the indices $n_u$ and $n_d$. Based on the assumption that the path loss is significant and the SI channel is strong, we can infer that the maximum absolute value of the SI channel coefficients $\max_{n_u,n_d} |h_{n_u,n_d}|$ is very small. Consequently, we can approximate the expression as follows:

$$|\mathbf{F}_U^T \mathbf{H}_{SI} \mathbf{F}_D| \approx \mathbf{0}. \quad (10.22)$$

This approximation is valid when SI channel is strong, the path loss is significant, and the array dimensions $M_U$ and $M_D$ are sufficiently large. It suggests that the design of DL and UL RF beamforming stages, represented by $\mathbf{F}_D$ and $\mathbf{F}_U$ respectively, can effectively eliminate the interference caused by the SI channel. Hence, we conclude that for a given UL-DL angle-pair $\theta_D, \theta_U$ in large array structures, the approximation $\mathbf{F}_U^T \mathbf{H}_{SI} \mathbf{F}_D \approx \mathbf{0}$ holds, indicating the suppression of SI interference.

## References


1. F. Chen, H.H. Lee, R. Morawski, T. Le-Ngoc, RF/Analog self-interference canceller for 2×2 MIMO full-duplex transceiver, in *2017 IEEE International Conference on Communications (ICC)* (IEEE, 2017), pp. 1–6
2. T. Le-Ngoc, Y. Gong, M. Mahmood, A. Koc, R. Morawski, J.G. Griffiths, P. Guillemette, J. Zaid, P. Wang, Full-duplex in massive multiple-input multiple-output. IEEE Open J. Veh. Technol. **5**, 560–576 (2024)
3. E. Everett, A. Sahai, A. Sabharwal, Passive self-interference suppression for full-duplex infrastructure nodes. IEEE Trans. Wirel. Commun. **13**(2), 680–694 (2014)
4. A. Koc, T. Le-Ngoc, Intelligent non-orthogonal beamforming with large self-interference cancellation capability for full-duplex multiuser massive MIMO systems. IEEE Access **10**, 51771–51791 (2022)
5. M. Rihan, T.A. Soliman, C. Xu, L. Huang, M.I. Dessouky, Taxonomy and performance evaluation of hybrid beamforming for 5G and beyond systems. IEEE Access **8**, 74605–74626 (2020)
6. S. Huberman, T. Le-Ngoc, MIMO full-duplex precoding: A joint beamforming and self-interference cancellation structure. IEEE Trans. Wirel. Commun. **14**(4), 2205–2217 (2014)
7. K. Satyanarayana, M. El-Hajjar, P.-H. Kuo, A. Mourad, L. Hanzo, Hybrid beamforming design for full-duplex millimeter wave communication. IEEE Trans. Veh. Technol. **68**(2), 1394–1404 (2018)
8. A. Koc, T. Le-Ngoc, Full-duplex mmWave massive MIMO systems: a joint hybrid precoding/combining and self-interference cancellation design. IEEE Open J. Commun. Soc. **2**, 754–774 (2021)
9. J.M.B. da Silva, A. Sabharwal, G. Fodor, C. Fischione, 1-bit phase shifters for large-antenna full-duplex mmwave communications. IEEE Trans. Wirel. Commun. **19**(10), 6916–6931 (2020)
10. K. Satyanarayana, M. El-Hajjar, A.A. Mourad, L. Hanzo, Multi-user full duplex transceiver design for mmwave systems using learning-aided channel prediction. IEEE Access **7**, 66068–66083 (2019)
11. M. Darabi, A.C. Cirik, L. Lampe, Transceiver design in millimeter wave full-duplex multi-user massive MIMO communication systems. IEEE Access **9**, 165394–165408 (2021)
12. Y. Zhang, M. Xiao, S. Han, M. Skoglund, W. Meng, On precoding and energy efficiency of full-duplex millimeter-wave relays. IEEE Trans. Wireless Commun. **18**(3), 1943–1956 (2019)
13. Z. Luo, L. Zhao, L. Tonghui, H. Liu, R. Zhang, Robust hybrid precoding/combining designs for full-duplex millimeter wave relay systems. IEEE Trans. Veh. Technol. **70**(9), 9577–9582 (2021)
14. D.H. Johnson, D.E. Dudgeon, Array signal processing: concepts and techniques (Simon & Schuster, Inc., 1992)
15. O.L. Frost, An algorithm for linearly constrained adaptive array processing. Proc. IEEE **60**(8), 926–935 (1972)
16. J. Capon, High-resolution frequency-wavenumber spectrum analysis. Proc. IEEE **57**(8), 1408–1418 (1969)
17. I.P. Gravas, Z.D. Zaharis, T.V. Yioultsis, P.I. Lazaridis, T.D. Xenos, Adaptive beamforming with sidelobe suppression by placing extra radiation pattern nulls. IEEE Trans. Antennas Propag. **67**(6), 3853–3862 (2019)

18. T. Le-Ngoc, A. Masmoudi, Full-duplex wireless communications systems. Wirel. Netw. (Springer, 2017)
19. K.E. Kolodziej, B.T. Perry, J.S. Herd, In-band full-duplex technology: techniques and systems survey. IEEE Trans. Microw. Theory Technol. **67**(7), 3025–3041 (2019)
20. Y.G.T. Zhang, T. Le-Ngoc, Supervised learning for nulling angle prediction with measured self-interference channels for full-duplex isolation enhancement, in *IEEE International Conference on Machine Learning for Communication and Networking (ICMLCN), Barcelona, Spain* (2025), pp. 1–6

# Chapter 11
# Metamaterial Absorber Structure for Tx-Rx Antenna Isolation Improvement in FD mMIMO

## 11.1 Introduction

FD operation is attracting growing interest in wireless communications due to its ability to transmit and receive signals simultaneously in the same frequency band, offering the potential for considerably higher spectral efficiency [1–3]. To achieve the theoretically doubled spectral efficiency and avoid significant degradation in the SINR caused by SI, a combination of Tx-Rx isolation and SIC techniques is crucial in FD systems [1, 4–6]. Previous chapters have introduced two primary methods for achieving Tx–Rx isolation: (1) leveraging cross-polarized antenna elements with strategic spacing, and (2) applying joint uplink/downlink beamformer design. Simulation and experimental findings reveal that this two-stage isolation design effectively prevents the residual SI from saturating the receiver's ADC or pushing the LNA into nonlinear operation. Further SI mitigation of over 56 dB is attainable through fractionally spaced (FS) SIC, which can reduce residual SI to levels comparable to thermal noise [7].

However, as presented in Chap. 10, beam-level isolation remains highly dependent on the relative positions of uplink and downlink users. When the uplink and downlink beams interact minimally, such as when they are steered away from each other, beam-level design techniques yield only moderate improvements. Additionally, in real-world conditions, the performance of SIC methods can be restricted by hardware impairments and the accuracy of SI channel estimation. Enhancing RF isolation by mitigating near-field coupling components ensures a robust overall Tx-Rx isolation level, even when far-field beam-level isolation design is less effective, thereby providing sufficient margin for total SI suppression.

Among the potential solutions, absorbers stand out as a promising technology to further improve Tx-Rx isolation in FD systems without altering the original antenna configuration. As shown in [8], using RF absorbers alongside directional antennas and carefully chosen separation can achieve up to 73.8 dB in total Tx–Rx isolation, offering more than 10 dB of improvement by blocking and absorbing LOS

 
T. Le-Ngoc et al., *Half-Duplex/Full-Duplex MU-mMIMO*, Wireless Networks,
https://doi.org/10.1007/978-3-032-09195-6_11

coupling waves. Nonetheless, for conventional RF absorbers, there is often a trade-off between thickness, height, and absorption efficiency. Metamaterial absorbers have grown in popularity because of their adjustable effective permittivity and permeability, low profile, easy fabrication, and low reflectivity [9]. As reviewed in Chap. 2, a typical metamaterial absorber consists of periodic elements smaller than $0.2\lambda_0 \times 0.2\lambda_0$ and a thickness of less than $0.04\lambda_0$ [10–13]. This compact size and flexible design allow easy integration into antenna systems without increasing the overall dimensions. Fabrication typically leverages PCB processes, as metamaterial absorbers consist of periodic metallic patches stacked on dielectric substrates.

When the effective permittivity ($\widetilde{\epsilon}$) and effective permeability ($\widetilde{\mu}$) of the homogenized structure are designed to be equal to each other, i.e., $\widetilde{\epsilon}/\widetilde{\mu} = 1$, the input impedance of the absorber approximates that of air, thus minimizing reflections at the air-absorber boundary. As a result, a substantial portion of the coupling waves penetrates the absorber layer and is dissipated within the dielectric substrate, offering power absorption rates above 90% [10–12]. Isolation improvements greater than 10 dB have been presented through different simulation or experimental studies [13–17]. Nonetheless, few existing studies address metamaterial absorbers in FD scenarios specifically. Researchers in [12] report an 8 dB SI reduction from a double-layer metamaterial absorber used in a four-element FD array, leading to a total isolation of 22 dB. However, very limited research exists on how metamaterial absorbers can benefit large-scale FD mMIMO antenna arrays and enhance beam-level isolation performance. Most absorber studies, such as [10–16], focus on array setup having no more than four elements. Although [18] shows an 8 dB MC reduction between sub-arrays in a 3×8 array using an orthogonally placed metamaterial absorber, the impact of absorbers on beam patterns and beamforming efficiency remains insufficiently explored.

Large-scale arrays hold considerable potential in FD system design by offering more directional and reconfigurable beams for enhanced isolation. Chapter 10, for instance, shows that steering antenna arrays toward interference nulls can effectively suppress SI. Nevertheless, near-field coupling remains a major hurdle in achieving uniformly high isolation across all uplink and downlink user locations. By mitigating near-field coupling components, metamaterial absorbers can strengthen overall Tx–Rx isolation, even when the beamforming isolation technique is less effective. Integrating metamaterial absorbers into large-scale FD arrays could therefore significantly boost total isolation. Despite this promise, there is still a shortage of experimental studies showing how metamaterial absorbers reduce near-field SI in tandem with beam-level isolation methods in FD mMIMO systems. In addition, experimental research on how different isolation techniques vary with Tx/Rx beam steering is also lacking. Examining these factors empirically could provide valuable insights into the interactions between near-field and far-field SI components. Such knowledge would also be instrumental in advancing multi-stage isolation strategies and boosting overall Tx–Rx isolation performance for various uplink and downlink user locations.

Based on the above observations, this chapter focuses on exploring and developing a metamaterial absorber structure, designed to enhance Tx-Rx isolation in large-scale arrays capable of FD mMIMO applications, without necessitating any alterations to the existing antenna designs. Utilizing realistic simulations and experimental measurements, our goal is to present a comprehensive methodology for designing the metamaterial absorber structures and to showcase its potential to significantly reduce near-field SI with negligible impact on the far-field beamforming performance. Furthermore, a simple and strategic Tx-Rx joint beamforming isolation improvement mechanism is implemented to further mitigate the far-field SI in a very low near-field SI scenario. It serves to highlight the importance of the absorber structure in overcoming the challenges of near-field coupling, as well as in enhancing overall beam-level isolation performance. To our knowledge, this work is the first to investigate the metamaterial absorber with beam-level isolation in FD mMIMO. The main contributions of this chapter are summarized as follows:

- **Metamaterial absorber for FD mMIMO:** A dual-layer metamaterial absorber is developed for our $8\times8$Tx/$8\times8$Rx large-scale antenna arrays capable of the FD mMIMO. The realistic simulation illustrates a power absorption ratio of more than 99.8%. In experimental measurements conducted with a 20 MHz modulated signal in the anechoic chamber, the absorber significantly enhanced Tx-Rx isolation. On average,[1] an antenna element isolation improvement of 17.5 dB is observed, with the closest Tx-Rx element pairs achieving an isolation of 65.4 dB. Moreover, the absorber has contributed to an average improvement of 18.6 dB in Tx-Rx beam-level isolation in near-field dominant SI scenarios, with the largest isolation improvement of 42.0 dB. An average beam isolation of 79.2 dB with the best beam isolation of 104.4 dB is achieved, underscoring the absorber's ability to mitigate the near-field SI.
- **Experimental studies of absorber-assisted angular-perturbation beamforming isolation:** Experimental investigations on the correlation between the near-field and far-field dominant SI scenarios and the downlink and uplink beam directions are conducted. The results highlight the critical role of absorber structures in overcoming the bottleneck presented by near-field dominant SI in beam-level isolation design. To enhance the total isolation performance in scenarios with very low near-field SI, we implemented an angular-based approach for Tx-Rx beam isolation improvement. Incorporating a simple steering angle perturbation of 2° in a 3dB-beamwidth search range with the proposed absorber structure, an average beam-level isolation of 90.8 dB is achieved within a downlink/uplink steering range between − 48° to 48°. Across 49 test cases, all the beam-level isolation is better than 70.8 dB and more than half the amount of the Tx/Rx beam pairs can achieve an isolation level better than 90 dB.
- **Metamaterial absorber impacts on the beamforming performance:** Experimental measurements were conducted to assess the impacts of the introduction of the metamaterial absorber on the far-field beam patterns. These measurements align well with our simulation findings. Within the steering range of − 50° to

[1] This study utilizes geometric means to characterize the average mutual coupling or isolation level, as the mutual coupling value represents a ratio of the received power to the transmitted power.

40°, we observed marginal differences in beam patterns. The average variations in beam directivity, 3dB-beamwidth, 10dB-beamwidth, and normalized sidelobe level were found to be 0.2 dB, 0.8°, 0.9°, and 0.9 dB, respectively. These results suggest that the addition of an absorber, positioned $\lambda_0$ away from the Tx antenna array, exerts a negligible impact on the far-field beam pattern in the beamforming. This finding is essential for ensuring the effectiveness of the absorber on near-field coupling mitigation without compromising the integrity of the far-field beam pattern in the beamforming process.

The rest of the chapter is structured as follows. Section 11.2 delves into the system models and SI channels employed in our study, providing the theoretical foundation for our research. Section 11.3 outlines the design process of the metamaterial absorber element. It includes simulation results demonstrating the absorber's effectiveness relative to its location and mounting position. Section 11.4 introduces the fabricated absorber prototype integrated into our FD mMIMO testbed. We present experimental measurements conducted in an anechoic chamber, focusing on isolation improvement in near-field dominant SI scenarios. This section also explores the impact of the absorber structure on the beamforming performance of a sample 1×8 sub-array. Section 11.5 presents our investigation into combining absorber technology with angular-perturbation-based methods to enhance total beam isolation. We discuss the effectiveness of this approach in scenarios dominated by near-field and far-field SI components, highlighting the necessity of the absorber structure in guaranteeing the total beam isolation performance. Section 11.6 concludes the chapter, summarizing our findings and suggesting future work in this area.

## 11.2 System Model

In this chapter, we adopt the FD mMIMO system model from Sect. 10.2, where a single-cell FD mMIMO system supports simultaneous DL and UL transmissions. The BS includes Tx and Rx antenna arrays with $M_{\text{Tx}}$ and $M_{\text{Rx}}$ elements, respectively, and serves $K_D$ DL and $K_U$ UL users simultaneously in FD mode. Meanwhile, users typically operate in HD mode due to limited power and signal-processing capabilities. To exploit beamforming technologies in large-scale arrays without incurring excessive hardware complexity, a two-stage HBF design is employed. The first stage, RF Tx/Rx beamforming, manages the Tx/Rx beam directions; the second stage, digital baseband precoding/combining, processes signals in the digital domain to maximize quality and efficiency. A low number of RF chains are used to connect the two stages, which can significantly reduce hardware complexity and cost compared with the fully digital approach.

In the FD mMIMO system, the DL signal undergoes processing through a baseband precoder $\mathbf{B}_D \in \mathbb{C}^{N_D \times K_D}$ and an RF beamformer $\mathbf{F}_D \in \mathbb{C}^{M_{\text{Tx}} \times N_D}$. The number of RF chains, $N_D$, satisfies $K_D \leq N_D \ll M_{\text{Tx}}$. Similarly, the UL signal received at the BS passes through an RF beamformer $\mathbf{F}_U \in \mathbb{C}^{M_{\text{Rx}} \times N_U}$ and a baseband combiner $\mathbf{B}_U \in \mathbb{C}^{N_U \times K_U}$, utilizing $K_U \leq N_U \ll M_{\text{Rx}}$ RF chains. The SI channel matrix $\mathbf{H}_{\text{SI}} \in \mathbb{C}^{M_{\text{Rx}} \times M_{\text{Tx}}}$ arises due to FD transmission, coupling between Tx and Rx antennas at the BS.

The beam steering directions for DL (UL) are defined by Azimuth and Elevation angles, denoted as $\theta_D(\theta_U)$ and $\phi_D(\phi_U)$, respectively. For this study, we focus on a single UL user and a single DL user scenario, i.e., $K_D = K_U = 1$, with two uniform linear sub-arrays. The downlink and uplink beamformer, $\mathbf{F}_D$ and $\mathbf{F}_U$, are constructed with phase shifters and uniform magnitudes, which focus on beam steering in Azimuth directions $\theta_D$ and $\theta_U$ with constant constant elevation angles $\phi_D$ and $\phi_U$.

Because the BS transmits and receives concurrently, the UL received signal suffers from significant SI caused by leakage from the local Tx. Before baseband processing, the residual SI power for serving the UL user at $\theta_U$ and the DL user at $\theta_D$ can be expressed as

$$\text{SI}(\theta_D, \theta_U) = \left|\mathbf{F}_U^T(\theta_U)\mathbf{H}_{\text{SI}}\mathbf{F}_D(\theta_D)\right|^2. \tag{11.1}$$

To attain UL detection performance in FD systems comparable to HD systems, minimizing SI, ideally to levels below thermal noise, is necessary. The SI channel encompasses both near-field and far-field components and can be expressed as

$$\mathbf{H}_{\text{SI}} = \underbrace{\mathbf{H}_{\text{Los}-\text{NF}}}_{\text{Near-field Coupling}} + \underbrace{\mathbf{H}_{\text{Los}-\text{FF}} + \mathbf{H}_{\text{NLos}-\text{FF}}}_{\text{Far-field Coupling}}. \tag{11.2}$$

Given the proximity of the Tx and Rx antennas, the predominant near-field SI components typically follow a LoS path and are denoted by $\mathbf{H}_{\text{LoS}-\text{NF}} \in \mathbb{C}^{M_{\text{Rx}} \times M_{\text{Tx}}}$. To minimize the $\mathbf{H}_{\text{LoS}-\text{NF}}$, antenna isolation design is implemented as the first line aiming to prevent coupling waves generated in potential interference directions from being captured by the local Rx. In our sample 8×8Tx/8×8Rx FD mMIMO prototype, Tx-Rx antenna isolation is initially achieved through cross-polarization and physical separation between the Tx and Rx antennas. To further improve the isolation performance, in this study, a metamaterial absorber is proposed and strategically placed within a distance of $\lambda_0$ from the transmitter. The absorber is specifically focused on absorbing and dissipating potential coupling waves, to reduce the strength of the near-field LoS component, $\mathbf{H}_{\text{LoS}-\text{NF}}$, and consequently, suppressing the $\mathbf{H}_{\text{SI}}$.

In contrast, the far-field coupling includes both LoS and NLoS components, represented by $\mathbf{H}_{\text{LoS}-\text{FF}} \in \mathbb{C}^{M_{\text{Rx}} \times M_{\text{Tx}}}$ and $\mathbf{H}_{\text{NLoS}-\text{FF}} \in \mathbb{C}^{M_{\text{Rx}} \times M_{\text{Tx}}}$, respectively. The $\mathbf{H}_{\text{NLoS}-\text{FF}}$ typically results from environmental reflections of radiation waves. In environments characterized by low reflection properties, the influence of $\mathbf{H}_{\text{NLoS}-\text{FF}}$ is considerably diminished. Such environments include simulations conducted under perfect electric (perfect-E) boundary conditions, experiments within anechoic chambers outfitted with efficient RF absorbers, and outdoor settings like rural areas

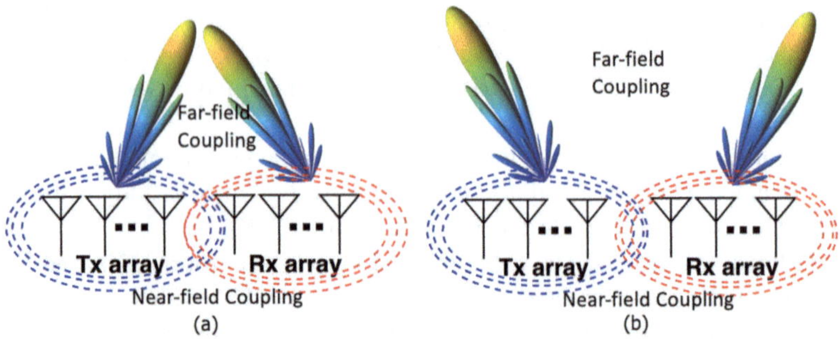


**Fig. 11.1** Far-field/Near-field coupling when Tx/Rx beams steering (**a**) towards each other (**b**) away from each other

or open fields. In these settings, the reduced presence of reflective surfaces and obstacles significantly lessens the impact of NLoS components in far-field coupling.

Additionally, due to the separation between our sample 8×8Tx and 8×8Rx, which is approximately $2.5\lambda_0$ edge-to-edge, the far-field beam interaction also contributes to the total SI. Therefore, by confining and controlling the wave propagation, beamforming plays a pivotal role in mitigating the LoS far-field component, $\mathbf{H}_{\mathrm{LoS-FF}}$, in the FD communications. Specifically, the direct interaction between Tx and Rx beams, represented by the Tx and Rx gains in specific directions, can be strategically managed through far-field beam steering techniques. As demonstrated in Chap. 10, steering the Rx beam towards the radiation nulls or low gain directions of the Tx beam is instrumental in reducing Tx-Rx MC. The approach entails introducing beam angle perturbations to align the Rx beam angles with the Tx radiation nulls. Such strategic alignment focuses on mitigating the beamforming-modified far-field LoS SI, $\mathbf{F}_U^T\mathbf{H}_{\mathrm{LoS-FF}}\mathbf{F}_D$. Consequently, this leads to a decrease in the magnitude of $\mathbf{F}_U^T\mathbf{H}_{\mathrm{SI}}\mathbf{F}_D$, thereby enhancing the overall performance of the FD system.

Two representative scenarios of the Tx/Rx beam directions against the MC scenarios are illustrated in Fig. 11.1. In the first scenario, where the Tx and Rx beams are directed across each other, i.e., $\theta_D > 0°$ and $\theta_U < 0°$, a relatively stronger far-field interaction between the two beams can be observed. Conversely, in the second scenario, the beams are steered away from each other, i.e., $\theta_D < 0°$ and $\theta_U > 0°$, and avoid a strong interaction between the Tx and Rx main lobes. Here, the far-field interaction between the Tx/Rx main lobes is comparatively weaker, which leads to a near-field dominant SI scenario. The detailed discussions of the isolation improvement strategies in different beam interaction scenarios will be presented in Sect. 11.5.

## 11.3 Metamaterial Absorber Design

### *11.3.1 Metamaterial Absorber Element Structure*

The proposed dual-layer metamaterial absorber unit cell is depicted in Fig. 11.2. Two identical Split Resonator Dual-Ring (SRDR) patch circuits are printed symmetrically on the bottom side of the lower FR-4 dielectric substrate and on the top side of the higher FR-4 dielectric substrate. The employed FR-4 dielectric material has a dielectric constant $\epsilon_r = 4.6$. The SRDR structure is strategically designed to be integrated with our previously proposed $8\times8$Tx/$8\times8$Rx array prototype for FD mMIMO operation. The copper ground plane in the middle of the sandwiched construction is shared by the two SRDR circuits. With concentric split resonators, the SRDR structure affords greater flexibility in manipulating the effective permittivity and the effective permeability. This is achieved by introducing adjustable gaps between the central and outer split rings and extending the length of metallic strips. The surface capacitance controlled by the gaps between and inside the metallic circuits and the dielectric material between the top circuits and ground plane contributes to the modification of the metamaterial's ability to store and respond to electric fields. Consequently, the effective permittivity of the metamaterial can be finely tuned. For instance, reducing the gap between conductors leads to increased capacitance, thereby enhancing the structure's ability to interact with and store electrical energy. This enhanced interaction, due to increased electrical energy storage capacity, facilitates an apparent adjustment in the metamaterial's effective permittivity. In addition, the inductance, determined by the length and geometry of the conducting strips in the meta-cell and generated by the current flowing through these adjacent cells, controls the metamaterial magnetic response. An increase in inductance, achieved through longer and twisted conducting paths that enhance magnetic field linkage, results in a higher effective permeability.

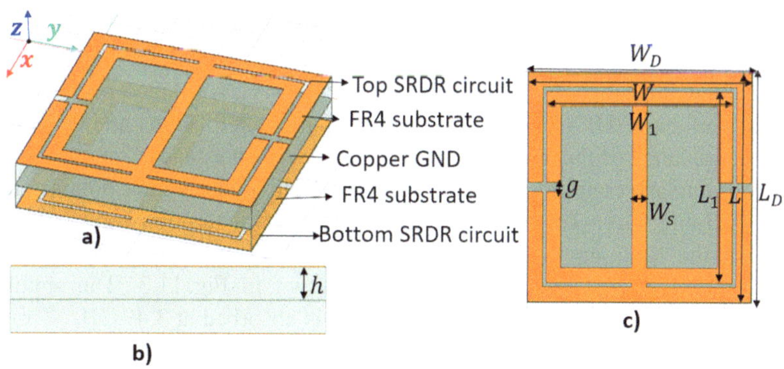


**Fig. 11.2** Metamaterial absorber element structure (**a**) layer diagram; (**b**) side view; (**c**) top view

**Table 11.1** Dimensions of the proposed metamaterial absorber element

| Label | Description | Value (mm) |
|---|---|---|
| $W$ | Outer layer circuit width | 16.9 |
| $L$ | Outer layer circuit length | 16.9 |
| $W_D$ | FR4 dielectric substrate width | 17 |
| $L_D$ | FR4 dielectric substrate length | 17 |
| $h$ | FR-4 substrate thickness | 1.7 |
| $W_1$ | Inner layer circuit width | 14.1 |
| $L_1$ | Inner layer circuit length | 14.1 |
| $W_s$ | Copper conductor circuit width | 1.1 |
| $g$ | Circuit gap width | 0.7 |

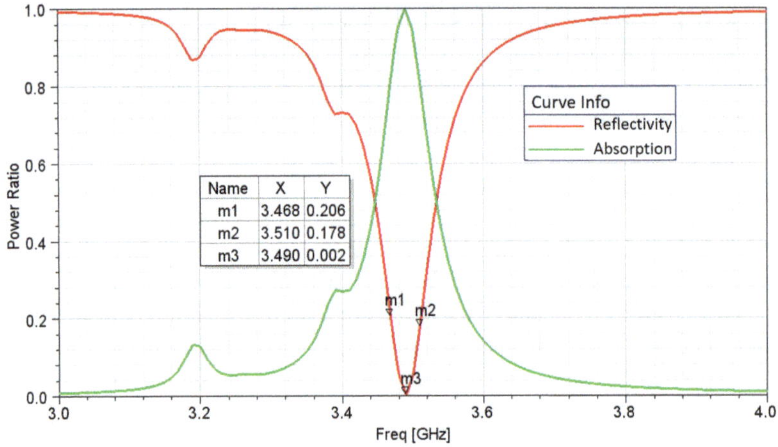


**Fig. 11.3** Power absorption and reflection of the proposed metamaterial absorber structure (simulation results)

The metamaterial absorber cells shown in Fig. 11.2 are designed to connect edge-to-edge, forming a two-dimensional array that extends along both the $x$- and $y$-axes. The overall size of the absorber structure is determined by the count of SRDR absorber elements along the $x$-axis ($N_x$) and the $y$-axis ($N_y$). Ansys high-frequency structure simulator (HFSS) is used in the design process to adjust and optimize the surface impedance of the absorber units using Floquet port and primary/secondary periodic boundary conditions. The suggested SRDR unit element dimensions are shown in Table 11.1. The absorber structure power absorption ratio ($\rho$) can be found via the S-parameter simulation, where $\rho = 1 - |S_{11}|^2$. The simulated power reflectivity and absorption ratio are shown in Fig. 11.3. The simulation results indicate that more than 99.8% of power is absorbed at 3.49 GHz. This high absorption power ratio ensures effective mitigation of near-field coupling waves, with a considerable amount of incident coupling wave travelling from the Tx to the Rx antennas being absorbed and dissipated in the absorber structure. Moreover, the absorber surface power reflection of less than 0.2% also contributes to reduced

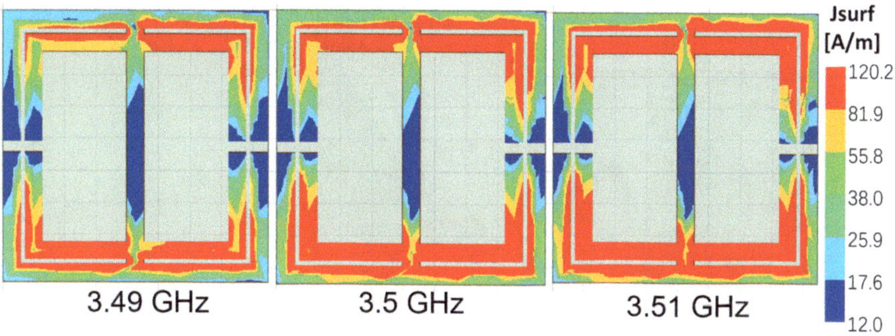


**Fig. 11.4** Surface current distribution of the metamaterial absorber at 3.49, 3.5, and 3.51 GHz (simulation results)

interference to the transmitting signals, preserving the integrity of the original beam patterns. Details of the absorber impact on the array beamforming will be further investigated via simulations and measurements.

The distribution of surface current on metamaterial absorbers also serves as an important indicator of the interaction strength between coupling waves and the metamaterial structure. The surface current distribution at three sample operating frequencies 3.49, 3.5, and 3.51 GHz are illustrated in Fig. 11.4. Across the target frequency bandwidth, a consistent distribution of high-strength surface currents can be observed. This consistent induction of strong surface currents indicates that a substantial portion of the incident EM waves effectively interacts with and couples to the periodic SRDR structures. The notable excitation of surface current within these frequencies indicates that the metamaterial is in strong resonance with the incoming EM waves, facilitating an effective absorption of the coupling waves.

### *11.3.2 Absorber Location Effects on Isolation Improvement*

The size of the absorber structure determines the path of the undesired leakage waves interacting with the absorber structures. A larger absorber structure typically exhibits stronger interactions with leakage waves, due to the increased number of pathways available for absorption. Therefore, a higher isolation enhancement can be achieved. For effective absorption and isolation, the absorber's dimensions should be comparable to the wavelength of the electromagnetic waves it is designed to interact with. Typically, effective absorbers are at least half a wavelength to several wavelengths in size in the dimension of interest. For an affordable

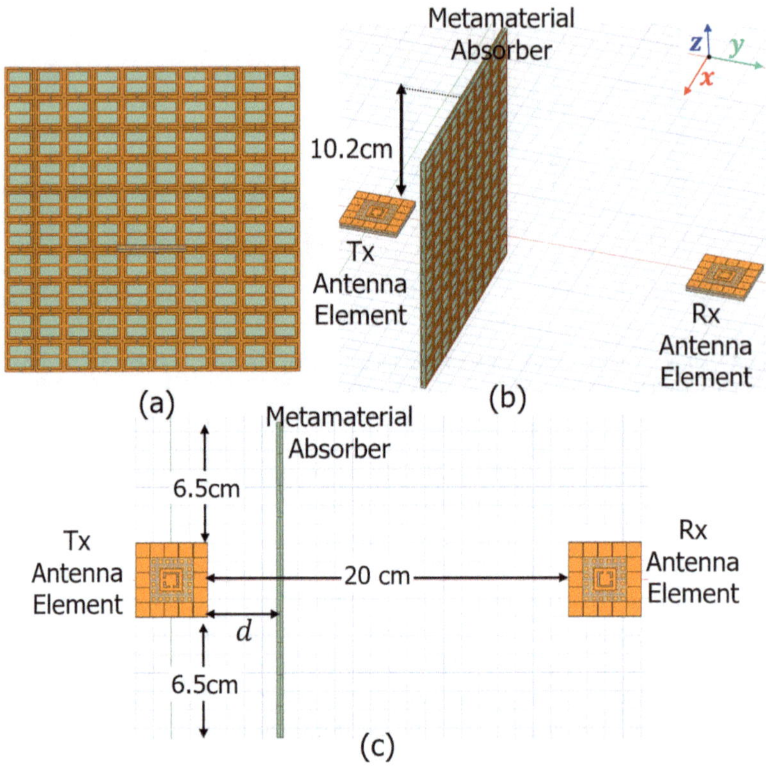


**Fig. 11.5** (**a**) 10×10 Metamaterial absorber, (**b**) perspective view, and (**c**) top view of the MC simulation setup with Tx/Rx antenna elements and absorber

iterative simulation time[2] and the effectiveness of the isolation enhancement, a metamaterial absorber with 10×10 (17 × 17 cm) proposed meta-cell structures, shown in Fig. 11.5a, is used in the HFSS simulation. The absorber has a width of 17 cm, which is approximately equivalent to two wavelengths, ensuring sufficient interaction with the leakage near-field coupling waves. As depicted in Fig. 11.5b and c, the absorber is vertically positioned between the Tx and Rx antennas, which are spaced 20 cm apart, edge-to-edge. The absorber extends 6.5 cm beyond the edge of the antennas in both $+x$ and $-x$ directions and is 10.2 cm above the surface of the patch antenna. The 6.8 cm absorber structure lower than the antenna surface can help mitigate the back leakage waves from the Tx element toward the Rx element. Only the Tx antenna (on the left) is excited at 3.5 GHz. The separation distance

[2] The realistic simulations used our proposed intricate dual-layer EBG-slotted antenna element demands high computational resources. The separation distance between the Tx-Rx antennas is kept at 20 cm which is consistent with our 8×8Tx/8×8Rx antenna array prototype.

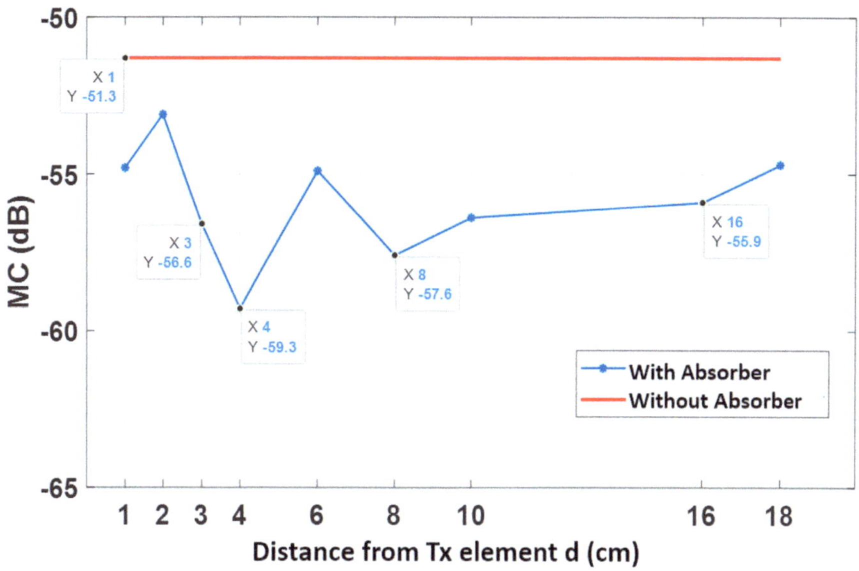


**Fig. 11.6** MC between Tx and Rx antennas versus $d$ (simulation results)

between the absorber and the Tx antenna ($d$) varies from 1 to 18 cm to find the best location of the absorber.

The simulation results are shown in Fig. 11.6. The findings show that the absorption effectiveness can be impacted by the absorber location. When 2 cm < $d$ <18 cm, a Tx-Rx element MC level lower (better) than −55 dB can be achieved. When comparing the Tx-Rx element MC with the scenario without the absorber structure, an 8 dB lower MC level can be attained when the absorber is at a distance of 4 cm ($d \approx \lambda_0/2$) and 6.3 dB isolation improvement is attained when the absorber is at 8 cm distance ($d \approx \lambda_0$). The effectiveness of the absorber diminishes when its distance from the transmitter exceeds the absorber's optimal operation range, which is determined by its resonant wavelength. Beyond these distances, a notable decline in isolation enhancement performance can be observed due to weaker resonant effects at the absorber surface and reduced interactions between the near-field coupling paths and the absorber structure.

To further investigate the absorber structure enhancement on the Tx-Rx element isolation level, two 16-element antenna arrays (4×4 or 8×2) with the proposed Tx and Rx antenna elements separated by 20 cm edge-to-edge are simulated. The absorber structure is extended along the $\pm x$-axis, to keep 6.5 cm longer than the antenna array on each side. The results are summarized in Table 11.2.

Placing an absorber at 4 and 8 cm can lower (improve) the Tx-Rx element MC by 9.2 and 7.2 dB on average, respectively, when two 4×4 arrays are used, and by 8.7 and 8.1 dB on average, respectively, when two 8×2 arrays are used. The average difference between the Tx-Rx element MC is less than 2 dB when placing

**Table 11.2** Tx/Rx MC without and with absorber at different *d* (simulation results)

| Tx/Rx arrays | Absorber to Tx distance (*d*) | Avg. Elem. Tx-Rx MC(dB) | Avg. MC diff. (dB) | Max Elem. Tx-Rx MC(dB) | Max MC diff. (dB) |
|---|---|---|---|---|---|
| 4×4 | No ABS | −50.9 | | −38.1 | |
| | 4 cm | −60.1 | 9.2 | −48.4 | 10.3 |
| | 8 cm | −58.1 | 7.2 | −46.6 | 8.5 |
| 8×2 | No ABS | −46.8 | | −38.0 | |
| | 4 cm | −55.5 | 8.7 | −45.9 | 7.9 |
| | 8 cm | −54.9 | 8.1 | −44.3 | 6.3 |

the absorber at 4 and 8 cm from the Tx array. When comparing the worst MC pairs, adding the absorber 4 or 8 cm from the Tx array leads to an improvement difference of 1.8 and 1.6 dB, respectively, when two 4×4 arrays or 8×2 arrays are used.

In the FD mMIMO design, the absorber's impact on array beamforming performance is also essential to be considered. To optimize beamforming and beam-level isolation without compromising beam steerability, the absorber must not obstruct the beam's path within its effective steering range. Consequently, we chose $d$=8 cm for the absorber in our design to prevent beam propagation blockage in the beamforming.

### *11.3.3 Absorber Mounting Effects on Tx-Rx Element Isolation Improvement*

Given the computational resource limitation, particularly in terms of Random Access Memory (RAM), and processing time of the simulation of the whole 8×8Tx/8×8Rx arrays, we use two sample orthogonally-polarized 1×8 arrays with the proposed Tx and Rx antenna element structures, operating at 3.5 GHz and separated by 20 cm edge-to-edge. The simulation setups with the antenna element indices are shown in Fig. 11.7a and b for vertically and horizontally mounted absorbers, respectively. Our investigation focuses on studying the Tx-Rx element/beam MC and beamforming performance under two different absorber mounting scenarios.

#### 11.3.3.1 Horizontally-mounted Absorber

By comparing Fig. 11.8a and b, the average Tx-Rx element MC is lowered (improved) from −47.3 to −57.4 dB, where an average 10.1 dB lower (better) MC level is achieved by adding the absorber. Figure 11.8b demonstrates that the horizontal absorber's effectiveness increases as the Tx and Rx elements move further apart. This enhancement, particularly evident in the lower-left quadrant

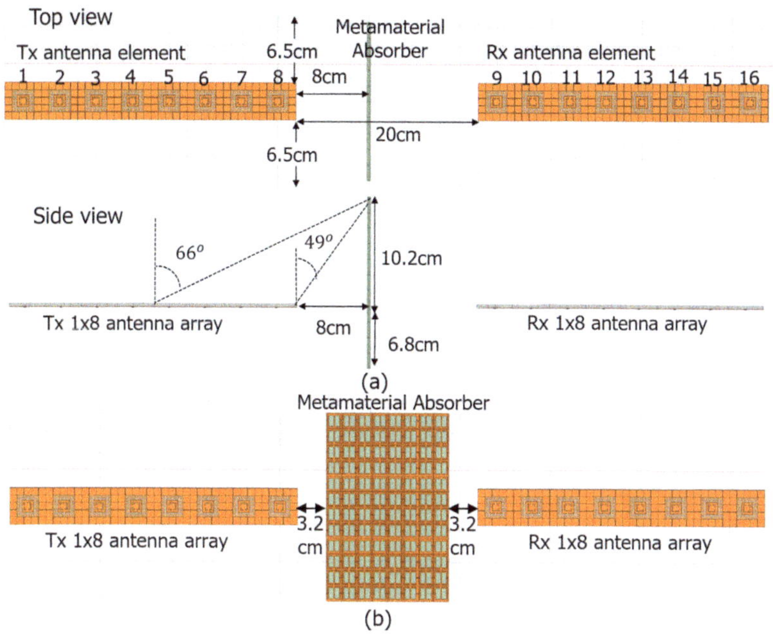


**Fig. 11.7** Simulation setups and antenna element indices with (**a**) vertically-mounted absorber (top and side views); (**b**) horizontally-mounted absorber (top view)

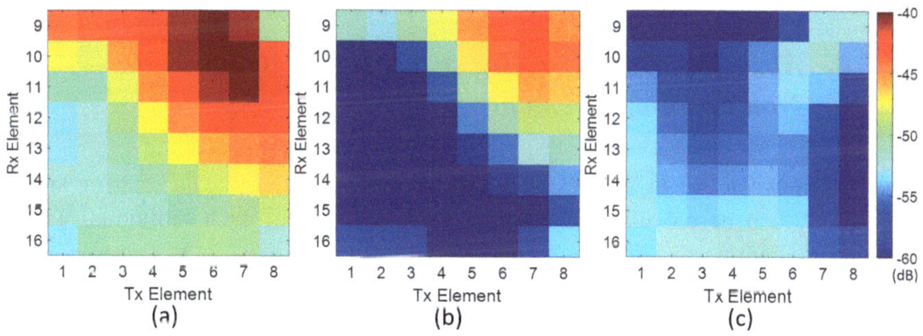


**Fig. 11.8** Tx-Rx antenna element MC magnitude maps in two 1×8 arrays at 3.5 GHz (**a**) without absorber; (**b**) with horizontally-mounted absorber; (**c**) with vertically-mounted absorber (simulation results)

of the plot, is attributed to stronger surface wave coupling interactions with the horizontally mounted absorber at greater distances (order of a half-wavelength or more). The rise in path loss attributed to the increased distance between elements also contributes to the reduction in MC between the Tx and Rx. The highest (worst) Tx-Rx element pair MC level is reduced from −38.9 to −42.7 dB, equivalent to a 3.8 dB improvement, which can be a potential limitation in the beam MC and isolation effectiveness when the antenna arrays are employed.

#### 11.3.3.2 Vertically-mounted Absorber

Comparison between Fig. 11.8a and c reveals a notable Tx-Rx isolation improvement. An average reduction of 9.2 dB in the Tx-Rx element MC level can be achieved, and an average MC level of −56.5 dB can be achieved. Across all the 64 Tx-Rx element pairs, the MC level is consistently lower than −51.9 dB. Furthermore, the worst-case (highest) MC level is reduced (improved) by 13.0 dB compared to the scenario without an absorber structure. This improvement is 9.2 dB greater than the case with a horizontally mounted absorber. As the Tx elements and Rx elements move closer to each other, a higher near-field MC intensity is suffering. Therefore, it can also be observed that as the Tx and Rx element pairs move close to the absorber structure (e.g., Tx element 8 and Rx element 9), a more considerable isolation improvement can be achieved. An average isolation improvement of 14.8 and 19.4 dB can be achieved between Tx-elem-8 and any Rx antenna elements and Rx-elem-9 and any Tx antenna elements, respectively. Furthermore, the effective attenuation of near-field LoS coupling components between the Tx and Rx significantly reduces the correlation between their distance and the MC level.

### *11.3.4 Absorber Mounting Effects on Tx-Rx Beam-Level Isolation Improvement*

The MC between the Tx and Rx arrays at various Tx and Rx beam steering angles $\theta_D$ and $\theta_U$ can be calculated through a weighted sum of each individual Tx-Rx antenna element pair's complex MC values and can be expressed as

$$S_b(\theta_D, \theta_U) = \sum_{j=9}^{16} \sum_{i=1}^{8} w_i(\theta_D) w_j(\theta_U) S_{ji}, \tag{11.3}$$

where $S_{ji}$ is the complex S-parameter that represents the coupling between the Tx antenna element $i$ and the Rx antenna element $j$ in the arrays. $w_i(\theta_D)$ and $w_j(\theta_U)$ are the complex steering weights for the Tx element $i$ and Rx element $j$ at the Tx beamforming angle $\theta_D$ and Rx beamforming angle $\theta_U$ in the Azimuth plane. The estimated MC values between Tx 1×8 and Rx 1×8 arrays for the three cases at 3.5 GHz are shown in Fig. 11.9. Both the Tx and Rx beams are steering from − 50° to 50° with a 2° step size. Therefore, a total of 51 × 51 = 2601 beamforming MC combinations are investigated.

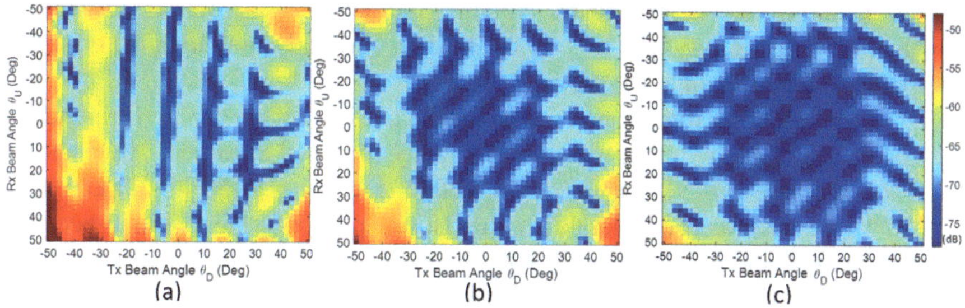


**Fig. 11.9** Tx-Rx beam MC magnitude maps with 1×8 antenna arrays at 3.5 Hz (**a**) without absorber; (**b**) with horizontally-mounted absorber; (**c**) with vertically-mounted absorber (simulation results)

#### 11.3.4.1 Horizontally-mounted Absorber

Without the absorber structure, the average beamforming MC level across the 2601 beam combinations is −64.2 dB as shown in Fig. 11.9a. Incorporating the horizontally-mounted absorber reduces the average beam MC to −68.1 dB, representing an improvement of 3.9 dB on average. This modest enhancement can be attributed to the reduced contribution of surface wave coupling to the total SI intensity when the elements are configured into directional beams.

It can be observed that when beams are steering away from each other (left-bottom quarter of the plot, i.e., $\theta_D \in [-50°, -2°]$ and $\theta_U \in [2°, 50°]$), the absorber structure is more effective, with an average improvement of 6 dB. When the Tx beam steers towards the Rx array, the absorber structure becomes less effective, and an average of 3.1 dB (right-top quarter of the plot, i.e., $\theta_D \in [2°, 50°]$ and $\theta_U \in [-50°, -2°]$).

#### 11.3.4.2 Vertically-mounted Absorber

As shown in Fig. 11.9c, with the vertically-mounted absorber structure, an average beam MC of −71.1 dB can be achieved, which is 6.9 dB lower (better) than the no-absorber case. A more dense amount of high-isolation Tx/Rx beam angle pairs, which results in a higher fraction of high-isolation $(\theta_D, \theta_U)$ pairs in the steering range, can be observed in Fig. 11.9c. All 2601 beamforming combinations have an MC lower than −55 dB. The result indicates that the vertically mounted absorber reduces (improves) the highest beam MC by 11.5 dB. A larger isolation improvement can be observed when steering Tx and Rx beams away from each other (i.e., $\theta_D \in [-50°, -2°]$ and $\theta_U \in [2°, 50°]$), with an average MC of 10.8 dB lower.

When comparing Fig. 11.9b and c, vertically mounting the absorber leads to an average of 3 dB higher isolation improvement compared with the horizontal case. Additionally, the highest (worst) beam-level MC is constrained by the most severe

Tx-Rx antenna element MC level. Notably, the worst beam MC is reduced by 6.6 dB, dropping from −48.4 to −55.0 dB when vertically mounting the absorber compared to the horizontal position.

The beam steering directions have a significant impact on the performance of the absorber. For both vertically and horizontally mounted absorbers, when the Tx and Rx beams are steered away from each other (i.e., $\theta_D \in [-50°, -2°]$ and $\theta_U \in [2°, 50°]$), we observe a considerably higher isolation improvement. This is because, in such configurations, the far-field beam interaction diminishes, making the near-field coupling components more dominant and the absorber structure is specifically designed to address these near-field coupling waves. To get better beamforming isolation across all the $(\theta_D, \theta_U)$ pairs, vertically mounting is used in our following simulations and measurements.

### *11.3.5 Absorber Impact on Beam Pattern: Simulation Results*

The Tx 1×8 array with the configuration depicted in Fig. 11.7a is simulated to assess the impact of a vertically-mounted absorber structure on beamforming performance. The simulated beam patterns with and without the vertically-mounted absorber structure at 3.5 GHz are illustrated in Fig. 11.10. The Tx 1×8 array is steering within the range $\theta_D \in [-50°, 50°]$. Without the absorber structure, an average directivity of 13.3 dB is achieved in the steering range. The 3dB- and 10dB-beamwidths are

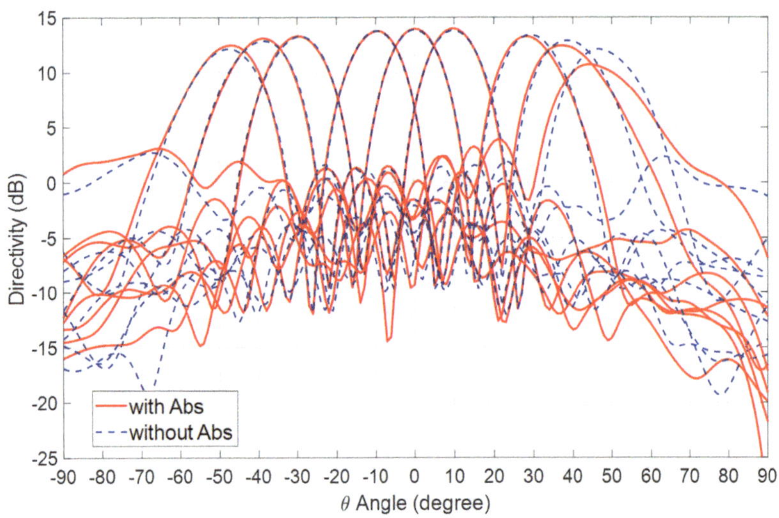


**Fig. 11.10** 1×8-array Tx beam pattern with and without the absorber at steering angle − 50°, − 40°, − 30°, − 10°, 0°, 10°, 30°, 40°, and 50° (simulation results)

consistently narrower than 18° and 32°, respectively. The normalized sidelobe level is consistently lower than the main lobe peak level by 9.5 dB.

With the incorporation of the absorber, the resulting beam from the Tx 1×8 array exhibits slight variations. An average directivity difference of 0.3 dB, an average 3dB-beamwidth difference of 0.6°, and an average beam peak direction variation of 0.9° can be observed over the steering range from − 50° to 50°.

The main lobes of the Tx 1×8 array with and without absorbers overlap with each other, and the difference between directivity and beamwidth is smaller than 0.4 dB, and 1° respectively, in the range − 50° to 40°. The slight beam pattern variation suggests that the introduction of the absorber, which is vertically mounted 8 cm from the Tx array, has a negligible effect on the array's beamforming performance and far-field beam patterns. The observation also indicates that the interaction between the absorber structure and the Tx radiation wave primarily occurs with the near-field coupling components. When $\theta_D \geq 50°$, the difference between the mainlobe peak and the sidelobe level decreases by 1.2 dB. As shown in Fig. 11.7a, at these larger steering angles, the absorber starts to impede the beam's propagation path, leading to a beam pattern degradation. The absorber intercepts the desired wave radiating in this direction that would otherwise contribute to the main lobe, diminishing antenna gain. This inadvertent interaction leads to an increase in the normalized side lobes and the main lobe beamwidth, diminishing the array directivity. Consequently, this can distort the intended radiation patterns of the antenna array.

## 11.4 Experiments with the Metamaterial Absorber

### *11.4.1 An 8×8Tx/8×8Rx Antenna Array Prototype with Metamaterial Absorber*

The 8×8Tx/8×8Rx antenna array prototype with the metamaterial absorber for FD mMIMO is shown in Fig. 11.11. The 8×8 Tx/Rx antenna arrays are positioned 20 cm apart. The entire Tx/Rx antenna-array prototype spans a total length of 84 cm and a width of 32 cm. The 10×27-element absorber prototype (Fig. 11.12) is vertically mounted between Tx and Rx antenna arrays. Additionally, the metamaterial absorber prototype has relative dimensions similar to the absorber utilized in the simulation. Specifically, the absorber structure extends approximately one wavelength longer, or 8.95 cm, relative to the center of the edge element in the antenna array, or it is 6.95 cm longer than the array prototype at both the top and bottom sides. To effectively block and absorb the LoS leakage waves without hindering wave propagation within the array's effective steering range, the absorber includes a section that is 10.2 cm above the array surface. Furthermore, a part of the absorber extends 6.8 cm below the antenna surface, facilitating not only an easier installation but also ensuring efficient absorption and interaction with potential back leakage waves. Additionally, the absorber is positioned 8 cm away from the Tx

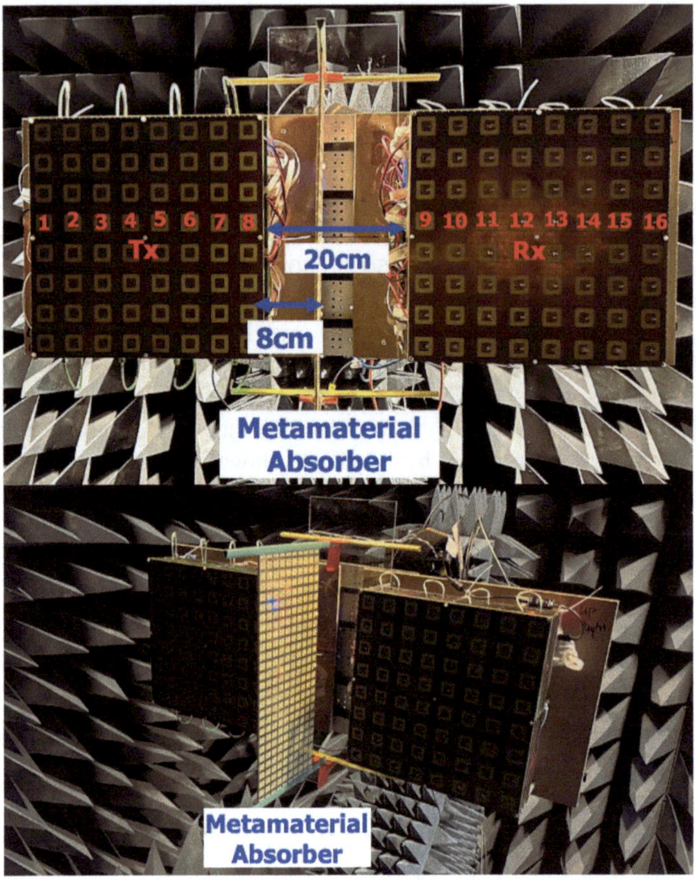


**Fig. 11.11** 8×8Tx/8×8Rx FD mMIMO array prototype with the metamaterial absorber (top: front-view with element and sub-array indices, bottom: perspective view)

array, ensuring optimal functionality in mitigating interference. Further details of the prototypes are briefly described in the following sections.

#### 11.4.1.1 Metamaterial Absorber Prototype

An absorber prototype consisting of 10×27 SRDR metamaterial elements is fabricated, as shown in Fig. 11.12. The absorber has a total dimension of 45.9 × 17 × 0.35 cm (length × width × height). The prototype is 13.9 cm longer than the antenna array. FR-4 material with a dielectric constant ($\epsilon_r$) 4.6 is used in the PCB manufacture.

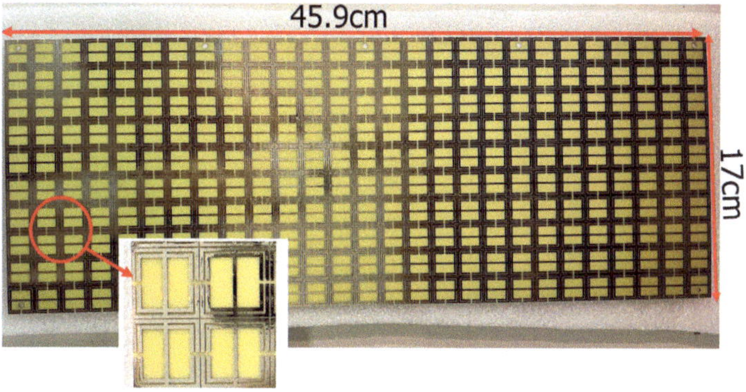


**Fig. 11.12** The 10×27-element metamaterial absorber prototype

#### 11.4.1.2 Cross-polarized 8×8Tx/8×8Rx Antenna Arrays

To investigate the Tx-Rx antenna element and sub-array beamforming isolation performance with and without the absorber structure, the FD mMIMO prototype consists of an RF feeding network and a pair of cross-polarized 8×8Tx and 8×8Rx antenna arrays, as presented in Chap. 10. The antenna element is designed to operate over a 250 MHz bandwidth, covering the frequency range from 3.35 to 3.6 GHz. As illustrated in Chap. 6, within the frequency range of 3.35–3.6 GHz, the antenna element return loss is consistently better than 10 dB and the inter-element mutual coupling is better than −14.3 dB. To avoid the grating lobes in the beamforming, the separation distance between neighbouring antenna elements is set to 2.78 cm (0.32 $\lambda_0$ at 3.5 GHz) edge-to-edge or 4.00 cm (0.47 $\lambda_0$) center-to-center. The direction of the circular polarization is controlled by the relative location between the feeding point and the slot direction on the central radiating patch. The polarization of the Tx elements is adjusted to be RHCP. Then, the Rx patch antenna is created by mirroring the Tx element structure, resulting in a LHCP.

#### 11.4.1.3 Measurement Setup

The placement of the array prototype in the anechoic chamber is shown in Fig. 11.13. The chamber is equipped with the C-RAM SFC-48 absorber to minimize reflections and external interference and has a dimension of 6.1 × 2.4 × 2.4 m (length × width × height). For both results presented in Sections IV and V, the Spectrum Analyzer is used to take measurements of the receiving signal at the Rx antenna element/sub-array when a modulated NR-FR1-TM3.1 test signal with a

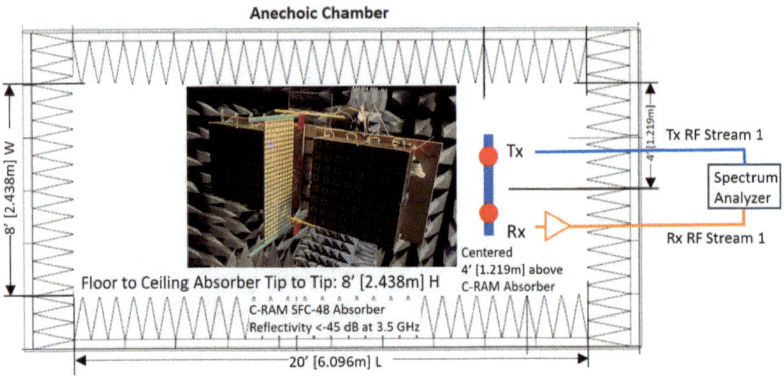


**Fig. 11.13** Measurement setup in the anechoic chamber

20 MHz bandwidth[3] (between 3.49 and 3.51 GHz) and an output power of 10 dBm generated from a Vector Signal Generator (VSG) is sent from the Tx antenna element/sub-array. In the Tx-Rx element mutual coupling measurements, only a single Tx antenna element is used for transmitting the signal and only a single Rx antenna element is used as receiver in each measurement. We manually switched the cable between different Tx-Rx element pairs.

In the sub-array beam-level experiments, two sample sub-arrays, Tx/Rx $1 \times 8$, located at the center (fourth row of the Tx/Rx arrays) are used for the beamforming and isolation measurements as labelled in Fig. 11.11. On the Tx side, the signal undergoes initial splitting via a 1-to-8 power splitter before entering the beamforming network. Subsequently, the eight paths will be modulated by 8 digitally controlled 6-bit phase shifters. These phase shifters, controlled by a pre-calculated steering weight vector, facilitate the manipulation of the phase across 360° with a fine step size of 5.625°, ensuring precise beam direction control. Before the signal is fed to the antenna, the final stage RF amplifiers are employed to maximize transmitted power and compensate for the insertion loss of the system.

On the Rx side, each sub-array is directly connected to Low-Noise Amplifiers (LNAs) and then 8 digitally controlled 6-bit phase shifters. The high Tx-Rx isolation which contributes to a low received coupling signal necessitates the use of LNAs to enhance measurement sensitivity by boosting the SNR. The signals from these eight paths are then combined through an 8-to-1 power combiner and fed into the Spectrum Analyzer.

Short coaxial cables and compact JSC connectors are used for interconnecting the components, all of which are arranged behind the antenna arrays for efficiency and convenience. The entire feeding network is controlled through two customized

[3] This study concentrates on narrow-band scenarios, with a 20 MHz bandwidth, to circumvent issues related to frequency-dependent beam patterns and avoid the complexity in calculating frequency-specific weight factors.

SPI modules, one for the Tx beamformer and the other one for the Rx beamformer, which are integrated into Raspberry Pi 4. The beamforming vectors $\mathbf{F}_D(\theta_D)$ and $\mathbf{F}_U(\theta_U)$ for the Tx and Rx antenna sub-arrays are generated by a computer running a MATLAB-based beamforming program, which utilizes Eqs. (10.6) and (10.7). The computer interfaces with the SPI modules for control purposes and the Spectrum Analyzer for analysis purposes via USB cables. During each beam-pair measurement, the control signals are first updated on the SPI modules based on the $\mathbf{F}_D(\theta_D)$ and $\mathbf{F}_U(\theta_U)$. Subsequently, the readings from the Spectrum Analyzer, indicating the mutual coupling levels between the Tx and Rx antennas, are recorded.

### *11.4.2 Measured Mutual Coupling between Tx and Rx Antenna Elements*

The measurement results of Tx-Rx antenna elements, both with and without the proposed metamaterial absorber structure, are illustrated in Fig. 11.14. The study focuses on Tx antenna elements 7 and 8, as well as Rx elements 9 and 10, which are the closest Tx/Rx antenna element pairs and most affected by the near-field MC, to assess the efficacy of the proposed absorber in isolation improvement, as labelled in Fig. 11.11. Only one of the Tx antenna elements is excited in each measurement.

As shown in Fig. 11.14a, without the absorber, the Tx-Rx antenna element MC level is −47.9 dB on average with an MC range from −49.7 to −45.8 dB. As the distance between Tx and Rx elements decreases, higher MC values are observed. For example, the MC between Tx-8 and Rx-9 is 3.9 dB higher than the farther element pair Tx-7 and Rx-10.

As illustrated in Fig. 11.14b, adding the absorber structure introduces an average reduction (improvement) of 17.5 dB in Tx-Rx antenna element MC, resulting

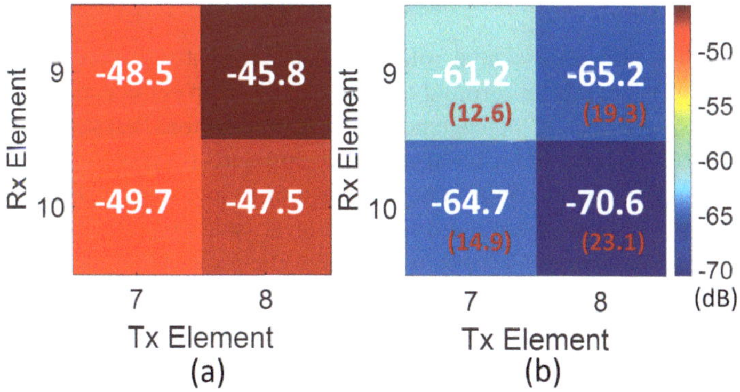


**Fig. 11.14** Tx-Rx antenna element MC magnitude maps in (dB) (**a**) without the absorber; (**b**) with the absorber (measured results)

in an average element MC of −65.4 dB. The Tx-Rx antenna element isolation improvement of the four Tx-Rx element pairs with and without the absorber is highlighted by the red values in Fig. 11.14b. The isolation enhancement is consistent across all four test cases, greater than 12.6 dB, with the highest improvement reaching 23.1 dB. The highest (worst) Tx-Rx element MC pair is −61.2 dB, with a reduction of 15.4 dB thanks to the absorber. Due to the effective suppression of near-field coupling waves by the absorber structure, the correlation between antenna element separation distance and MC levels is notably weakened.

The measured results show a similar absorber effectiveness change as the simulation results at similar relative locations. A significant isolation improvement can be seen as the Tx and Rx antenna element pairs become closer to the absorber structure. For instance, an MC improvement of −21.2 dB is obtained on average between Tx element 8 and the Rx element 9 and 10.

The EBG patches in the array surrounding the tested sample Tx and Rx elements create a high surface impedance, which works as a band-stop filter. This impedance effectively mitigates and blocks the propagation of coupling waves in the direction parallel to the absorber structure, along the $x$-axis. Consequently, the near-field coupling along the $y$-axis, perpendicular to the absorber, becomes the predominant factor contributing to SI, leading to a more considerable isolation improvement.

As the distance between the Tx element and the absorber increases, a decrease in isolation improvement can be observed. This reduction is attributed to the diminishing influence and contribution of near-field MC on the total SI. Furthermore, when the distance exceeds the absorber's optimal range, defined by its resonant wavelength, the interaction between the incident wave and the absorber surface becomes weaker, further diminishing its effectiveness in isolation improvement.

### 11.4.3 Measured Beam Mutual Coupling with/Without the Absorber

Antenna arrays play an essential role in enhancing spatial isolation in FD communications. Through beamforming techniques, these arrays concentrate radiation power, enabling effective management of the interactions between Tx and Rx beams. The proposed metamaterial absorber structure focuses on mitigating the near-field coupling wave components. Therefore, to evaluate the effectiveness of the metamaterial absorber structure in beam-level isolation, we first focus on the near-field dominant beam MC scenarios, where the Tx and Rx beams are steering away from each other to minimize the interaction between the far-field beams as shown in Fig. 11.1b. A study of both near-field and far-field beam-level isolation will be presented in Sect. 11.5.

To evaluate the absorber improvement in the beam-level isolation, we consider the Tx-Rx beam pairs $\theta_D \in [-50° : 2° : -2°]$ and $\theta_U \in [2° : 2° : 50°]$. Therefore total of $25 \times 25 = 625$ pairs of Tx/Rx beam pairs are measured. The measurement

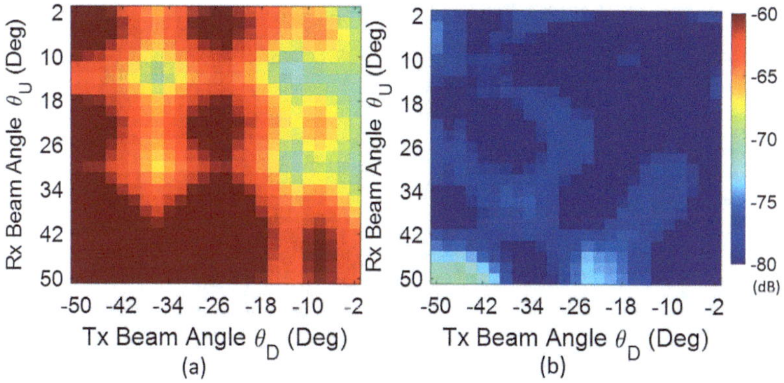


**Fig. 11.15** Tx-Rx 1×8 arrays' beam MC magnitude maps in (dB) (**a**) without the absorber; (**b**) with the absorber (measured results)

results without and with the proposed metamaterial absorber structure are presented in Fig. 11.15.

The addition of a strategically placed absorber structure significantly reduces the MC between the Tx and Rx beams. Without the metamaterial absorber, an average beam MC of −60.6 dB, ranging from −53.4 to −72.1 dB can observed. When the absorber structure is vertically inserted between the Tx and Rx arrays, an average beam MC of −79.2 dB can be achieved, which indicates an average beam isolation improvement of 18.6 dB. Notably, all 625 beam pairs measured achieve isolation better than 70.3 dB with the absorber, representing a 16.9 dB improvement in the worst (highest) beam MC. The most substantial isolation improvement occurs at $(\theta_D, \theta_U) = (-48°, 38°)$ and an isolation improvement of 42.0 dB can be achieved.

The highest (best) isolation of 104.4 dB can be achieved, which indicates a significant 32.3 dB increase compared to cases without the absorber shown in Fig. 11.15a. Out of the 625 tested beam pairs, 609 pairs (approximately 97%) exhibit an isolation improvement exceeding 10 dB, 490 pairs (around 78%) show more than 15 dB improvement, and 234 pairs (about 37%) achieve an increase in isolation beyond 20 dB. These results underscore the efficacy of the absorber structure in enhancing beam isolation across a wide range of near-field coupling dominant test cases.

## *11.4.4 Absorber Impacts on the Beamforming Performance*

### 11.4.4.1 Tx/Rx 1x8 Sub-array Beamforming Without the Absorber

The measured normalized 3D radiation pattern of the Tx 1×8 and Rx 1×8 sub-arrays with uniform feeding power and zero phase shift at 3.5 GHz in the anechoic chamber are shown in Fig. 11.16. Both sub-arrays have a very similar plate-shaped

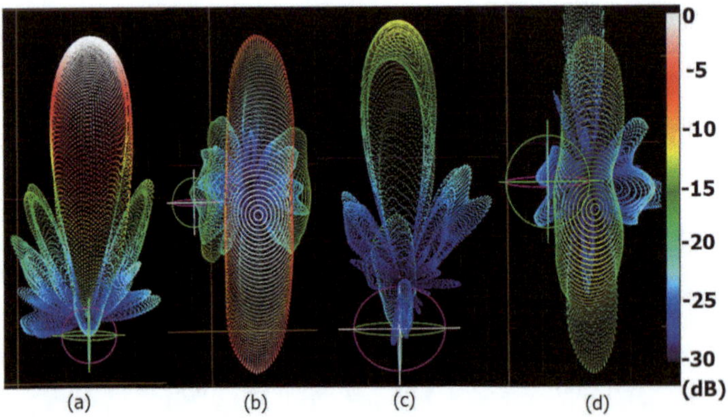


**Fig. 11.16** 3D beam patterns of the Tx/Rx 1×8 sub-arrays: (**a**) Tx side view, (**b**) Tx top view, (**c**) Rx side view, (**d**) Rx top view (measured results)

beam, with two sidelobes on the two sides of the main lobe. The directivity of the Tx 1×8 sub-array is 14.0 dB, and the directivity of the Rx 1×8 sub-array is 14.9 dB. The Tx and Rx sub-arrays have 3dB-beamwidths of 14.2° and 13.4° in the Azimuth direction, respectively. The normalized side lobe level is −13.6 dB for the Tx sub-array and −13.0 dB for the Rx sub-array.

The Tx 1×8 sub-array is used for the experimental beam steering investigation. The beam characteristics without absorber are measured within the steering range − 60° to 60° at 3.5 GHz. An effective beam peak steering range from − 58.1° to 59.3° is achieved. Within the steering range from −50° to 50°, an average directivity of 13.4 dB, an average 3dB-beamwidth of 16.3°, and an average 10dB-beamwidth of 27.8° can be achieved. Within this range, the 3dB- and 10dB-beamwidths are narrower than 19.2° and 34°, respectively. The simulation and experimental results are in good agreement, with average directivity and 3dB-beamwidth difference of 0.1 dB and 0.8°, respectively. The sidelobe level is lower than the main lobe by −9.3 dB in the steering range of − 40° to 50°. The sidelobe increases quickly when the steering angle is larger than 50°. The observed beam patterns indicate that the array possesses an effective steering range extending from − 50° to + 50°. Steering the sub-array beyond this angular range results in a notable expansion of the main lobe's 3dB- and 10dB-beamwidths. Such expansions lead to a rapid degradation in beam resolution. As the steering angle increases, the effective spacing between elements, in terms of wavelength, also increases due to the projected element spacing along the beam direction. This can result in element spacing effectively exceeding half a wavelength, leading to the formation of grating lobes, which leads to a faster sidelobe increase.

#### 11.4.4.2 Absorber Effects on the Sub-array Far-field Pattern

To examine the impact of the absorber on the beam radiation patterns, the Tx $1\times8$ sub-array is steered from $-50°$ to $+50°$ with a $10°$ step size. The vertically-mounted absorber is 8 cm from the Tx array. The beam-pattern azimuth-cut comparison at eight sample steering angles is shown in Fig. 11.17. It can be observed that the main lobes with and without the absorber overlap each other within the steering range $\theta_D \in [-50°, 40°]$. The measurement results are in line with the simulation findings. Within this range, the average difference in the directivity, the 3dB-beamwidth, the 10dB-beamwidth, and the normalized sidelobe level are 0.2 dB, 0.8°, 0.9°, and 0.9 dB, respectively. In other words, adding the absorber (located 8 cm away from the Tx array) has a negligible influence on the Tx $1\times8$ sub-arrays far-field beam pattern during the beamforming process. This observation also illustrates that the absorber, designed to improve near-field isolation, has limited impacts on the far-field waves.

When the steering angle $\theta_D \geq 50°$, the beam sidelobe level increases quickly. As illustrated in Fig. 11.17h, although the shape of the main lobe remains similar, the beam sidelobe level grows by 2 dB when $\theta_D \geq 50°$, due to the absorber blocking the path of the steering beam's propagation. When the blockage happens, the absorber inadvertently dissipates waves that are radiating in the target directions, which would otherwise contribute to the main lobe, thus reducing the antenna's directivity. This unintended interaction results in an elevation of the side lobes and an expansion of the main lobe's beamwidth, causing the energy to scatter in undesired directions and reducing the array's directivity.

## 11.5 Metamaterial Absorber with Beamforming Isolation Optimization

In Sect. 11.4.3, our study focused on the efficacy of the metamaterial absorber in improving isolation, particularly in conditions where Tx-Rx far-field beam interactions are minimized. This section aims to demonstrate the performance of a combined two-stage isolation design, where using a metamaterial absorber to suppress near-field coupling, complemented by beam angle perturbations to address far-field components. The experimental findings highlight two insights:

1. The necessity of a metamaterial absorber for managing near-field MC components in guaranteeing the overall beamforming isolation design within the steering range. Otherwise, the isolation performance can be predominantly constrained by the near-field SI, especially when the beams are oriented away from each other and the Tx-Rx beam interaction is minimal.
2. Incorporating an absorber structure streamlines the beam-level optimization process. By employing the absorber structure, we can achieve a significant average isolation level of 90 dB, with a simple beam angle perturbation technique.

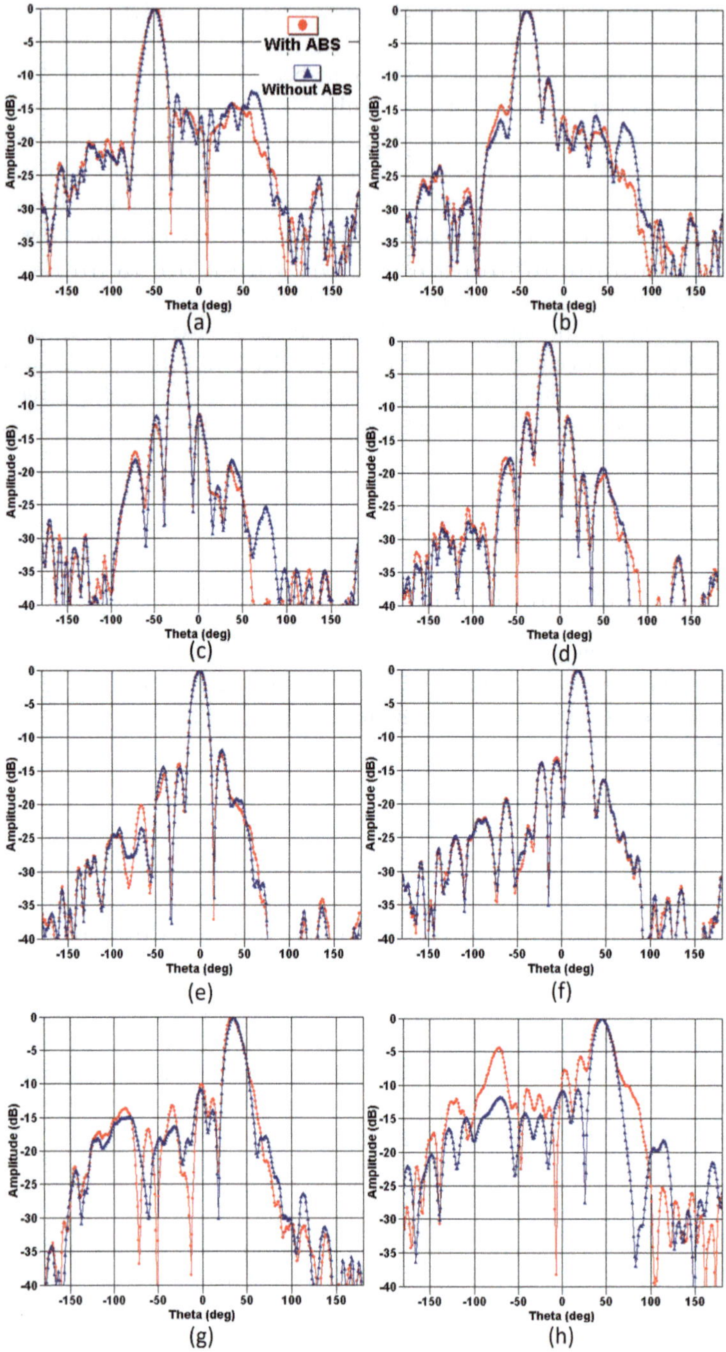


**Fig. 11.17** Tx 1×8 sub-array beams (with and without absorber) at different steering angles at (**a**) $\theta_D = -50°$, (**b**) $\theta_D = -40°$, (**c**) $\theta_D = -20°$, (**d**) $\theta_D = -10°$, (**e**) $\theta_D = 0°$, (**f**) $\theta_D = 10°$, (**g**) $\theta_D = 40°$, (**h**) $\theta_D = 50°$ (measured results)

In Fig. 11.18, we present the experimental results of Tx-Rx beam isolation, encompassing four distinct scenarios:

- Scenario 1: without the absorber and beam steering angle at exact $(\theta_D, \theta_U)$. This scenario serves as a baseline with no extra isolation improvement.
- Scenario 2: without the absorber, but employing beam angle perturbation for isolation optimization technique, thereby steering the beams to adjusted angles $(\hat{\theta}_D, \hat{\theta}_U)$ and targeting on far-field isolation improvement.
- Scenario 3: with the metamaterial absorber while steering the beams at exact $(\theta_D, \theta_U)$ angles, leading to an enhancement in near-field isolation.
- Scenario 4: the combination of the metamaterial absorber with adjusted beam steering angles at $(\hat{\theta}_D, \hat{\theta}_U)$, aiming to optimize both near-field and far-field isolation.

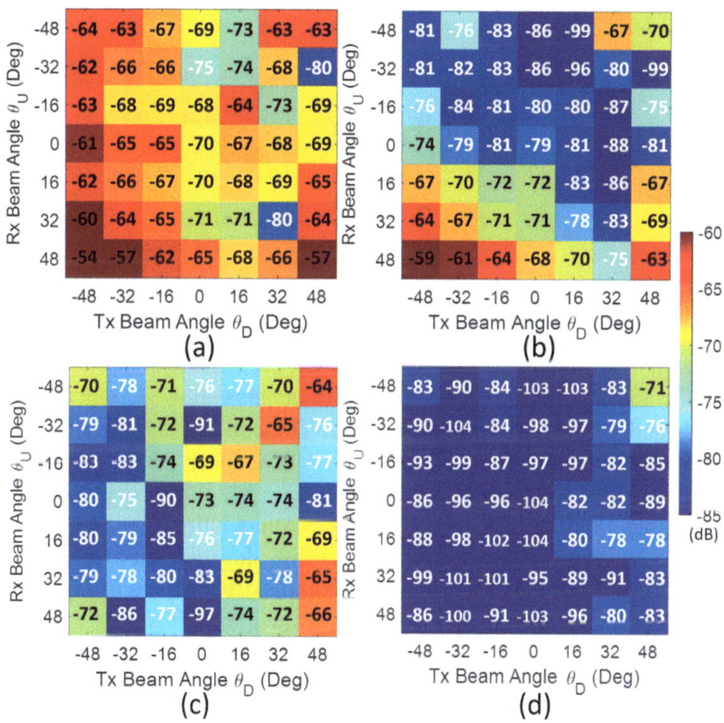


**Fig. 11.18** Beam MC between the sample Tx-1×8 and Rx-1×8 sub-arrays (**a**) without the absorber and steering at $(\theta_D, \theta_U)$, (**b**) without the absorber and steering at $(\hat{\theta}_D, \hat{\theta}_U)$, (**c**) with the absorber and steering at $(\theta_D, \theta_U)$, (**d**) with the absorber and steering at $(\hat{\theta}_D, \hat{\theta}_U)$ (measured results)

### 11.5.1 Angular-Perturbation Far-Field Isolation Improvement

Beamforming plays a crucial role in enhancing spatial isolation by (1) generating directional beams that confine radiation power towards the intended user, minimizing power leakage into potential interference paths; (2) focusing the Rx beams to the directions where interference radiation is nullified, reducing the receiver's detection sensitivity to interference signals.

Therefore, isolation improvement in FD mMIMO systems can be effectively achieved through beam angle perturbation techniques. As presented in Chap. 10 Algorithm 10.1, rather than aligning the Tx and Rx beams precisely at the target angles $\theta_D$ and $\theta_U$, which maximize beam directivity, we allow for a minor variation in the beam steering angles. This approach involves searching for new steering angles $(\hat{\theta}_D, \hat{\theta}_U)$ that prioritize achieving the highest possible isolation, albeit with a minor compromise in directivity. For an FD mMIMO system equipped with DL and UL RF beamformers $\mathbf{F}_D$ and $\mathbf{F}_U$, we can formulate the optimization problem for Tx-Rx isolation as follows

$$\begin{aligned} \min_{\{\hat{\theta}_D,\hat{\theta}_U\}} \quad & \left|\mathbf{F}_U^T(\hat{\theta}_U)\mathbf{H}_{\text{SI}}\mathbf{F}_D(\hat{\theta}_D)\right|^2 \\ \text{s.t.} \quad & C_1: \ |\hat{\theta}_D - \theta_D| \leq \epsilon_{DL}, \\ & C_2: \ |\hat{\theta}_U - \theta_U| \leq \epsilon_{UL}, \end{aligned} \tag{11.4}$$

where $C_1$ and $C_2$ refer to the beam angle perturbation constraints in desired DL and UL beam directions, respectively. In our experiments, a maximum DL and UL beam angle variation within a 3dB-beamwidth range is allowed, with a center at the desired DL and UL beam angles $\theta_D$ and $\theta_U$. The constraints indicate a maximum 3 dB directivity degradation. The proposed 1×8 sub-arrays have an average 3dB-beamwidth around 16°. Therefore, $\epsilon_{DL} = \epsilon_{UL} = 8°$. When an exhaustive search scheme is used in finding the $(\hat{\theta}_D, \hat{\theta}_U)$, the computational complexity and the number of measurement iterations increase with the resolution of the searching angle steps ($\Delta\theta$). Therefore, to mitigate the complexity involved in angle searching, our experiments adopt a large search step of 2°, i.e., $\hat{\theta}_D \in \{\theta_D - 8° : 2° : \theta_D + 8°\}$ and $\hat{\theta}_U \in \{\theta_U - 8° : 2° : \theta_U + 8°\}$.

#### 11.5.1.1 Experimental Measurements

In comparing beam-level MC as shown in Fig. 11.18a and b, it can be observed that implementing beam angle perturbations for Tx/Rx can effectively lower (improve) the beam MC by 10.5 dB, reducing the average MC level from −66.5 to −77.0 dB. The effectiveness of this isolation enhancement, however, is notably influenced by the directional orientation of the Tx and Rx beams. For instance, when Tx and Rx beams are oriented towards each other (i.e., $\theta_D > 0°$ and $\theta_U < 0°$), an average isolation improvement of 13.9 dB, with a maximum of 25.6 dB, can be observed. The strong interaction between the Tx and Rx beams results in the high effectiveness of the approach when there is a change in the mainlobe direction.

On the other hand, as demonstrated in the lower left quadrant of Fig. 11.18b, the performance of the beamforming isolation is limited when $\theta_D < 0°$ and $\theta_U > 0°$. An average MC of −66.0 dB and the highest (worst) MC of −58.8 dB can be observed. This scenario only allows for a marginal isolation improvement of 4.2 dB through far-field beam angle perturbation. Such constraints are primarily attributed to reduced Tx/Rx beam interaction and the dominance of near-field coupling waves, due to the Tx and Rx beams' main lobes directing away from each other. Consequently, effectively mitigating near-field SI components is essential for enhancing overall beam-level isolation in the FD systems.

### *11.5.2 Metamaterial Absorber Near-Field Isolation Improvement*

When comparing the beam-level isolation shown in Fig. 11.18a and c, the introduction of the metamaterial absorber structure results in a notable isolation improvement of 9.6 dB, reducing the average MC from −66.5 to −76.1 dB. Additionally, the orientation of the Tx and Rx beams which control the far-field beam interaction still impacts the effectiveness of this improvement. When the Tx and Rx beams are steered away from each other (i.e., $\theta_D < 0°$ and $\theta_U > 0°$), a more considerable isolation enhancement can be observed with an average isolation improvement of 17.5 dB and the maximum enhancement reaching up to 28.9 dB in the tested cases. An average beam MC of −79.4 dB can be achieved.

Furthermore, as the Tx beams are steered towards the Rx sub-array, i.e., when $(\theta_D, \theta_U)$ pair moves towards the right half of Fig. 11.18c, the level of isolation improvement decreases. In beam pairs that $\theta_D > 0°$ and $\theta_U < 0°$, an average MC of −71.3 dB is observed, which is 8.1 dB higher (worse) than the case when $\theta_D < 0°$ and $\theta_U > 0°$. This reduction in isolation performance is mainly attributed to the stronger interaction between the far-field components of the Tx and Rx beams, where far-field coupling becomes increasingly dominant in the total SI.

### *11.5.3 Combined Near-Field and Far-Field Isolation Improvement*

From the experimental results in Fig. 11.18b and c, following observations can be made

- The Angular-Based Beamforming Isolation Technique, primarily aimed at mitigating far-field coupling, demonstrates its effectiveness in scenarios charac-

terized by strong interactions between the Tx and Rx beams. However, its performance tends to be limited in situations where the interaction between Tx and Rx beams is weak. In such cases, the near-field coupling component becomes the dominant factor, thus constraining the overall isolation performance.
- The Metamaterial Absorber Technique can effectively mitigate the near-field coupling. At the same time, it is important to note that in scenarios where far-field coupling is dominant, additional isolation techniques are important to enhance the overall isolation performance.

As a result, combining these two techniques could potentially yield high isolation performance and eliminate the limitations of both methods in FD communications. The corresponding experimental results are depicted in Fig. 11.18d. Across 49 tested cases, an average Tx-Rx beam-level MC of −90.8 dB can be achieved, indicating an average improvement of 24.2 dB compared to scenarios without the absorber and beam-angle optimization. The highest recorded isolation level is 104.4 dB. All 49 beam pairs achieve a beam isolation level better than 70.8 dB.

Moreover, among the 49 Tx-Rx beam pairs, except for 5 cases, all others show an MC better (lower) than −80 dB. Notably, more than half of the tested cases (26 out of 49) achieve a beam MC lower than −90 dB. When there is no absorber and the beam angle-perturbation isolation technique, only two Tx/Rx beam pairs reach the MC level lower (better) than −80 dB, and 39 angle pairs (or 80% of the measured beam pairs) exhibit an MC worse (higher) than −70 dB, as shown in Fig. 11.18a).

Additionally, the beam-level isolation diminishes as the $\theta_D$ becomes larger, i.e., the Tx beam steering more towards the Rx sub-array. Further far-field isolation can be achieved by either decreasing angular-perturbation step $\Delta\theta$ or implementing arbitrary beamforming nulling techniques in these beam pairs.

## 11.6 Concluding Remarks

This chapter presents our design and study of a metamaterial absorber structure for FD mMIMO Tx-Rx antenna isolation improvement. The design processes of the absorber unit element, absorber placement, and influence on the beamforming radiation patterns are investigated. Working with our 8×8Tx/8×8Rx FD mMIMO antenna array prototype, an average Tx-Rx antenna element mutual coupling level of −65.4 dB is achieved between the closest Tx-Rx element pairs, which shows an improvement of 17.5 dB isolation by adding the absorber. Illustrative results using Tx/Rx 1×8 sub-arrays indicate an average 18.6 dB isolation improvement in near-field dominant SI scenarios. Experimental investigations on the correlation between the near-field and far-field dominant SI scenarios and the downlink and uplink beam directions are conducted. The critical role of the absorber structure in managing near-field self-interference, thereby ensuring the efficacy of the overall beamforming isolation performance, is demonstrated. Combined with a simple angular perturbation beam isolation technique, an average Tx-Rx beam-level isolation of 90.8 dB with a maximum beam isolation level of 104.4 dB can be achieved within a Tx/Rx steering range $-48°$ to $48°$ with 20 MHz modulated signals.

## References


1. A. Sabharwal, P. Schniter, D. Guo, D.W. Bliss, S. Rangarajan, R. Wichman, In-band full-duplex wireless: challenges and opportunities. IEEE J. Sel. Arcas Commun. **32**(9), 1637–1652 (2014)
2. Z. Zhang, X. Chai, K. Long, A.V. Vasilakos, L. Hanzo, Full duplex techniques for 5G networks: self-interference cancellation, protocol design, and relay selection. IEEE Commun. Mag. **53**(5), 128–137 (2015)
3. B. Smida, A. Sabharwal, G. Fodor, G.C. Alexandropoulos, H.A. Suraweera, C.-B. Chae, Full-duplex wireless for 6G: Progress brings new opportunities and challenges. IEEE J. Sel. Areas Commun. **41**(9), 2729–2750 (2023)
4. M. Duarte, C. Dick, A. Sabharwal, Experiment-driven characterization of full-duplex wireless systems. IEEE Trans. Wirel. Commun. **11**(12), 4296–4307 (2012)
5. B. Debaillie, D.-J. van den Broek, C. Lavin, B. van Liempd, E.A. Klumperink, C. Palacios, J. Craninckx, B. Nauta, A. Pärssinen, Analog/RF solutions enabling compact full-duplex radios. IEEE J. Sel. Areas Commun. **32**(9), 1662–1673 (2014)
6. M. Chung, M.S. Sim, J. Kim, D.K. Kim, C.-B. Chae, Prototyping real-time full duplex radios. IEEE Commun. Mag. **53**(9), 56–63 (2015)
7. T. Le-Ngoc, Y. Gong, M. Mahmood, A. Koc, R. Morawski, J.G. Griffiths, P. Guillemette, J. Zaid, P. Wang, Full-duplex in massive multiple-input multiple-output. IEEE Open J. Veh. Technol. **5**, 560–576 (2024)
8. E. Everett, A. Sahai, A. Sabharwal, Passive self-interference suppression for full-duplex infrastructure nodes. IEEE Trans. Wirel. Commun. **13**(2), 680–694 (2014)
9. N.I. Landy, S. Sajuyigbe, J.J. Mock, D.R. Smith, W.J. Padilla, Perfect metamaterial absorber. Phys. Rev. Lett. **100**(20), 207402 (2008)
10. F. Tofigh, M. Amiri, N. Shariati, J. Lipman, M. Abolhasan, Polarization-insensitive metamaterial absorber for crowd estimation based on electromagnetic energy measurements. IEEE Trans. Antennas Propag. **68**(3), 1458–1467 (2019)
11. A. Jafargholi, A. Jafargholi, J.H. Choi, Mutual coupling reduction in an array of patch antennas using CLL metamaterial superstrate for MIMO applications. IEEE Trans. Antennas Propag. **67**(1), 179–189 (2018)
12. J. Zhang, J. Li, J. Chen, Mutual coupling reduction of a circularly polarized four-element antenna array using metamaterial absorber for unmanned vehicles. IEEE Access **7**, 57469–57475 (2019)
13. G. Zhai, Z.N. Chen, X. Qing, Enhanced isolation of a closely spaced four-element mimo antenna system using metamaterial mushroom. IEEE Trans. Antennas Propag. **63**(8), 3362–3370 (2015)
14. Z. Qamar, U. Naeem, S.A. Khan, M. Chongcheawchamnan, M.F. Shafique, Mutual coupling reduction for high-performance densely packed patch antenna arrays on finite substrate. IEEE Trans. Antennas Propag. **64**(5), 1653–1660 (2016)
15. Q.-L. Zhang, Y.-T. Jin, J.-Q. Feng, X. Lv, L.-M. Si, Mutual coupling reduction of microstrip antenna array using metamaterial absorber, in *2015 IEEE MTT-S International Microwave Workshop Series on Advanced Materials and Processes for RF and THz Applications (IMWS-AMP)* (IEEE, 2015), pp. 1–3
16. M. Agarwal, M.K. Meshram, Isolation improvement of 5 GHz WLAN antenna array using metamaterial absorber, in *2016 URSI Asia-Pacific Radio Science Conference (URSI AP-RASC)* (IEEE, 2016), pp. 1050–1053

17. C. Wang, X.-S. Yang, B.-Z. Wang, A metamaterial-based compact broadband planar monopole MIMO antenna with high isolation. Microwave Opt. Technol. Lett. **62**(9), 2965–2970 (2020)
18. D.H.T. Hoang, V.X.T. Nguyen, T.T. Do, D. N. Nguyen, T.A. Vu, Q.N. Tran, Mutual coupling reduction of MIMO antenna using metamaterial absorber for 5G base stations, in *2022 Asia-Pacific Microwave Conference (APMC)* (IEEE, 2022), pp. 767–769

# Chapter 12
# Conclusions and Future Works

In this chapter, we provide a detailed overview of the monograph's main contributions, followed by a discussion of potential research topics for future work.

## 12.1 Summary

Massive multiple-input multiple-output (mMIMO) technology is a pivotal innovation for emerging wireless systems, which face rapidly increasing demands for higher data rates, improved spectral efficiency, greater capacity, and more reliable channels. The deployment of large-scale arrays marks a major step forward, enabling networks to manage soaring data traffic more efficiently. By incorporating numerous elements, the antenna arrays can be programmed to focus power in specific directions, thereby improving coverage and enhancing signal strength. Dividing large arrays into sub-arrays allows base stations (BS) to serve multiple users in different locations simultaneously, significantly increasing overall throughput. Additionally, metamaterial-based structures, composed of periodic sub-wavelength unit cells, have also garnered attention for their ability to enhance antenna array design. These artificial materials aid in boosting isolation, refining radiation characteristics, and enabling antenna element miniaturization—capabilities that help address potential bottlenecks in large-scale antenna array design and provide new opportunities in interference suppression.

In this monograph, we proposed and developed metamaterial-based large-scale antenna arrays alongside various beamforming optimization techniques for both half-duplex (HD) multi-user mMIMO in Part I (Chaps. 4–8) and full-duplex (FD) mMIMO systems in Part II (Chaps. 9–11). Full-wave simulations and experimental measurements have been conducted to evaluate and illustrate the performance of these designs under practical considerations. Collectively, these findings underscore the potential of metamaterial-based large-scale arrays to meet evolving needs and



T. Le-Ngoc et al., *Half-Duplex/Full-Duplex MU-mMIMO*, Wireless Networks,
https://doi.org/10.1007/978-3-032-09195-6_12

effectively boost the desired signal gain while minimizing the interference of future wireless networks.

In Chap. 3, two antenna element structures operating around 3.5 GHz are proposed. The first design utilizes a dual-layer metamaterial-based design that achieves patch antenna miniaturization through a slow-wave effect and enhances inter-element isolation via an electromagnetic bandgap (EBG) structure. A strategically designed L-slotted radiating patch enables circular polarization using a simple probe-feeding mechanism. The second design features a dielectric resonator antenna (DRA) structure with dual-polarization, achieved through two distinct feeding mechanisms. Full-wave simulations are performed to optimize and validate the radiation characteristics, including operating impedance bandwidth, element gain, and sample array beamforming performance. Due to its ease of fabrication, flexibility, and scalability in array configurations, the dual-layer metamaterial-based patch design is selected for experimental measurements in subsequent chapters.

Chapter 4 presents the design and evaluation of three hybrid beamforming (HBF) architectures aimed at enhancing base station capacity in mMIMO systems while reducing hardware cost and CSI overhead. The first uses full-resolution phase shifters (PSs) and DACs with a constant-modulus RF beamformer; the second employs b-bit PSs and full-resolution DACs, leveraging an OMP-based algorithm to approach optimal unconstrained RF performance; and the third integrates b-bit PSs with q-bit DACs, explicitly accounting for DAC quantization noise. The proposed HBF schemes reduce hardware cost and complexity by employing a limited number of RF chains, while also minimizing the channel state information (CSI) overhead through RF-stage design based on slow time-varying CSI (i.e., angular information). The resulting reduced-size effective CSI is then used for baseband-stage design. Performance comparisons across different antenna array configurations show that the proposed schemes consistently outperform other HBF benchmarks in terms of spatial, spectral, and energy efficiency, achieving notably higher sum-rate and energy efficiency in the illustrative results.

Then, Chap. 5 evaluates the performance of different 3D array structures for HBF in mmWave MU-mMIMO systems. The proposed HBF framework operates in two stages: an RF beamforming stage designed via singular value decomposition (SVD) of the channel matrix, followed by baseband multi-user (MU) precoding using the instantaneous effective baseband channel, where MU interference is mitigated through a regularized zero-forcing (RZF) approach. Illustrative results demonstrate that, compared to 2D arrays, 3D array structures offer higher spectral and energy efficiencies and maintain a stable achievable rate regardless of the user's angular position.

The proposed metamaterial-based patch antenna structure in the previous chapter is implemented in a large-scale patch antenna array in Chap. 6. Prior to fabrication, a quantitative analysis was conducted to assess the impact of array length on beam radiation patterns under various element radiation pattern (ERP) considerations, resulting in theoretical expressions for directivity and 3dB- and 10dB-beamwidths. The study also examined how sub-array locations and surrounding environments influence beam steering characteristics, revealing potential deviations in beam steering angles, directivity, and sidelobe characteristics caused by asymmetric sub-array placement and environmental factors. Based on these findings, an 8×8 metamaterial-based patch antenna array was designed and fabricated for mMIMO operation in the 3.35 to 3.6 GHz band. Measurements of element return loss (RL), inter-element mutual coupling (MC), and sub-array beamforming performance showed close agreement with simulation results, validating the array design. A detailed evaluation of ERPs across all 64 elements demonstrated how symmetry and distortion in ERP variations can potentially guide more effective beamforming optimizations. Furthermore, beamforming experiments conducted in an anechoic chamber with quantized phase shifters and attenuators confirmed the array's steering capabilities, investigated the influence of array length on steerability, and highlighted the potential impact of environmental factors on performance.

Then, in Chap. 7, a Perturbation-Based Nulling Control Beamforming (PNCB) scheme is proposed to effectively suppress multi-user interference (MUI) in MU-mMIMO systems by creating deep and wide radiation nulls precisely aligned with interference directions. The scheme leverages symmetry information from ERPs to streamline optimization by reusing optimized beamforming vectors for symmetric sub-arrays. The PNCB approach begins with weights generated by the Linearly Constrained Minimum Variance (LCMV) method, which are subsequently refined through perturbations to identify optimal beamforming vectors that account for variations caused by heterogeneous ERPs. Particle Swarm Optimization (PSO) is employed to enhance search efficiency. Simulation results demonstrate the scheme's ability to achieve effective interference mitigation in multi-user scenarios with minimal impact on the main lobe pattern. The scheme is further extended to control nulling width, allowing consistent MUI suppression even when interference user locations deviate or move. Experimental validations conducted in an anechoic chamber, involving two users with distinct modulation signals, confirm the PNCB's effectiveness. Results show considerable improvements in Error Vector Magnitude (EVM) and Signal-to-Interference Ratio (SIR), which successfully bring the MUI to the noise floor. Additionally, the study examines the effects of attenuator and phase-shifter quantization on nulling depth, providing valuable insights into achievable performance under different hardware configurations.

As the array aperture size continues to grow, Chap. 8 explores the opportunities provided by extra-large (XL) arrays for near-field beamfocusing in MU-mMIMO, accounting for the spherical wavefront characteristics introduced by the large aperture. Comprehensive full-wave simulation analyses are conducted to investigate near-field beam radiation pattern transitions with changing distances, incorporating realistic EM wave characteristics, heterogeneous ERPs, and interactions among array elements. To harness the advantages of distance-dependent radiation patterns in the near-field, two Nulling Control Beamforming (NCB) algorithms are developed to effectively mitigate MUI in mMIMO systems. The approaches demonstrate their effectiveness by achieving significant focusing gain differences between desired and interfering locations in both colinear and non-colinear user scenarios. The first algorithm employs an LCMV-based NCB scheme, which adjusts array feeding magnitudes and phase shifters. The second algorithm introduces a constant-modulus beamfocusing scheme, based on a perturbation-based NCB algorithm that utilizes Particle Swarm Optimization (PSO) and relies solely on phase shifters. Iterative full-wave simulations further explore variations in beamfocusing gain differences across diverse colinear desired and interfering user locations, providing valuable insights into the performance and adaptability of these algorithms.

To further enhance spectral efficiency, full-duplex (FD) technologies have emerged as a promising advancement, enabling simultaneous transmission and reception over the same bandwidth. In Chap. 9, a cross-polarized $8\times8$Tx/$8\times8$Rx large-scale array prototype was designed, fabricated, and characterized to evaluate its effectiveness in achieving Tx-Rx antenna isolation for FD mMIMO applications around 3.5 GHz. Building on the dual-layer metamaterial-based antenna design introduced in Chap. 3, the prototype employs mirrored L-slotted radiating patches to form cross-polarized CP Tx/Rx pairs with comparable radiation characteristics. Full-wave simulations conducted in HFSS illustrate the efficacy of the cross-polarized configuration and strategic element separation in achieving high Tx-Rx isolation. Experimental validation, involving the measurement of 64-Tx-to-64-Rx self-interference (SI) channels over a 1 GHz bandwidth (3–4 GHz), revealed an average element isolation of 59.1 dB within a 20 MHz band centered at 3.5 GHz. Additionally, the spatial isolation between various Tx and Rx element positions was investigated. The comprehensive dataset of measured SI channels offers a realistic and robust foundation for the development and evaluation of beam-level isolation and cancellation algorithms, enabling further advancements in FD mMIMO systems.

Moreover, in Chap. 10, joint uplink/downlink beamforming optimization schemes are implemented to further suppress SI by optimizing beam-level spatial isolation while maintaining gain toward desired directions. The chapter highlights the necessity and advantages of multi-stage SI isolation and cancellation techniques in FD mMIMO systems. The relationship between array size, beamwidth, and beam-level MC is analyzed. Results demonstrate that focusing Tx/Rx beam radiation patterns significantly enhances Tx-Rx beam-level isolation. Building on this insight, two beam-level isolation optimization algorithms are proposed. The first algorithm employs an angular perturbation search to identify optimal uplink and downlink steering angles within an acceptable power degradation limit for desired users. The second algorithm integrates a Null Steering Beamforming (NSB) approach with a PSO algorithm to determine optimal nulling angles for enhanced isolation. The effectiveness of these beam-level optimization algorithms in achieving robust SI mitigation is studied using the $8\times8$Tx/$8\times8$Rx antenna array prototype with quantized beamforming networks in the anechoic chamber. Measurement results also illustrate a strong correlation between Tx/Rx beam radiation patterns and achieved isolation, offering valuable insights for the development of advanced beam-level isolation algorithms. By combining Tx-Rx antenna isolation with a joint uplink/downlink beamformer design, the system effectively prevents residual SI from saturating the receiver's ADC or driving the LNA into nonlinear operation. With additional SI mitigation through SI cancellation (SIC) techniques, the residual SI is successfully reduced to levels comparable to thermal noise.

To achieve more uniform Tx-Rx isolation performance across all uplink and downlink user locations, Chap. 11 presents the design and evaluation of a metamaterial absorber structure aimed at enhancing Tx-Rx RF isolation in FD mMIMO systems without requiring modifications to the existing antenna array design. The chapter details the absorber unit element design, its placement, and its influence on beamforming radiation patterns. Using the $8\times8$Tx/$8\times8$Rx FD mMIMO antenna array prototype, the addition of the absorber significantly improves isolation between the closest Tx-Rx element pairs. Experiments with $1\times8$ Tx/Rx sub-arrays demonstrate an average isolation improvement of 18.6 dB in near-field dominant SI scenarios, with negligible impact on the original beam radiation patterns. The study further explores the correlation between near-field and far-field dominant SI scenarios and the alignment of downlink and uplink beam directions. Findings emphasize the critical role of the absorber in mitigating near-field SI, which is essential for maintaining effective beamforming isolation performance. By mitigating near-field coupling components, the absorber enhances RF isolation, ensuring a robust overall Tx-Rx isolation level. This approach provides sufficient margin for total SI suppression, even when far-field beam-level isolation techniques are less effective, thereby contributing to a more reliable FD mMIMO system performance.

## 12.2 Potential Future Works

In this monograph, we proposed and developed metamaterial-based large-scale antenna arrays, integrating various beamforming optimization techniques for both MU-mMIMO and FD mMIMO systems. These developments were supported by comprehensive full-wave simulations and experimental measurement studies. As we conclude, we would like to outline potential avenues for future research directions:

- **Artificial Intelligence-Based Wireless Communication Optimization:** In an era of rapidly increasing data traffic and ever-stringent latency and reliability requirements, the computational complexity of extra-large MIMO (XL-MIMO) systems presents significant challenges for beamforming and resource allocation optimization. This monograph has demonstrated that nature-inspired optimization algorithms can effectively and efficiently tackle interference, such as multi-user interference (MUI) and self-interference (SI). These optimized datasets, derived from experimental measurements, can be leveraged to train intelligent learning-based models capable of predicting and managing beamforming networks in real time. As we move toward 6G networks, artificial intelligence (AI) will play a pivotal role in enhancing network throughput and automating system management. A variety of AI-driven approaches, including deep learning, Ensemble Methods, reinforcement learning–based approaches, and Generative Adversarial Networks (GANs), can be integrated into our large-scale antenna array testbed to achieve more robust, adaptive, and efficient wireless communication.

- **Centimeter-Wave Communications and XL-mMIMO Systems:** Wireless service providers are navigating a rapidly evolving technological landscape, driven by increasing demands for faster data speeds, enhanced spectrum efficiency, and improved energy efficiency. In this context, centimeter-wave (cmWave) communication technology emerges as a promising candidate for future wireless systems [1]. The cmWave band strikes a practical balance between wide coverage and ample bandwidth, offering favourable propagation characteristics along with manageable hardware complexity. Compared to sub-6 GHz designs, cmWave technology provides significantly broader spectrum resources, resulting in higher data throughput. Moreover, the higher operating frequency leads to a shorter wavelength, enabling more compact antenna designs and allowing a larger number of elements to be placed within a given array aperture. At these frequencies, XL-MIMO arrays become feasible for various applications. By deploying larger-scale arrays, XL-MIMO technology can substantially increase system capacity, nearly in proportion to the number of transmitters and receivers [2]. Such arrays also enhance beamforming capabilities, delivering improved spatial resolution, tighter energy focusing, and higher array gain.

  Meanwhile, unlike millimeter-wave (mmWave) systems, cmWave operates at wavelengths that offer more robust and reliable propagation characteristics. Signals experience lower loss from reflection, scattering, and diffraction, allowing better diffraction around obstacles when direct line-of-sight paths are obstructed [3, 4]. Additionally, the cmWave design aligns well with standard PCB manufacturing processes, making it less sensitive to surface roughness or material imperfections. Although this monograph focuses primarily on sub-6 GHz structures, the proposed designs can be adapted to cmWave mMIMO systems and extended to XL-mMIMO deployments operating in this appealing frequency range.
- **Antenna and Sub-array Selection Optimization in Metamaterial-Based Large-Scale Arrays:** Metamaterial technology facilitates the realization of electrically small, high-isolation antenna elements with sub-wavelength dimensions and intrinsically suppressed inter-element coupling. By selectively exciting sub-arrays, these compact elements enable ultra-dense array layouts that adapt electromagnetic performance and tailor wireless-channel characteristics to specific user locations and operating conditions. These reconfigurable arrays emulate quantized-movable or semi-fluid array configurations, thereby expanding design freedom for managing both desired signals and interference channels. The effects of element spacing and overall sub-array topology on signal quality, MUI, and SI can be studied via full-wave simulations and experimental measurements. A cohesive hardware-software framework can then optimally schedule antenna elements or sub-arrays and refine beamformers across near- and far-field regimes. Such a framework might leverage nature-inspired heuristics, advanced optimization schemes, or machine-learning approaches to efficiently explore the high-dimensional selection and beamforming space, and accelerate convergence. Beyond communications, these programmable, ultra-dense arrays offer unique capabilities for advanced sensing, such as real-time imaging, object

detection, and non-invasive health monitoring, by dynamically sampling the electromagnetic environment with sub-wavelength spatial resolution.

- **Metamaterial Reconfigurable Intelligent Surface:** The recent advent of reconfigurable intelligent surfaces (RIS) has opened up new possibilities for creating favourable propagation environments in wireless communications [5, 6]. Conventional phased-array beamforming typically relies on multiple RF chains-comprising amplifiers, mixers, and filters-which can lead to high power consumption and hardware complexity. In contrast, metamaterial-based large-scale reflectarrays comprise nearly passive components, offering a high-flexibility, low-profile, and energy-efficient alternative. By embedding tunable circuit components and reconfigurable metamaterials, the array's radiation pattern, polarization, and frequency response can be dynamically adapted in real time. By controlling surface impedance and outgoing wave phase, these arrays can effectively concentrate reflective beams toward desired directions. A frequent challenge in wireless systems arises from the lack of line-of-sight (LOS) channels, where obstacles block or deflect signals, potentially causing service disruptions. The adaptability and energy efficiency of metamaterial reflectarrays make them a promising approach to both sensing and improving the radio environment. The beamforming algorithms introduced in this monograph can also be extended to RIS control by dynamically adjusting the structure electromagnetic properties in response to environmental or application requirements. Such adaptability can further strengthen the radio channel, improve signal quality, and enhance overall wireless performance.

## References

1. E. Semaan, E. Tejedor, R. Kumar-Kochhar, S. Magnusson, S. Parkvall, 6G spectrum-enabling the future mobile life beyond 2030. Ericsson AB, White Paper (2023)
2. P. Xingdong, H. Wei, Y. Tianyang, L. Linsheng, Design and implementation of an active multibeam antenna system with 64 RF channels and 256 antenna elements for massive MIMO application in 5G wireless communications. China Commun. **11**(11), 16–23 (2014)
3. T.S. Rappaport, S. Sun, R. Mayzus, H. Zhao, Y. Azar, K. Wang, G.N. Wong, J.K. Schulz, M. Samimi, F. Gutierrez, Millimeter wave mobile communications for 5G cellular: it will work!. IEEE Access **1**, 335–349 (2013)
4. H.C. Nguyen, I. Rodriguez, T.B. Sorensen, L.L. Sanchez, I. Kovacs, P. Mogensen, An empirical study of urban macro propagation at 10, 18 and 28 GHz, in *2016 IEEE 83rd Vehicular Technology Conference (VTC Spring)* (IEEE, 2016), pp. 1–5
5. E. Basar, M. Di Renzo, J. De Rosny, M. Debbah, M.-S. Alouini, R. Zhang, Wireless communications through reconfigurable intelligent surfaces. IEEE Access **7**, 116753–116773 (2019)
6. I. Yildirim, A. Koc, E. Basar, T. Le-Ngoc, RIS-aided angular-based hybrid beamforming design in mmWave massive MIMO systems, in *GLOBECOM 2022–2022 IEEE Global Communications Conference* (IEEE, 2022), pp. 5267–5272

MIX
Papier aus verantwortungsvollen Quellen
Paper from responsible sources
FSC® C105338

If you have any concerns about our products,
you can contact us on
**ProductSafety@springernature.com**

In case Publisher is established outside the EU,
the EU authorized representative is:
**Springer Nature Customer Service Center GmbH**
**Europaplatz 3, 69115 Heidelberg, Germany**

Printed by Libri Plureos GmbH
in Hamburg, Germany